MOLECULAR ANATOMY AND PHYSIOLOGY OF PROTEINS

PHOSPHOGLYCERATE KINASE

A HINGE-BENDING ENZYME

MOLECULAR ANATOMY AND PHYSIOLOGY OF PROTEINS

VLADIMIR N. UVERSKY - SERIES EDITOR -

(INDIANA UNIVERSITY OF MEDICINE, INDIANA, USA)

Alpha -Lactalbumin
Eugene A. Permyakov
2005. ISBN: 1-59454-107-8

Creatine Kinase
Christian Vial
2006. ISBN: 1-59454-715-7

Cellulosome
Irina A. Kataeva (Editor)
2006. ISBN: 1-59454-950-8

Neuronal Calcium Sensor Proteins
Pavel P. Philippov and Karl-Wilhelm Koch (Editors)
2006. ISBN: 1-59454-978-8

Human Stefins and Cystatins
Eva Zerovnik, Nataša Kopitar-Jerala and Vladimir Uversky (Editors)
2006. ISBN: 1-60021-233-6

Parvalbumin (Volume in Molecular Anatomy and Physiology of Proteins Series)
Eugene A. Permyakov
2006. ISBN: 1-60021-337-5

Methods in Protein Structure and Stability Analysis: Luminescence Spectroscopy and Circular Dichroism
Vladimir N. Uversky and Eugene A. Permyakov (Editors)
2007. ISBN: 1-60021-404-5

Methods in Protein Structure and Stability Analysis: Vibrational Spectroscopy
Vladimir N. Uversky and Eugene A. Permyakov (Editors)
2007. ISBN: 1-60021-703-6

Methods in Protein Structure and Stability Analysis: NMR and EPR Spectroscopies, Mass-Spectrometry and Protein Imaging
Vladimir N. Uversky and Eugene A. Permyakov (Editors)
2008. ISBN: 1-60021-705-2

Protein Adaptation in Extremophiles
Khawar Sohail Siddiqui, Torsten Thomas, and Vladimir N. Uversky (Editors)
2008. ISBN: 978-1-60456-019-0

Ribonucleotide Reductase
Kristoffer Andersson (Editor)
2008. ISBN: 978-1-60456-199-9

Foldases: Enzymes Catalyzing Protein Folding
Natalya K. Nagradova
2008. ISBN: 978-1-60456-389-4

Methods in Protein Structure and Stability Analysis: Conformational Stability, Size, Shape and Surface of Protein Molecules
Vladimir N. Uversky and Eugene A. Permyakov (Editors)
2011. ISBN: 978-1-60021-704-3

Biochemistry and Function of Antifreeze Proteins
Steffen P. Graether (Editor)
2010. ISBN: 978-1-61668-265-1
2012. ISBN: 978-1-62100-435-6 (Softcover)

Beta-Lactamases
Jean-Marie Frère (Editor)
2012. ISBN: 978-1-61324-638-2

Calcium Binding Lysozymes
Eugene A. Permyakov, Ludmilla A. Morozova-Roche and Lawrence J. Berliner
2012. ISBN: 978-1-62081-755-1

PHOSPHOGLYCERATE KINASE: A Hinge-Bending Enzyme
Mária Vas (Editor)
2014. ISBN: 978-1-62808-836-6

MOLECULAR ANATOMY AND PHYSIOLOGY OF PROTEINS

PHOSPHOGLYCERATE KINASE

A HINGE-BENDING ENZYME

MÁRIA VAS
EDITOR

New York

For permission to use material from this book please contact us:
Telephone 631-231-7269; Fax 631-231-8175
Web Site: http://www.novapublishers.com

Library of Congress Cataloging-in-Publication Data

Library of Congress Control Number: 2013946507

ISBN: 978-1-62808-836-6

Published by Nova Science Publishers, Inc. † New York

Contents

Acknowledgments

Here I would like to express my sincere thanks to everybody who supported and helped me by any means or contributed in preparing this book. First of all, many thanks to my previous (Andrea Szilágyi, Beáta Flachner) and present (Andrea Varga, Éva Gráczer) co-workers, including graduate and Ph.D. students who participated enthusiastically in various lines of the research work devoted to uncovering the structure-function relationship of phosphoglycerate kinase, PGK. In addition, I thank the inspiring atmosphere that has been existing and I could enjoy for many years in Institute of Enzymology under the leadership of the previous (Péter Friedrich, Péter Závodszky) and the present (László Buday) directors. Many thanks are also due to the collaborating laboratories. Among them, first of all, Colin Blake's laboratory (Laboratory of Molecular Biophysics, Oxford University, UK) has to be mentioned, where one of the first X-ray structures of PGK from horse muscle has been determined and from where the "hinge-bending hypothesis" was originated. Unfolding-refolding studies of pig muscle PGK have been carried out in collaboration with Oleg Borisovich Ptitsyn and Gennady Semisotnov (Institute of Protein Research, Russian Academy of Sciences, Pushchino, Russia). Structural works (X-ray and NMR) with both wild-type and mutant human PGKs were done by contribution by Matthew Bowler (Structural Biology Group, ESRF, Grenoble Cedex, France) and by Jonathan Waltho's group (The Krebs Institute, University of Sheffield, UK). The functional aspects of crystalline PGK have been tested by contribution of Angelo Merli and Gian Luigi Rossi (Department of Molecular Biology, University of Parma, Italy). The work devoted to the drug-candidate nucleotide substrate analogues were done in collaboration with Corinne Lionne, Laurent Chaloin and Tom Barman (Research Centre on Pathogens and Biotechnologies for Health, CNRS, University of Montpellier, France). Finally, the valuable contribution of the Co-authors of this book in preparing particular chapters are greatly appreciated and thanked.

Mária Vas
Institute of Enzymology
Research Centre for Natural Sciences
Hungarian Academy of Sciences
Budapest, Hungary

Introduction

3-Phosphoglycerate kinase (PGK), among the abundant and well soluble glycolytic enzymes, has been discovered at the early stage of enzymology. Isolation of 1,3-bisphosphoglycerate (bPG) as an intermediate in the oxidation of glyceraldehyde-3-phosphate to 3-phosphoglycerate (PG) was the first indication that two separate enzymes (GAPDH and PGK) were involved at this stage of the glycolytic pathway [1]. A little later the enzyme PGK was isolated from yeast and was shown to catalyse the high energy phospho-transfer reaction from the acid anhydride bond of 3-phosphoglyceroyl-phosphate (bPG) to the β-phosphate of ADP [2]. The reaction requires a divalent metal ion, commonly magnesium (Mg^{2+}) [3], complexed to the phosphates of the nucleotide, as proved by later crystallographic works [4-7]. The purified yeast PGK has solely served for at least the following 20 years for structure-function relationship studies, but later on PGK was also isolated (or produced by genetic expression) from various sources (e.g., higher plants [8] human erythrocytes [9], muscles of chicken breast [10], rabbit [11, 12], horse [13], fish [14], pig [15] and human [16, 17] as well as bacteria of *B. stearothermophilus* [18], *Thermotoga maritima* [19], *Thermus caldophilus* [20], the extreme thermophile *Thermus thermophilus* [21], archaea [22] and a single cell parasite [23]) and found to possess high degree of similarity both in the primary sequence and in the tertiary fold [24-26].

PGK is a fundamentally important enzyme in most living cells, where it is required for ATP generation in the glycolytic pathway of aerobes and in the fermentation of anaerobes as well as in many plant tissues for carbon fixation. Recently, PGK was also shown to possess other type of functions such as specific thiol-reductase activity, important in regulation of tumour angiogenesis [16, 17, 27], phosphorylating reactions important in antiviral and antitumour therapy [28-33] and regulated by CaM binding [34].

Due to its simple monomeric and highly conserved structure as well as the relatively small molecular size (cf. reviews [35, 36]) PGK can be considered as an attractive model to study various general aspects of enzymology and protein chemistry.

First of all, the chemical mechanism of *enzyme-catalysed phospho-transfer* is only poorly understood, in spite of its wide occurrence and general importance in several biochemical pathways and regulatory mechanisms. In particular, little is known about the nature of the transition state, the role of metal (i.e., shielding the negative charges on the nucleotide substrate, help in development of negative charge on the leaving group, stabilization of the transition state or optimal orientation of the reacting groups) and whether the reaction is associative or dissociative [37-46] A closely related problem is the paucity of structural information about the binding details of the metal-complex of ATP (the universal phosphate donor) to proteins, especially in the functionally competent complexes (e.g., [47, 48]).

As a second, PGK molecule with its two domains can also be considered as a simple model of multidomain enzymes. Accumulated experimental data (cf. reviews [49-52]) indicate that *relative domain movements* (mediated by protein structural flexibility) are essential in functioning of such enzymes. Domain closure over the active sites of multidomain enzymes can often create an optimal environment for catalysis, orient properly the substrate reactive groups and impedes side-reactions. The special way of co-operation between domains and the detailed molecular mechanism of relative domain movements, however, are not yet delineated in terms of the structure. Apart of the mechanism, we are still far from understanding the role of such motions in catalysis and regulation in terms of the atomic details. The mechanism of domain closure and its immediate consequences have to be clarified in individual cases (such as PGK) in order to reach more general conclusions. On the basis of accumulated structural and functional studies, carried out with PGK, a fairly detailed mechanism of domain closure has been suggested and outlined below.

In addition to achievement of biological function, another general question is related to understanding *the basic rules of self-organization (folding) process* of native three-dimensional structure from the various known protein building blocks (modules or domains). PGK molecule with its two separated structural domains (each of them binds one of the two substrates) has also served for a long time as a promising model structure in numerous folding studies. From these studies not only the possible role of domains in the self-organisation process has been exemplified, but some characteristics of the sophisticated *protein misfolding* mechanism has also been described. These separate aspects are also overviewed and the respective results are summarized in this volume.

Besides, this book is attempted to provide a historical overview of the researches, especially of the structural-functional studies. It is remarkable that originally the detailed structural basis of domain motion could be successfully derived by the time consuming comparison of the open and closed PGK conformations determined from sequentially different species. Finally, this mechanism of domain closure has been clearly approved by the much later solved crystal structure of the sequentially identical closed conformation of the mammalian PGK, the open form of which was determined at least 30 years earlier.

Abbreviations

AMP-PNP	β,γ-imido-adenosine-5'-triphosphate
AMP-PCP	β,γ-methylene-adenosine-5'-triphosphate
ANS	1-anilinonaphthalene-8-sulfonic acid
AP_4	adenosine 5'-tetraphosphate
Ap5A	P^1,P^5-di(adenosine-5') pentaphosphate
bPG	1,3-bisphosphoglycerate
BPTI	bovine pancreatic trypsin inhibitor
BsPGK	*Bacillus stearothermophilus* PGK
CaM	calmodulin
CFR	contact forming residue
CIDNP	chemically induced dynamic nuclear polarisation
CK	creatine kinase (EC 2.7.3.2)
CM-	carboxymetil-
CXCR4	CXC-motif receptor
D-dFdC (gemcitabine)	2',2'-difluorodeoxycytidine
DSC	differential scanning calorimetry
DEN	deformable elastic network
DTT	dithiothreitol
EPR	electron paramagnetic resonance
FDA	US Food and Drug Administration
G-3-P	DL-glycerol-3-phosphate
GAPDH	D-glyceraldehyde-3-phosphate dehydrogenase (EC 1.2.1.12)
GuHCl	guanidine hydrochloride
HsPGK	human PGK
IAA	iodoacetamide
IAEDANS	5-[2-(2-iodoacetamido)ethylamino]-1-naphthalenesulfonic acid
IAF	5-iodoacetamidofluorescein
IAOH	iodoacetate
L-3TC or L-SddC (lamivudine)	β-L-2',3'-dideoxy-3'-thiacytidine
L-FMAU	2'-fluoro-5-methyl-β-L-arabinofuranosyluracil
L-OddC	β-L-(-)-dioxolanecytidine
MET	2-mercapto-ethanol
Nbs_2	5,5'-ditiobis(2-nitrobenzoát)

NDPK	nucleoside diphosphate kinase (EC 2.7.4.6)
NEM	N-ethyl-maleimide
NMR	nuclear magnetic resonance
NOE	nuclear Overhauser effect
P_i	inorganic orthophosphate
PG	3-phospho-D-glycerate
PGK	3-phospho-D-glycerate kinase, or ATP 3-phospho-D-glycerate 1-phospho-transferase (EC 2.7.2.3)
PK	pyruvate kinase (EC 2.7.1.40)
PPP_i	tripolyphosphate
R_g	radius of gyration
RMSD	root main square deviation
SANS	small angle neutron scattering
SAXS	small angle X-ray scattering
TBP	trigonal bipyramid
TbPGK	*Trypanosoma brucei* PGK
TmPGK	*Thermotoga maritima* PGK
TNP-ATP	2'3'-O-(2,4,6-trinitrophenyl)ATP
TRNOESY	transfer nuclear Overhauser effect spectroscopy
TSA	transition state analogue
uPA	urokinase plasminogen activator
uPAR	urokinase plasminogen receptor
YPGK	yeast PGK

Part I

Normal Anatomy: Structural Features

In: Phosphoglycerate Kinase: A Hinge-Bending Enzyme
Editor: Mária Vas
ISBN: 978-1-62808-836-6

Chapter I

Three-Dimensional Structure as Determined by X-Ray Studies

Mária Vas
Institute of Enzymology, Research Centre for Natural Sciences,
Hungarian Academy of Sciences, Budapest, Hungary

As for other proteins, the three-dimensional fold of the polypeptide chain of PGK at high resolution has been determined by X-ray crystallography, and later NMR investigations have provided consistent results. Crystal structure of PGK has been solved already at the very early stage of protein crystallography and at the time of writing this review more than a dozen of independent PGK structures have been published.

1.1. History of Early X-ray Studies Leading to Structure Determination of PGK

Because of deficiencies in X-ray sources and measuring equipment in the 1950/1960 period, the first X-ray diffraction studies on proteins were restricted to relatively small, stable molecules from which large single crystals could be readily grown. The biological roles of the proteins were of secondary importance. The consequence of these selection criteria was that nearly all the proteins initially studied were extracellular proteins with robust molecular structures, often rich in disulphide bridges, required for their harsh functional environment. Typical of this self-selected group of proteins studied in this first stage of X-ray analysis were ribonuclease, lysozyme, chymotrypsin, and insulin. The result was of course that after quite a number of successful X-ray determinations we had a body of information on the molecular architecture and catalytic activities of a number of small extracellular enzymes, mostly proteinases, glycosidases and nucleases, but we had virtually no knowledge of these characters in the larger intracellular proteins which exhibited more interesting biological activities.

It was not until the "second wave" of X-ray examinations that intracellular proteins began to be considered. Not only was clear that structural studies of intracellular proteins would be of greater importance because of their central roles in a wide range of metabolic processes, but it was also known that extracellular and intracellular proteins differed from one another in certain basic structural aspects. In fact, the first intracellular proteins began to be studied soon after the initial studies proved successful. For example, work on glyceraldehyde phosphate dehydrogenase was started in Cambridge by Watson [53] soon after the first protein of all, myoglobin, had been solved there in 1958 by Kendrew's group [54]. In addition, lactate dehydrogenase was being studied by Rossmann at Purdue [55]. However the first concerted examination of intracellular proteins was on the enzymes of the glycolytic pathway mostly carried out by protein crystallographers in Oxford and Bristol. Indeed it used to be said that the glycolytic pathway ran between Oxford and Bristol in England.

The reason for this concentration of effort was the presence in Bristol of Robert Scopes, an Australian biochemist who was intent on purifying and crystallising as many of the enzymes of glycolysis as possible for X-ray analysis. He was very successful in this enterprise [56], and produced enough protein to satisfy both the small X-ray group in Bristol and the larger group in Oxford, which had been successful with lysozyme. Members of these groups worked independently on individual glycolytic enzymes from about 1970 onwards. The first of these enzymes to be solved was PGK. The structure of the horse muscle enzyme, and the enzyme from yeast were solved independently but almost simultaneously by Blake and Evans [57, 58], and by Wendell, Bryant and Watson [59, 60], respectively.

1.2. The Tertiary Fold of PGK Molecule

The initial views of the molecular structure uncovered one of the enzyme's major characteristics and one that dominates considerations of its enzymatic activities to this very day. The low resolution (6Å) image [57] showed the PGK molecule to have overall dimensions of 70Åx45Åx35Å, with its single polypeptide chain folded into two distinct folding domains of equal size widely separated by a comparatively thick 'neck' region. The linking region was marked by a strong rod of density, which was later confirmed to be an α-helix. The implications of this, for the time, remarkable bilobal structure for the activity of PGK were considerable. One possibility was that the enzyme might have two activities, different ones being associated with each domain. In fact, there were preliminary enzymological observations suggesting the presence of two binding sites for each of the two substrates. There was also biochemical evidence for both an active site and a strong anion-binding site. An alternative possibility was that in addition to the active site PGK also contained a binding site for another protein, or cell component. One of the substrates of PGK, 1,3-bisphosphoglyerate (bPG) is very unstable to hydrolysis and it was thought possible that PGK might form a molecular complex with glyceraldehyde-3-phosphate dehydrogenase, its immediate neighbour in the glycolytic pathway, to facilitate the rapid exchange of the unstable substrate. This complex, however, could not be detected by in vitro investigations [61, 62], but the question remained open and has been reappraised later [63, 64]. A third possibility was that the active site of PGK was split between the two domains even though they were widely separated in the native enzyme.

Extension of the crystallographic study to higher resolution [58] revealed more detail of the molecular structure, particularly when the complete amino acid sequence was incorporated for the horse enzyme [4]. It became apparent that the two domains corresponded closely to the N- and C-terminal halves of the 416-amino acid polypeptide chain. The correspondence was not quite exact because the last 10 amino acids at the C-terminus formed a short α-helix packed against the N-terminal domain. Thus, there were two connections between the two domains, one around residues 186-189 and the other around residues 404-406. The overall structure of the two domains was rather similar: each was based on a six-stranded parallel β-sheet core, flanked on both sides by α-helices. Both domains were typical of what became known as the α/β class of protein tertiary structure [65-67]. However in their details the domains differed to a considerable extent. For example, although both β-sheets have 6 parallel strands, the strand order was different. The order in the N-terminal domain was CDBAEF, but the C-terminal domain was CBADEF (corresponding to IHGJKL according to PGK nomenclature, cf. Figure 1A). The minimal difference could be assigned to the position of strand D that was placed second in the N-terminal sheet but fourth in the C-terminal sheet. In addition, the positions of the α-helices and connecting loops of chain were quite different in detail. Figure 1A illustrates the steric structure of PGK molecule with the positions and labelling of the secondary structural elements, as determined with the horse muscle enzyme.

One remarkable feature of the structure is the incorporation of the C-terminal helix and tail into the N-domain as illustrated by the ribbon diagram in Figure 1B. This type of "arm-domain" interaction is a special characteristic of the molecule and up to now several such examples are accumulated among other proteins. This structural organisation is thought to have general importance in enhancing domain-domain interactions by assisting in transmission of conformational effects from one domain to the other [68].

This molecular structure obtained with horse muscle PGK has been first confirmed by the X-ray study of yeast PGK [5] and later by crystal structures of PGK-s from the bacteria *Bacillus stearothermophilus*, Bs [71] and from pig muscle [15, 72]. The superposition of the Cα-traces of horse muscle (black), pig muscle (orange), yeast (grey) and *B. strearothermophilus,* Bs (pink) PGK structures is shown in Figure 1C. These structures are extremely similar, despite of various biological origin, different crystallization conditions (ammonium sulphate or PEG) and state of ligation: only very slight differences in the relative domain orientations can be observed. All of the later crystal structures of PGK (*Thermotoga maritima,* Tm [73]; *Trypanosoma brucei*, Tb [74] and human, Hs [75, 76]) essentially confirmed this fold, but in some cases (as outlined below) extremely different relative domain positions could be observed.

The next major step of PGK structure determination was to incorporate the amino-acid sequence into the molecular model. In the 1970's the determination of the complete sequence of a molecule the size of PGK containing 416 amino acids was a very considerable undertaking using the chemical methods then available. In fact, the sequence of the horse-muscle enzyme made use of the crystallographic model to facilitate the sequence determination [4]. Horse muscle PGK contains 13 methionine residues, which could be used to split the chain into 14 fragments using cyanogen bromide. These cyanogen bromide peptides varied in length between 104 residues for the longest and 6 residues for the shortest.

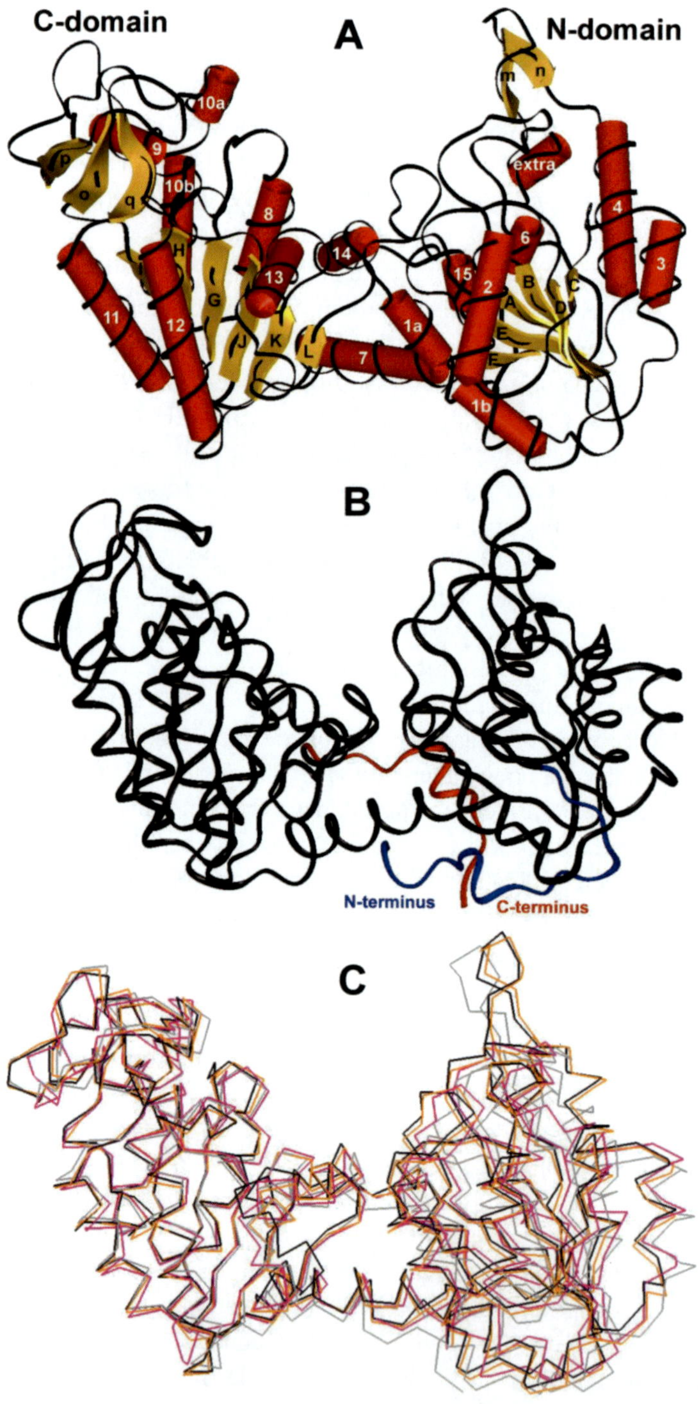

Figure 1. Three-dimensional crystal structure of PGK: (A) Arrangement and numbering of the secondary structural elements, (B) Ribbon diagram with the highlighted N- and C-terminus, (C) Traces of Cα atoms of horse muscle (black) [4], pig muscle (orange) [69], yeast (grey) [70] and *B. stearothermophilus,* Bs (pink) [71] PGK-s superimposed according to the C-domain β-strands.

Making use of the 2.5 Å map obtained by isomorphous replacement the sequenced cyanogen bromide peptides were fitted to the electron density one at a time, starting with the longest and proceeding down to the shortest. The task was facilitated by the fact that the peptides began and ended at methionine residues, which are usually marked by large, strong electron density features in the map. In addition, the location of six of the seven cysteinyl residues in horse muscle PGK were known from the sites of mercury binding from the isomorphous derivatives. Using these markers together with the normal process of the appropriate fitting of large and small, hydrophobic and hydrophilic residues, enabled all peptides to be fitted end-to-end on the map to produce a complete amino-acid sequence derived jointly from chemical and diffraction evidence. Thus, the chemically determined sequence [77, 78] was in complete agreement with that based on X-ray analysis. Confirmation of the sequence was also available from its overall homology with the sequence of yeast PGK determined throughout by conventional chemical methods [79].

1.3. Homology of the Primary Sequences

Since interpretation of several further aspects of the three-dimensional structure requires comparison of PGK crystal structures of various origins, it is needed in advance to consider briefly the differences in their primary structures.

High extent of similarity in PGK primary structures has been suggested even at the early time when only 6 PGK sequences were known: three from mammals (mouse, human and horse), one from yeast and two from *Trypanosomes* [24]. It has been claimed that the homology between the three mammalian PGKs is greater than 96% over the entire 416 amino acid residues. Yeast PGK (containing 415 residues) could be aligned by introducing two gaps at positions 73 and 130 with respect to the mammalian PGK sequences. The overall homology between mammalian and yeast PGKs is about 65%. In order to align the PGK sequences from trypanosomes with the other species several gaps had to be introduced: one between positions 66 and 67 in the mammalian sequence, the others between positions 136 and 152 in the trypanosome sequence. The overall homology is found to be only 44%, but if conservative substitutions are included the homology is increases to 65%. It is also recognised that most of the conserved positions fall into the twelve β-sheets and adjacent peptide region that form the inner loops between the two domains. The homology of this region between mammalian and yeast PGKs is greater than 94%. It will be shown below that this interdomain region constitutes the substrate binding cleft, thus the active site region is evolutionary highly conserved.

The evolutionary conservation of PGK sequences become more obvious as soon as the primary sequences of 22 PGKs, isolated from species ranging from human to bacteria and plant sources, have been determined (cf. Watson et al. [25] and references therein). The alignment of these sequences has been greatly helped by the information of the tertiary structures known at that time. The alignment of these sequences showed that 62 amino acid residues (i.e., approximately 15% of the average size of PGK sequence) are totally conserved. Approximately 30% of the remaining residues cannot be considered as invariant, but the changes involved are all of a conservative nature.

```
                                         βA                         α1A        α1B          βB
Pig(X-ray)      1  SLSNKLTLDK LDVKGKRVVM RVDFNVPMKN NQITNNQRIK AAIPSIKFCL DNGAKSVVLM  60
Pig(DNA)        1  SLSNKLTLDK LDVKGKRVVM RVDFNVPMKN NQITNNQRIG AAIPSIKFCL DNGAKSVVLM  60
Horse           1  SLSNKLTLDK LNVKGKRVVM RVDFNVPMKN NQITNNQRIK AAVPSIKFCL DDGAKSVVLM  60
Human(X-ray)    1   LSNKLTLDK LDVKGKRVVM RVDFNVPMKN NQITNNQRIK AAVPSIKFCL DNGAKSVVLM  60
Human(DNA)      1  SLSNKLTLDK LDVKGKRVVM RVDFNVPMKN NQITNNQRIK AAVPSIKFCL DNGAKSVVLM  60
T.brucei        5  ----EKKSINE CDLKGKRVLI RVDFNVPVKN GKITNDYRIR SALPTLKKVL TEGG-SCVLM  60
T.maritima      1    MEKMTIRD VDLKGKRVIM RVDFNVPVKD GVVQDDTRIR AALPTIKYAL EQGAK-VILL  57
B.stearo        1    MNKKTIRD VDLRGKRVFC RVDFNVPMEQ GAITDDTRIR PALPTIRYAL EHGAK-VILA  57
Yeast           1  SLSSKLSVQD LDLKDKRVFI RVDFNVPLDG KKITSNQRIV AALPTIKYVL EHHPRYVVLA  60
                                          * * * *         *

                                                             α2        βC        α3
Pig(X-ray)     61  SHLGRPDGIP M--------- ------PDKY SLEPVAVELK SLLGKDVLFL KDCVG-PEVE 104
Pig(DNA)       61  SHLGRPDGIP M--------- ------PDKY SLEPVAGELK SLLGKDVLFL KDCVG-PEVE 104
Horse          61  SHLGRPDVGP M--------- ------PDKY SLQPVAVELK SLLGKDVLFL KDCVG-PEVE 104
Human(X-ray)   61  SHLGRPDGVP M--------- ------PDKY SLEPVAVELK SLLGKDVLFL KDCVG-PEVE 104
Human(DNA)     61  SHLGRPDGVP M--------- ------PDKY SLEPVAVELK SLLGKDVLFL KDCVG-PEVE 104
T.brucei       61  SHLGRPKGKP MAQAGKIRST GGVPGFQQKA TLKPVAKRLS ELLLRPVTFA PDCLNAADV- 119
T.maritima     58  SHLGRPKGGP ---------- ------SPEF SLAPVAKRLS ELLGKEVKFV PAVVG-DEVK 100
B.stearo       58  SHLGRPKGKV ---------- ------VEEL RLDAVAKRLG ELLERPVAKT NEAVG-DEVK 100
Yeast          61  SHLGRPNGER ---------- ------NEKY SLAPVAKELQ SLLGKDVTFL NDCVG-PEVE 103
                    *

                   α3        βD      α(extra)        βm     βn         α4
Pig(X-ray)    105  KACADPAAGS VILLENLRFH VEEE------ -GKGKDASGS KVKADPAKIE AFRASLSKLG 157
Pig(DNA)      105  KACADPAAGS VILLENLRFH VEEE------ -GKGKDASGS KVKADPAKIE AFRASLSKLG 157
Horse         105  KACADPAAGS VILLENLRFH VEEE------ -GKGKDASGN KVKAEPAKIE TFRASLSKLG 157
Human(X-ray)  105  KACANPAAGS VILLENLRFH VEEE------ -GKGKDASGN KVKAEPAKIE AFRASLSKLG 157
Human(DNA)    105  KACANPAAGS VILLENLRFH VEEE------ -GKGKDASGN KVKAEPAKIE AFRASLSKLG 157
T.brucei      120  --VSKMSPGD VVLLENVRFY KEE-GSKKAK D--------- --------RE AMAKILASYG 159
T.maritima    101  KAVEELKEGE VLLLENTRFH PGE------- ---------- -----TKNDP ELAKFWASLA 138
B.stearo      101  AAVDRLNEGD VLLLENVRFY PGDE-----K N--------- --------DP ELAKAFAELA 138
Yeast         104  AAVKASAPGS VILLENLRYH IEEEGSRKV- D-------GQ KVKASKEDVQ KFRHELSSLA 155
                                   * *

                   βE      α5        α6              βF         α7                βG
Pig(X-ray)    158  DVYVNDAFGT AHRAHSSMVG V--NLPKKAG GFLMKKELNY FAKALESPER PFLAILGGAK 215
Pig(DNA)      158  DVYVNDAFGT AHRAHSSMVG V--NLPKKAG GFLMKKELNY FAKALESPER PFLAILGGAK 215
Horse         158  DVYVNDAFGT AHRAHSSMVG V--NLPQKAG GFLMKKELNY FAKALESPER PFLAILGGAK 215
Human(X-ray)  158  DVYVNDAFGT AHRAHSSMVG V--NLPQKAG GFLMKKELNY FAKALESPER PFLAILGGAK 215
Human(DNA)    158  DVYVNDAFGT AHRAHSSMVG V--NLPQKAG GFLMKKELNY FAKALESPER PFLAILGGAK 215
T.brucei      160  DVYISDAFGT AHRDSATMTG IPKILGNGAA GYLMEKEISY FAKVLGNPPR PLVAIVGGAK 219
T.maritima    139  DIHVNDAFGT AHRAHASNVG IAQFIP-SVA GFLMEKEIKF LSKVTYNPEK PYVVVLGGAK 197
B.stearo      139  DLYVNDAFGA AHRAHASTEG IAHYLP-AVA GFLMEKELEV LGKALSNPDR PFTAIIGGAK 197
Yeast         156  DVYINDAFGT AHRAHSSMVG F--DLPQRAA GFLLEKELKY FGKALENPTR PFLAILGGAK 213
                        * *    *                          *                 *     ** *

                      α8        βH        α9                           α10
Pig(X-ray)    216  VADKIQLINN MLDKVNEMII GGGMAFTFLK VLNNMEIGTS LFDEEGSKIV KDLMSKAEKN 275
Pig(DNA)      216  VADKIQLINN MLDKVNEMII GGGMAFTFLK VLNNMEIGTS LFDEEGSKIV KDLMSKAEKN 275
Horse         216  VADKIQLINN MLDKVNEMII GGGMAFTFLK VLNNMEIGTS LFDEEGAKIV KNLMSKAEKN 275
Human(X-ray)  216  VADKIQLINN MLDKVNEMII GGGMAFTFLK VLNNMEIGTS LFDEEGAKIV KDLMSKAEKN 275
Human(DNA)    216  VADKIQLINN MLDKVNEMII GGGMAFTFLK VLNNMEIGTS LFDEEGAKIV KDLMSKAEKN 275
T.brucei      220  VSDKIQLLDN MLQRIDYLLI GGAMAYTFLK AQGY-SIGKS KCEESKLEFA RSLLKKAEDR 278
T.maritima    198  VSDKIGVITN LMQKADRILI GGAMMFTFLK ALGK-EVGSS RVEEDKIDLA KELVEKAKEK 256
B.stearo      198  VKDKIGVIDN LLEKVDNLII GGGLAYTFVK ALGH-DVGKS LLEEDKIELA KSFMEKAKEK 256
Yeast         214  VADKIQLIDN LLDKVDSIII GGGMAFTFKK VLENTEIGDS IFDKAGAEIV PKLMEKAKAK 273
                     *                   *  *          *
```

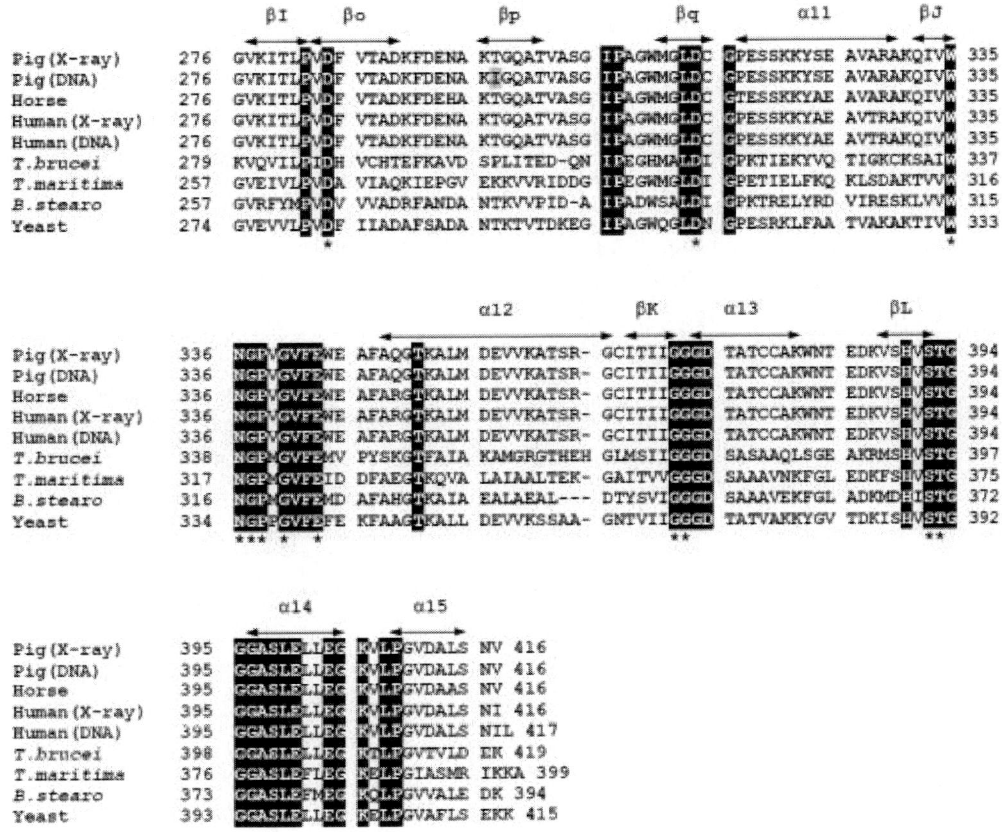

Figure 2. Sequence alignment of PGKs discussed in this volume: The identical residues are highlighted in black background and the completely conserved residues are labelled with black stars. As a comparison the DNA sequences of pig and human muscle PGKs are also aligned: the differences from the crystal structure are highlighted in grey background (only 3 places).

At the year of 2010 about 1100 PGK sequences are found in ExPASy Molecular Biology Server. After making alignments only 32 totally conserved residues are counted, but as many as 122 residues (about 30 % of the whole molecule) are found to be invariant in at least 90 % of the sequences. Thus, this large scale analysis has approved the conservative nature of PGK sequence.

Figure 2 shows the sequence alignment of those PGK species, crystal structures of which have been determined at high resolution up to 2010 and served the bases of further structural information described below. For pig muscle PGK the amino acid sequence was derived from its various X-ray structures [7, 15, 69, 72, 80]. This sequence deviates only in three positions (K40G, V82G and T297I) from the sequence derived from DNA sequence data. In case of the recently determined X-ray structure of HsPGK [76] the agreement with the DNA sequence is even more complete.

1.4. Binding Mode of the Nucleotide Substrates

Preliminary Studies

The availability of a high-resolution model of PGK, complete with side-chain information, enabled useful studies of substrate binding and catalytic activity to begin. The initial studies were on the binding of the true substrates MgADP and MgATP, and their active manganese homologues, since these ligands could be diffused into either the horse muscle [4] or yeast PGK [5] crystals without causing damages, although in the latter case no more than 20 % of occupancy had been reached. Thus, the binding experiments to horse muscle PGK crystals provided more definitive data about the way of nucleotide binding. Both nucleotide substrates bound to the same site on inner surface of the C-domain opposite the N-domain but some distance from it. The location of nucleotide binding will be illustrated below by the series of Figures 3A-3D, prepared from the data of later crystallographic studies. Part of the structures also contained bound PG, the other substrate, the binding characteristics of which will be described later (cf. Chapter I. 1.6.).

During the early studies with horse muscle [4] and yeast PGK-s [5] the difference maps of the bound MgADP and MgATP at 3 Å resolution and of the bound MnATP at 2.5 Å resolution allowed to determine the conformation of the bound nucleotides. In both cases the base is in the *anti* conformation with respect to the sugar (χ ~80°); the sugar pucker is C2'-*endo*; and the conformation about the exocyclic C4'-C5' bond is *gauche* (-) (or *trans-gauche*). This conformation was quite different from the conformation observed in crystals of ATP [81]. The adenine group of the nucleotides was located deep in a narrow crevice in the PGK molecule, its NH_2 group was deepest in the crevice, hydrogen bonded to a main chain carbonyl group (Gly 312:O according to the mammalian PGK sequence numbering[1]). The crevice itself was largely composed of glycine residues, making hydrophobic interactions with the adenine ring. These residues (Gly-s 212, 213, 237, 238, and 340, as well as Val 341) immediately follow β-strands G, H and J of the C-domain core, thus binding of adenine keeps together these β-strands and thereby possibly stabilizes the C-domain structure (for the effect of substrates on PGK conformation cf. Chapter II. 2.5-2.9.). Similar interactions with the equivalent residues of yeast PGK (Gly-s Y210, Y211, Y235, Y236 and Y338, as well as Val Y339) was also observed upon diffusing MnAMP-PNP (an analogue of MgATP [82]) into the crystal soaked in a mother liquor at pH 9 [5]. Later on it became also evident that these residues are all conserved (cf. the sequence alignment in Figure 2). The ribose moiety of the bound nucleotide was located in a shallow depression above the pirrolidine ring of the conserved Pro 338/Y336 in the enzyme's surface, with its free 2'- and 3'-hydroxyl groups each hydrogen-bonded to one of the two oxygen atoms of the carboxyl side-chain of the conserved Glu 343/Y341.

While the interactions with the adenosine parts of both MgADP and MgATP (or its analogue) with the enzyme have been found to be very similar, significant differences have been observed already in the abovementioned initial studies both the location and interactions of their phosphates as well as of the associated metal ion. The only similarity is the hydrogen-

---

[1] Numbering of residues throughout the text refers to the highly similar (98 % identity) mammalian (pig muscle, horse or human) PGK sequence, unless quoted differently, as e.g., Y (yeast), Bs (B*acillus. stearothermophilus*), Tb (*Trypanosoma brucei*) or Tm (*Thermotoga maritima*).

bonding of the α-phosphate of either MgADP or MgATP to the charged amino group of the extended side-chain of Lys 219/Y217. In Mg- and MnADP the metals are equidistant from the α- and β-phosphate and from the carboxyl side-chain of the conserved Asp 374/Y372, and the β-phosphate is located at the amino-terminal end of helix 13, where it may be able to interact with the helix dipole as proposed by Hol et al. [83]. In Mg- and MnATP the interactions are very similar, but different from MgADP: the metal has moved from its position linking the α- and β-phosphates to Asp 374/Y372, showing that although this site is occupied in ADP, it is empty in ATP [4]. Difference map suggests that in ATP the metal ion is associated with the γ-phosphate, but neither phosphate group nor the metal ion is strongly bound to the enzyme. The location of the metal ions in the Mg and MnATP binary complexes of horse muscle PGK is not entirely clear but it appears to differ significantly from either the ADP or the AMPPNP containing complexes described below. The weak electron-density at the γ-phosphate position suggests that this end of the substrate is partially disordered and may point away from the surface of the C-domain in the absence of the phosphoglycerate substrates.

Binding of MgADP

Later crystallographic works has essentially confirmed the initial picture about the details of nucleotide binding to PGK. At the time of starting writing this book the highest resolution (1.65 Å) structure was published about the MgADP binary complex of BsPGK crystallized from PEG [6] (coloured as pink in Figures 3A, 4A and 5A). It might be interesting to note that originally MgATP was included in the crystallization medium, the γ-phosphate of which was possibly split by hydrolysis during the long time of crystal growing and resulted in bound MgADP. This structure has revealed further details, meanwhile essentially supported the originally found binding mode of MgADP, showing the same type of interactions with adenine and ribose rings, as well as the same metal position and co-ordination as for horse muscle PGK.

One slight difference is the ribose ring conformation, namely, the ring adopts an O4'-*endo* conformation, in contrast to the C2'-*endo* conformation observed in the horse muscle PGK nucleotide complexes. The difference in energy between these two conformations is small, however, and the O4'-*endo* conformation is found in a number of small-molecular sugar structures [85, 86]. The Mg-ion is shown in a tetragonal distorted octahedral coordination (Figure 4A). The axial ligands are a single water molecule and one of the carboxylate O atoms of Asp Bs352/374/Tb377[2]. The equatorial ligands are provided by two water molecules and two O atoms from each of the α- and β-phosphates. Most of the coordinating atoms lay between 2.27 and 2.29 Å, characteristic of the interaction distances of the magnesium, which are shorter than the usual H-bond [87]. The aspartate carboxylate O atom lays only 1.97 Å from the Mg ion, reflecting the charged nature of this interaction. The details of interactions of BsPGK with MgADP (coloured as pink) [6] are shown in Figure 4A.

[2] The numbering of the equivalent residues in other relevant PGK structures are also indicated, although the actual statement is derived from BsPGK structure.

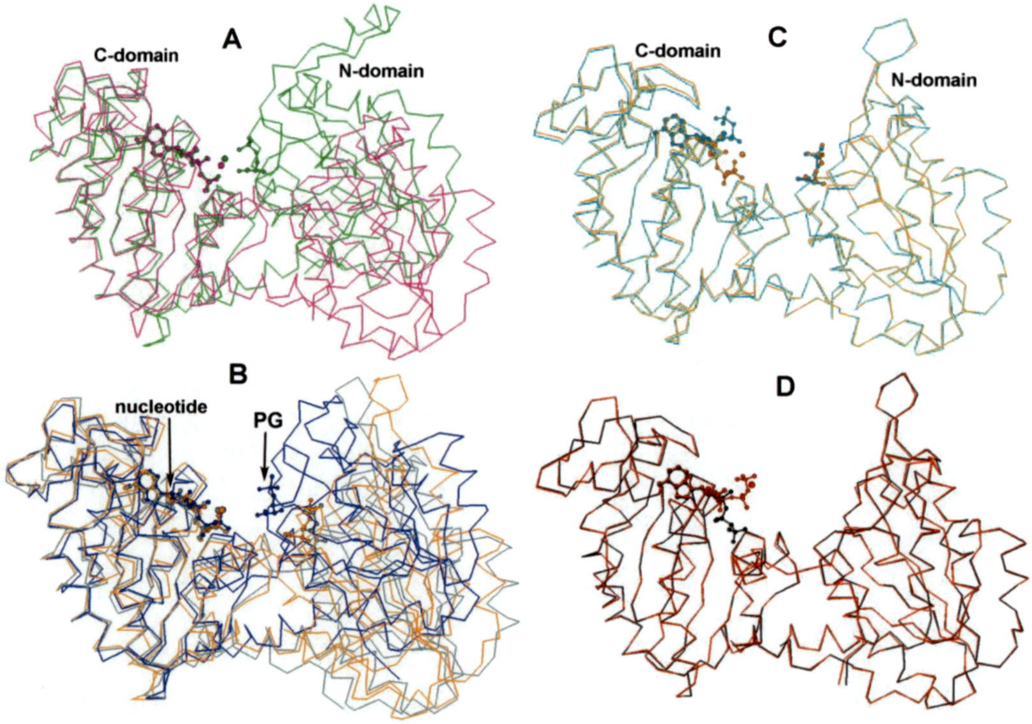

Figure 3. Comparison of various pairs of PGK structures containing bound nucleotides: The traces of Cα atoms are superimposed according to the C-domain β-strands. Ball and stick models represent the bound ligands. A: MgADP binary and MgADP·PG ternary complexes of Bs (pink) [71] and Tb (green) [84] PGKs, respectively, B. Mg(Mn)AMP-PNP·PG ternary complexes of Tm (blue) [73], pig muscle (orange) [69] and Y (grey) [70] PGKs, C: MnAMP-PNP·PG (orange)[69] and MgAMP-PCP·PG (cyan) [72] ternary complexes of pig muscle PGK and D: MgATP (red) and ATP (black) binary complexes [7]of pig muscle PGK.

For comparison, Figure 4B shows MgADP binding to TbPGK (coloured as green), determined later [74, 84]. In general, the interactions of the two nucleotides are strongly similar and only slight differences can be observed, although the latter one is a ternary complex containing additionally bound PG and exhibits largely different domain positions (cf. Figure 3A) that will be discussed later (cf. Chapter I. 1.8. -1.11.). Beside the permanent interaction of Lys Bs201/219/Tb223 side-chain with the nucleotide α-phosphate (noted even in the early studies), the β-phosphate of MgADP also interacts with this side-chain, although in case of TbPGK it is only a weak ionic interaction (Figure 4B). It may be notable that this Lys occupies a particular position in the protein molecule, namely it is situated at the beginning of helix 8, as shown below in Figure 5A. The importance of this location will be discussed later, in relation to the function. Furthermore, the main chain N-atom of Ala Bs196/214/Tb218 of the same protein region forms one identical H-bond with the ribose O-atom in the two structures, but in TbPGK two additional H-bonds exist with O-atoms of the nucleotide α-phosphate (compare Figures 4A and 4B). The other interacting residues are from different structural elements of the C-terminal domain.

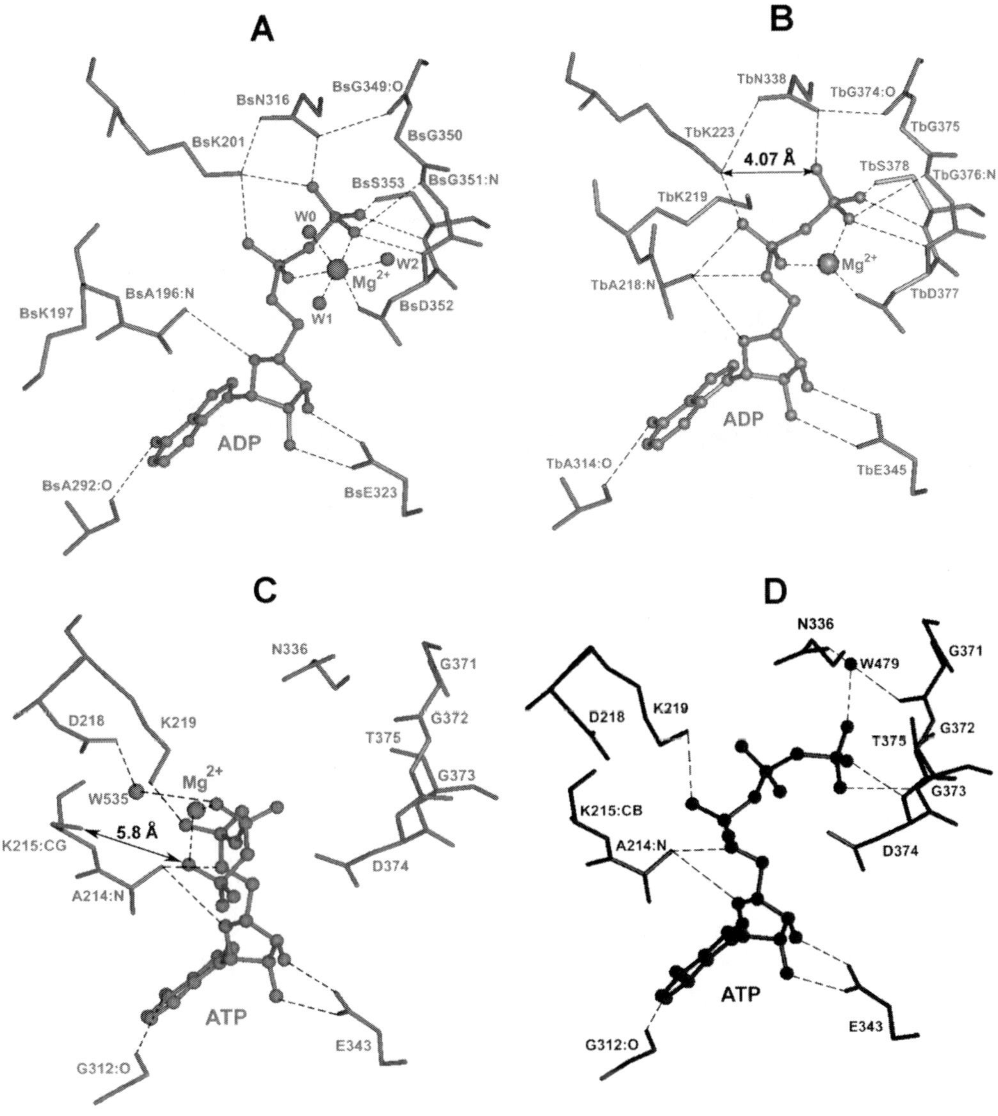

Figure 4. Binding details of the nucleotide substrates (analogues) as bound to various crystal structures: The bound ligands are represented by ball and stick models, while stick models label the interacting side-chains. Dashed lines connect the interacting atoms (the limiting distances are 3.5 and 4.5 Å-s for H-bonds and electrostatic interactions, respectively). A: MgADP in Bs [71], B: MgADP in Tb [84], C: MgATP in pig muscle [7] and D: ATP in pig muscle [7].

One of the β-phosphate O-atom is H-bonded to the amide N-atom of the conserved Asn Bs316/336/Tb338 from the β-strand J (it is not shown here), which in turn is H-bonded to a peptide O-atom of Gly Bs349/371/Tb374 at the N-terminus of helix 13 (cf. also Figure 5A). As a consequence of these interactions with MgADP, a further H-bond is also formed between the side-chains of Lys Bs201/219/Tb223 and the amide CO-group of

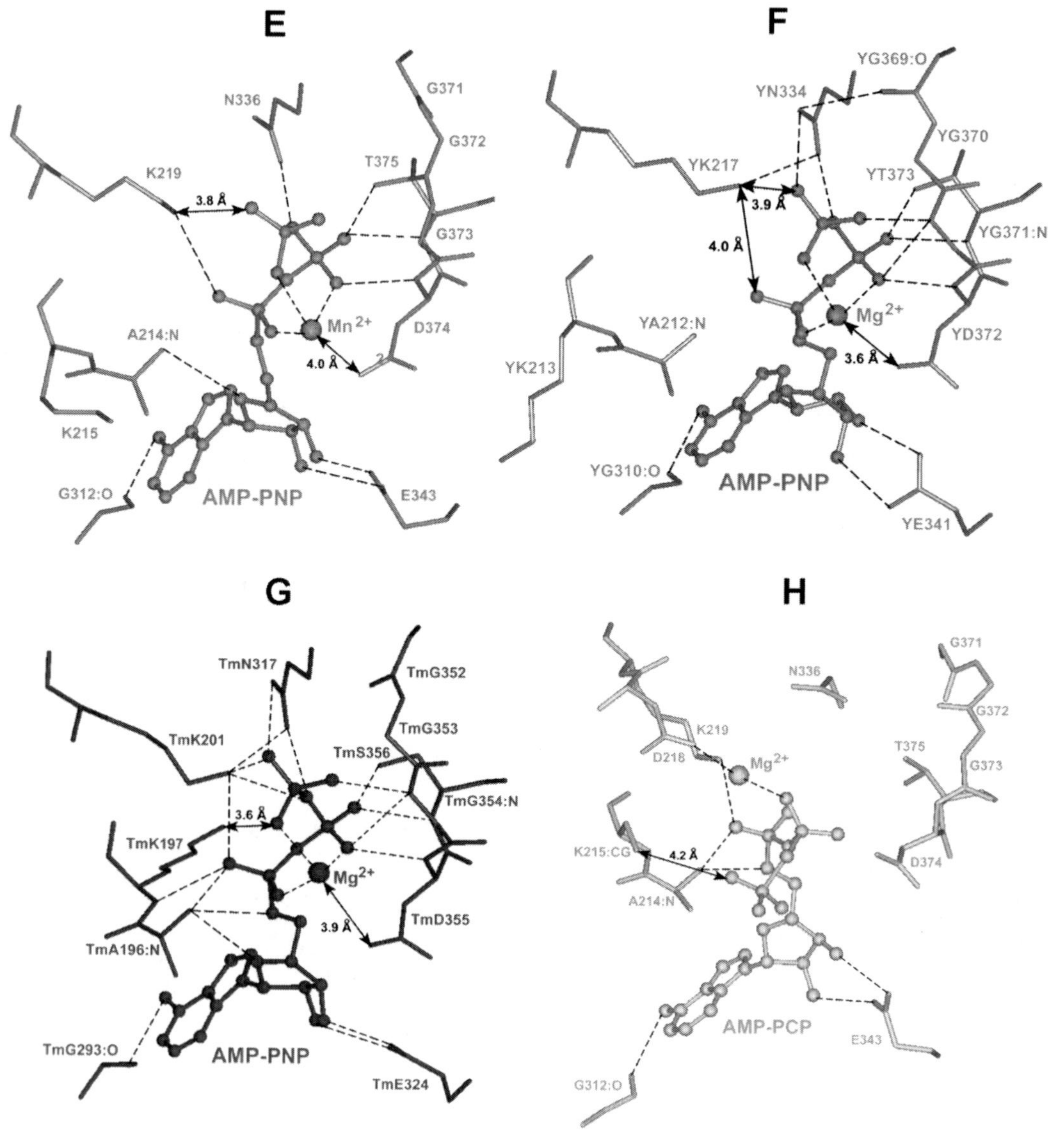

Figure 4. (continuation) Binding details of the nucleotide substrates (analogues) as bound to various crystal structures: The bound ligands are represented by ball and stick models, while stick models label the interacting side-chains. Dashed lines connect the interacting atoms (the limiting distances are 3.5 and 4.5 Å-s for H-bonds and electrostatic interactions, respectively). E: MnAMP-PNP in pig muscle [69], F: MgAMP-PNP in yeast [70], G: MgAMP-PNP in Tm [73] and H: MgAMP-PCP in pig muscle [72] PGK structures.

Asn Bs316/336/Tb338. These interactions among the above side-chains around the bound MgADP are absent in the substrate-free PGK. The hydroxyl group of Ser Bs353/Thr 375/Ser Tb378 from helix 13 makes also a H-bond with another O-atom of the β-phosphate. Finally, the β-phosphate lies close to the N-terminus of helix 13 and interacts with its characteristic Gly-Gly-Gly motif where it is presumably stabilized by the α-helix dipole [83, 88] as well as by H-bond interactions with the main chain amide N-atoms of Gly Bs351/373/Tb376, Asp Bs352/374/Tb377 and Ser Bs353/Thr 375/Ser Tb378.

In summary, the crystal structures of the MgADP binary complex of BsPGK and the ternary complex of TbPGK, containing also bound PG, have revealed a very similar binding modes and interactions of MgADP. The subtle differences in TbPGK, compared to BsPGK, are the stronger interaction (involved the α-phosphate) with Ala Tb218/214/Bs196 and a weaker (ionic) interaction of the β-phosphate with Lys Tb223/219/Bs201 [74, 84]. This binding mode of MgADP is essentially approved by the recently published 1.47 Å resolution crystal structure of HsPGK [76] that also provided all mechanistic details of the active site architecture in the transition state of the enzyme reaction (cf. Chapter IX.). It is also notable that the absolutely conserved, sequentially neighbouring side-chain Lys Tb219/215/Bs197 in either of these structures does not interact with MgADP (Figures 4A and 4B).

The multiple interactions of the N-terminus of helix 13 are characteristics of the structure containing only bound MgADP. This explains the observation that helix 13 is highly ordered in these structures as evidenced by the low crystallographic B-factor values of its backbone atoms (19.5 $Å^2$ in the structure of BsPGK [6]) compared to its relatively high value (53.0 $Å^2$) in the structure of substrate-free PGK (Kovári, Náray & Vas, unpublished). The corresponding crystallographic B-factor values in various PGK crystal structures are summarized in Table 1.

Table 1. List of the crystallographic B-factor values (in $Å^2$)

Substrate complex	Resolution (Å)	Reference	Source	Conformation	Helix 13 backbone	Nucleotide phosphates
No substrate	2.7	unpublished	pig muscle		53.0	-
MgADP	1.65	[6]	*Bacillus stearothermophilus*	open	19.5	16.7
MgATP	2.1	[7]	pig muscle		38.9	55.3
ATP	1.9	[7]	pig muscle		47.2	62.0
PG	2.0	[15]	pig muscle	open	22.3	-
PG·MgAMP-PNP	2.0	[69]	pig muscle		27.5	46.9
PG·MgAMP-PCP	2.5	[72]	pig muscle	(slightly closed)	5.9	53.1
PG·MgAMP-PNP	2.0	[73]	*Thermotoga maritima*	largely closed	27.1	32.6
PG·MgADP	2.5	[74]	*Trypanosoma brucei*	completely	31.1	30.8
PG·MgADP·MgF_3	1.47	[76]	human	closed	5.5	7.4

Binding of Mg(Mn)AMP-PNP

For a long time no structure could be obtained for the true crystalline binary MgATP complex produced by co-crystallization of PGK with this substrate. Thus, the binding mode of MgATP as suggested from the early soaking experiment with horse muscle PGK, could not be confirmed. On the other hand, three different PGK crystal structures have been determined containing the metal complexes of the bound analogue adenylylimidodiphosphate (AMP-PNP), which differs from ATP only in a single atom, namely at the connection between the β- and γ- phosphate it possesses an –NH- rather than –O- link. This link however is much less reactive to hydrolysis and phosphorolysis [82] and found to be no substrate, but inhibitor for PGK catalysis [72]. The various PGK structures obtained with Mg(Mn)AMP-PNP have yielded coherent results, independent of the enzyme source, ligation with the other substrate,

PG, and even of the conformational state (the relative domain positions) of the enzyme [69, 70, 73] (Figure 3B).

In the ternary complex structures (with bound PG) the bound MnAMP-PNP to pig muscle PGK [69] (orange in Figure 3B; Figure 4E), the bound MgAMPPNP to a mutant yeast PGK [70] (grey in Figure 3B; Figure 4F) and the bound MgAMPPNP to TmPGK [73] (blue in Figure 3B; Figure 4G) (2.0 Å, 2.4 Å and 2.0 Å resolutions, respectively) are found at the same site in the C-terminal domain as the nucleotide substrates in their binary complexes (Figures 3A and 3D). Before comparison these structural data, it may be notable that no differences can be expected solely due to substitution of Mg^{2+} into Mn^{2+}, since it is known that the bivalent metal ion, most commonly magnesium, required by PGK catalysis, can be replaced by other metals under physiological conditions, of which the most important one is manganase [3].

As in the complexes with the other nucleotides, the adenine ring of Mg(Mn)AMP-PNP is bound in a hydrophobic slot and the ribose is in a shallow depression in the enzyme surface with its 2'- and 3'-hydroxyls hydrogen bonded to the carboxylate group of Glu 343/Y341/Tb345/Bs323/Tm324 (Figures 4E, 4F and 4G). The ribose rings are reported to be in the energetically only slightly different O4'-*endo* and C2'-*endo* conformations in the pig muscle and yeast PGK structures, respectively. However, unlike the above-described observation of MgATP binding to horse muscle PGK, the metal-ion interacts with all three-phosphate groups of AMP-PNP. This type of α,β,γ-tridentate coordination of a divalent metal-ion by a triphosphate group has been observed in the crystal structures of the magnesium and manganese complexes of protein-free ATP [89, 90]. Furthermore, in the complex with PGK, in contrast to MgADP where the metal is bound strongly to the carboxylate group of Asp Bs352/Tb377/374/Tm355/Y372 (cf. Figures 4A and 4B), here the metal-ion makes no or only very loose contact with the equivalent residue of pig muscle (Asp 374) [69], Tm (Asp Tm355) [73] and of the yeast (Asp Y372) [70] PGK structures (Figures 4E, 4F and 4G, respectively). This change in co-ordination between MgADP and MgAMP-PNP involves a shift of about 1.5 Å in the position of the metal ion, as illustrated in Figure 5A, where the nucleotides are superimposed according to the ring atoms. Such movement might be relevant in respect of the catalysis and will be discussed later (cf. Chapter VIII.).

In comparison the details of interaction with AMP-PNP in the structures of pig muscle, yeast and TmPGK-s, the following observations can be made. The α-phosphate of Mg(Mn)AMP-PNP, similar to all nucleotides, makes more or less uniform ionic interactions with Lys 219/Y217/Tm201/Bs201/Tb223. The β-phosphate, similar to MgADP, forms multiple H-bonds with the main-chain N-atoms of Asp 374/Y372/Tm355/Bs352/Tb377 and Thr 375/Y373 or of the equivalent Ser Tm356/Bs353/Tb378 as well as the OH side-chain atom of the latter residues at the N-terminus of helix 13 (for the secondary structural elements in the surroundings of the bound nucleotide cf. Figure 5A). These interactions of AMP-PNP are possibly independent of the presence of the other substrate, PG, since the β-phosphate could be best assumed in a very similar position from the low (3 Å) resolution map of the binary complex of TbPGK with MgAMP-PNP [84].

The unique -NH-link between the β- and γ-phosphates of AMP-PNP in all pig muscle, yeast and TmPGK structures is noted to form H-bond with the amido CO-group of a conserved Asn (Asn 336/Y334/Tm317/Bs316/Tb338) in β-strand J. Namely, the NH-bridge *donates* a proton, which *is accepted* by the amido CO-group. One may expect a reverse-type of interaction with the natural substrate, MgATP (similar to the one with MgADP), i.e., the *acceptance* of a proton by the equivalent O-bridge of ATP from the *donating* -NH_2 of the amido group of the same Asn. A possible consequence of this change would be that, with an ATP substrate, the amido–CO of Asn 336 would be free to make hydrogen bond to the amino group of Lys 219/Y217/Tm201/Bs201/Tb223 that could help to stabilise its ion-pair interaction with the ATP α-phosphate. In this respect, however, the crystallographic data on AMP-PNP binding is not coherent. In fact, a much weaker interactions of the α-phosphate of AMP-PNP with this lysyl side-chain (characterized by 3.5 and 4.0 Å distances, respectively) has been observed in both pig and yeast PGK structures, although the interaction of the amido-CO group of Asn 336 with the ε-amino-group of Lys 219 is absent only in pig PGK [69] while in yeast PGK a strong H-bond does exist between the equivalent atoms of Asn Y334 and Lys Y217 [70] (Figures 4E and 4F). Thus, binding of Lys 219/Y217/Tm201 to the nucleotide α-phosphate cannot be simply considered as a consequence of the interaction of this residue with Asn 336/ Y334/Tm317. Furthermore, this criteria does not hold either in the crystal structure of Tm PGK, where Lys Tm201 forms strong interaction not only with the α-phosphate, but also with the γ-phosphate of AMP-PNP, in spite of the fact that the amido-CO group of Asn Tm317 is involved in interaction with the NH-group of AMP-PNP (Figure 4G). This interaction, however, similarly to yeast PGK, does not interfere with the simultaneous interaction of the amido CO-group of Asn Tm317 with the ε-amino-group of Lys Tm201. On the other hand, in both structures the free NH_2-group of the same Asn forms a good H-bond with one of the O-atom of the nucleotide γ-phosphate (Figures 4F and 4G). Thus, in the crystal structure of Tm PGK Lys Tm201 and Asn Tm317 are interacted both directly and indirectly, since each of them are involved in multiple interactions with the nucleotide phosphates [73] (Figure 4G). These interactions are partially also exist in yeast PGK [70] (Figure 4F). In contrast, only a single weak H-bond (3.48 Å) of the equivalent Lys 219 with an O-atom of the α-phosphates of AMP-PNP as well as a single H-bond of the amido CO-group of Asn 336 with the imido-group of the nucleotide analogue exist in pig muscle PGK structure, the interaction of Lys 219 with the γ-phosphate of the nucleotide is weaker (Figure 4E) [69].

There are also further differences among the binding modes of AMP-PNP to PGK. While the backbone N-atom of Ala Tm196/214/Y212 in the loop preceding helix 8 (Figure 5A) forms triple H-bonds with the nucleotide in the structure of TmPGK (Figure 4G), only a single bond is formed with the ribose O-atom in pig muscle PGK (Figure 4E) and completely absent in the yeast enzyme (Figure 4F). Another striking difference in the same loop region is the double interaction of Lys Tm197/215/Y213 with AMP-PNP in case of TmPGK (the side-chain of Lys Tm197 with the γ-phosphate O-atom as well as the peptide N-atom of this residue with the α-phosphate O-atom), which was observed exclusively only in this structure (Figure 4G). The same conserved Lys residue is surprisingly located far away and not involved in any interactions with AMP-PNP bound to either pig muscle (Figure 4E) or to yeast (Figure 4F) PGK-s.

Thus, besides the largely similar features there are striking differences in binding details of AMP-PNP to PGKs in different crystal structures. It will be pointed out later that these differences are not originated from the existing differences between the investigated PGK sources (e.g., pig *vs.* yeast or bacterium), but are rather correlated with the conformational state of the enzyme (Chapter I. 1.8.). Solely from these structural observations with AMP-PNP it would be difficult to derive the exact roles of Asn 336/Y334/Tm317, Lys 219/Y217/Tm201 and Lys 215/Y213/Tm197 in binding of ATP, when it replaces AMP-PNP. Thus, the question is still remained open, whether the roles and positions of the equivalent Asn 336/Bs316/Tb338 and Lys 219/Bs201/Tb223, as observed in MgADP-binding to Bs and TbPGKs, can be applicable in MgATP-binding [69].

Meanwhile, further structural information has been also accumulated concerning the binding of another ATP-analogue, AMP-PCP and about the binding of MgATP, itself.

Binding of MgAMP-PCP

Co-crystallization of pig muscle PGK with MgAMP-PCP and PG has yielded crystals suitable for X-ray studies. Crystallographic work on these crystals has provided a 2.5 Å resolution structure [72]. AMP-PCP, i.e., β,γ-methylene adenosine 5'-triphosphate, besides AMP-PNP, is a commercially available and widely used non-reactive ATP-analogue. Both analogues are equally applied to mimic the interactions of ATP with various proteins. While in AMP-PNP an imido-, in AMP-PCP a methylene-group replaces the O-atom that connect the β- and γ-phosphates of ATP. In spite of the great similarity of the structure of AMP-PCP to AMP-PNP, striking differences in their binding modes have been observed (Figures 3C, 4E and 4H). Although in one case MnAMP-PNP, while in other case MgAMP-PCP has been used, as argued above, substitution of Mg^{2+} into Mn^{2+}, itself, should have no influence the mode of nucleotide binding. On this basis the most obvious conclusion would be that the observed differences in the binding of AMP-PCP and AMP-PNP is most possibly due to the inability of the CH_2-group to be involved in H-bonding and electrostatic interactions. However, it will be pointed below, that this explanation may not be satisfactory.

The adenine ring and the ribose moiety binds as usual [4, 6, 69-71, 73, 84], but the phosphate chain of AMP-PCP folds back to the C-domain (Figure 3C), in contrast to the previous observations with either MgATP [4] or Mg(Mn)AMP-PNP [69, 70, 73]. Although, the averaged B-factor value of the whole AMP-PCP molecule is relatively high (40.1 $Å^2$), those of the adenine moiety and the ribose part are only 18.4 and 31.1 $Å^2$, respectively, which argue in favour of full occupation of the nucleotide site. Keeping this in mind, the much higher B-factor (53.1 $Å^2$) of the phosphate chain and the attached Mg^{2+} ion indicates an increased mobility of this part of the molecule. On the other hand, the electron density map shows clear orientation of the triphosphate chain. This orientation is surprisingly new and is not seen in any of the previously determined nucleotide containing complexes [4, 6, 69-71, 73, 84]. Details of MgAMP-PCP binding to pig muscle PGK are seen in Figure 4H.

The identification of Mg^{2+} in this structure requires some caution since its electron density is rather weak and its B-factor is rather high (54 $Å^2$). Furthermore, the coordination sphere of the postulated Mg^{2+} is only partially occupied by two ligands: the carboxylate O-atom of the conserved Asp 218/Y216/Tm200/Bs200/Tb222 and an O-atom of the nucleotide β-phosphate. Despite of these uncertainties, the assumed position of Mg^{2+}-ion is approved by its characteristic coordination distances [87]. In addition, the position of the two coordinating ligands corresponds rather to an octahedral coordination sphere (characteristic of Mg^{2+}), than to a tetrahedral one (characteristic to a water molecule).

Accordingly, the ligation of the nucleotide β-phosphate (through Mg^{2+}) to Asp 218/Y216/Tm200/Bs200/Tb222 represents a new type of nucleotide-metal-protein interaction of PGK (Figure 4H), compared to the previously observed binding modes. Interestingly, the completely conserved Lys 215/Y213/Tm197/Bs197/Tb219, which is part of the nucleotide-binding region (and shown above to interact with AMP-PNP bound to TmPGK), does not interact with this analogue, as no significant density could be observed around the terminal part of the side-chain indicating its mobility in this structure. The position of Lys 215 (the distance of its CG atom from an O-atom of the γ-phosphate is 4.2 Å), otherwise, would allow an interaction with the closely located nucleotide γ-phosphate, which does not form any contact with the protein in this structure. The other conserved lysine (Lys 219/Y217/Tm201/Bs201/Tb223) that affects nucleotide binding has a well-defined electron density map around its side-chain. Its ε-amino-group interacts with one of the O-atoms of the α-phosphate with a distance of 3.1 Å (Figure 4H) in a way similar to all other nucleotide complexes of PGK. The interacting residues are located close to the N-terminus of helix 8 as seen from Figure 5A showing also the secondary structural elements in the vicinity of the superimposed nucleotides.

While the phosphate chain interacts with helix 8, practically no interactions are observed with helix 13, in contrast to the other complexes containing bound Mg(Mn)AMP-PNP [69, 70, 73] or bound MgADP [6, 84]. The absence of interaction with helix 13 correlates well its elevated flexibility as evidenced by the relatively high average B-factor values of its backbone atoms (55.9 $Å^2$) in the MgAMP-PCP bound structure (Table 1). Several studies with PGK indicate the importance of helix 13, and support the observation that its mobility can be reduced by binding of substrates, as discussed above in case of MgADP binding. Similar to MgADP, but in contrast to MgAMP-PCP, binding of Mn(Mg)AMP-PNP to either pig muscle or TmPGKs is accompanied by interaction of nucleotide phosphate O-atoms with the N-terminus of helix 13 and thereby possibly extending and stabilizing the H-bond network of this helix. In accordance, the mobility of helix 13 in the AMP-PNP bound structures is relatively restricted as indicated by its lower B-factor value (27.5 and 27.1 $Å^2$, respectively, in case of pig muscle and Tm PGKs).

Besides the crystallographic B-factor, the flexibility of helix 13 can be also characterized by the reactivity of the thiol groups located in this helix. Pig (as well as horse and human) PGK are known to possess two fast reacting thiol groups, which have been identified as Cys 378 and Cys 379, i.e., their position is within helix 13 [91]. The reactivity of these groups is sensitively reduced by the presence of substrates [92, 93]. The electron density map of the pig enzymes shows that when the helix is ordered Cys 378 is partially, and Cys 379 is totally buried, within the molecule. Neither of these side-chains is accessible when the helix is ordered, although presumably both are accessible when the helix is disordered. Thus, a close relationship between flexibility of helix 13, orientation of the bound nucleotide phosphates and thiol-reactivity apparently holds in the different nucleotide-containing complexes of pig muscle PGK [7, 72] (the results of thiol-modification is discussed in Chapter II. 2.5.).

The most interesting feature of the AMP-PCP bound structure is the unprecedented orientation of the bound nucleotide phosphates. The strikingly different binding modes of the metal-complexes of AMP-PCP and AMP-PNP are apparently not the result of a difference in the crystallization protocols (co-crystallization *vs.* diffusion into the pre-grown crystal). At least, the binding mode of AMP-PNP depends neither on the way of crystal growing nor on the origin of the enzyme and similarly, it is independent of the conformational state of the enzyme as exemplified by comparing panels E, F and G in Figure 4.

To interpret the binding differences, at first one should consider that the phosphate chain of AMP-PCP contains a methylene group, while of AMP-PNP contains an imido-group in place of the O-atom between β- and γ-phosphates of ATP. These local changes influence differently both the electrostatic and H-bonding potential of the two analogues. These inherent differences may be reflected not only in different energies, but also in different modes of the binding of the two analogues. The latter can be facilitated by the inherent flexibility of their phosphate chain. In fact, the phosphate chain can occupy various positions relative to the adenine and ribose rings, as also shown by superimposing the analogues bound to various other proteins [72]. As for PGK, independent inhibitory and binding studies, carried out at different pH-s, indicated the importance of the electrostatic interactions in binding of the analogues. Namely, weaker binding of AMP-PCP, compared to AMP-PNP, has been detected at pH 7.5 (or lower), corresponding to the pH of X-ray data collection, which is entirely consistent with the higher pK_a of the last dissociating protons of AMP-PCP phosphates as compared to that of AMP-PNP. Thus, AMP-PCP phosphates may form a less favourable electrostatic interaction with the positively charged N-terminus of helix 13.

The above considerations, however, may not offer an entirely satisfactory explanation for the different orientation of AMP-PCP phosphates. In fact, solution experiments at saturating concentrations of the nucleotides suggested similar binding modes of MgAMP-PCP and MgAMP-PNP in their both binary complexes [72]. Namely, both nucleotide analogues exert similar extent of protection of the reactive thiol groups present in helix 13 of pig muscle PGK. The different binding mode of AMP-PCP phosphates, however, was observed in the crystal in the presence of the bound substrate, PG. Solution binding data in the presence of PG, in fact, are in good agreement with this: it is well conceivable that PG (as a result of substrate antagonism, discussed later, cf. Chapter VII.) abolishes more easily the weaker interactions of the phosphates of MgAMP-PCP with the protein than the stronger interactions of MgAMP-PNP. Thus, AMP-PCP phosphates may move with a higher probability to the novel site. Although, no structural data on MgAMP-PCP binding in its binary complex with PGK exist, the structure of the ternary complex indicates increased flexibility of AMP-PCP phosphates (the average B-factor is as high as 53.1 $Å^2$) and the solution binding data also argue in favour of the movement of the flexible phosphate chain of the nucleotide from one alternating binding site (at helix 13) to the other one (at helix 8), cf. Figure 5A.

The question arises whether the assumed fluctuation of the AMP-PCP phosphates between the two alternative sites has any possible relevance to the binding of the physiological substrate, MgATP, itself. The crystal structure of the binary complex with MgATP has allowed closer insight into this problem.

Binding of MgATP

The true crystal structure of MgATP binary complex has been determined at 2.1 Å resolution using crystals obtained by co-crystallisation of pig muscle PGK with MgATP. In order to clarify the role of Mg^{2+} in the binding, for comparison the structure of the binary complex with the metal-free ATP has also been solved at 1.9 Å resolution [7].

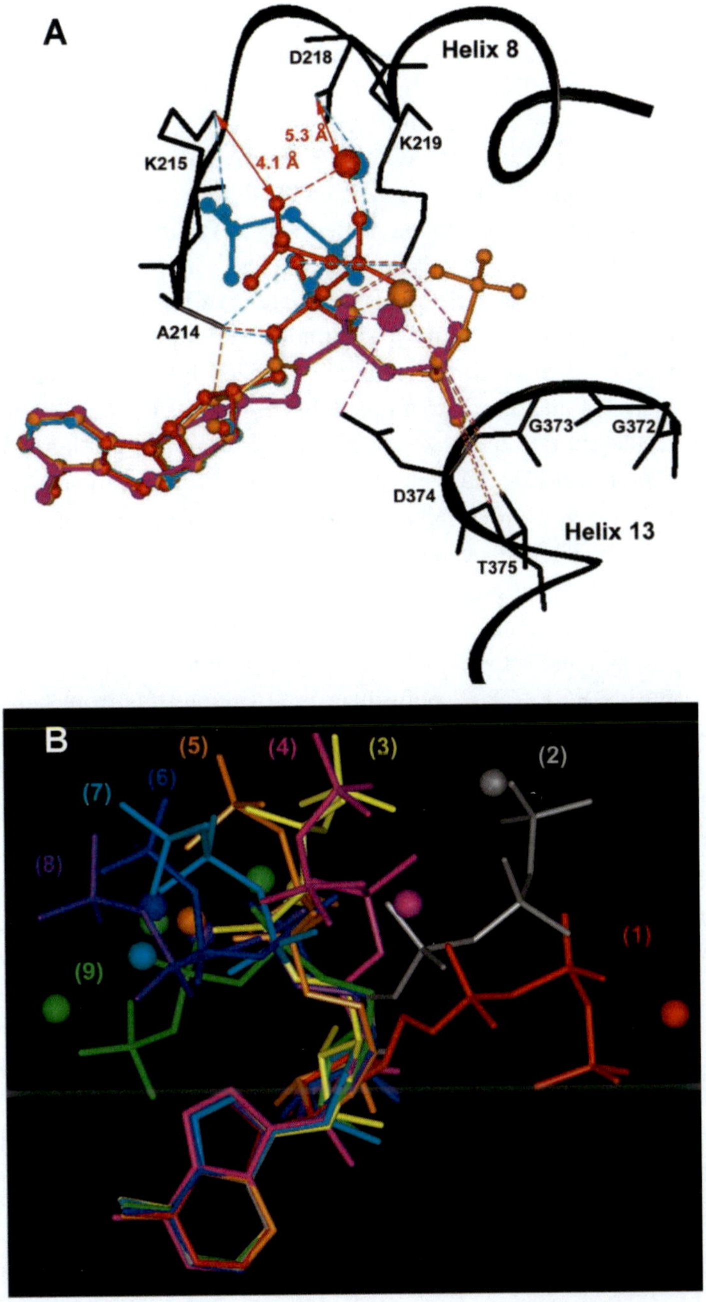

Figure 5. Superimposition of various nucleotides as bound to PGK (A) and of MgATP as bound to various other proteins (B): The nucleotides were superimposed according to all atoms of adenine and ribose rings. In (A) MgADP (violet) as bound in Bs [71], MgAMP-PCP (cyan) as bound in pig muscle [72] and MgATP (red) as bound in pig muscle [7] PGK-s are superimposed onto MnAMP-PNP (orange) as bound in pig muscle PGK [69]. Protein surroundings are shown by black ribbons and by the interacting side-chains (stick models). In (B) MgATP-s are superimposed as bound to (1, red) pig muscle PGK [7], (2, grey) an ABC transporter ATPase domain [94], (3, yellow) casein kinase [95], (4, violet) mevalonate kinase [96], (5, orange) cyclin-dependent protein kinase [97], (6, blue) tryptophanyl-tRNA synthetase [98], (7, cyan) a universal stress protein-like hypothetical protein [99], (8, purple) phosphopantetheine-adenylyl-transferase [100] and (9, green) lysyl-tRNA synthetase[101].

The nucleotide ATP or its complex with Mg^{2+}, like the previously described nucleotide-containing complexes, binds to the known adenosine-binding pocket on the C-terminal domain. However, the direction and geometry of their phosphates are strikingly and remarkably different, not only from each other (Figures 3D, 4C and 4D), but also from the known conformations of the bound MgADP (Figures 4A and 4B) or the metal complexes of AMP-PNP (Figures 4E, 4F, 4G) and AMP-PCP (Figure 4H). The difference is better illustrated by superimposition of the bound nucleotides in Figure 5A.

In the absence of Mg^{2+}, a highly ionic, possibly non-specific interaction of ATP dominates between the uncompensated negative charges of their phosphates. Details of ATP binding are shown in Figure 4D. Namely, the phosphate chain of ATP is directed towards helix 13 and the γ-phosphate oxygen atoms are stabilized by multiple H-bonds to the main chain N-atoms of Asp 374 and Thr 375 of this helix, as well as by the electrostatic interaction with the positively charged end of helix dipole at its N-terminus. Furthermore, an O-atom of the γ-phosphate also forms a H-bond with the NH_2-group of the amide of Asn 336, belonging to β-strand J, which is, in turn, H-bonded, through a water molecule (Wat 479) to the carbonyl O-atom of Gly 372. Since the latter residue is also situated at the N-terminus of helix 13, this H-bond network strengthens the link of the γ-phosphate here.

The position and interactions of the MgATP phosphates are remarkably different from those of ATP (Figure 4C) and roughly resemble to the peculiar binding mode found originally in case of horse muscle PGK. The γ-phosphate is positioned far from helix 13 and moved closer to the N-terminus of helix 8, although no interactions are formed with it; only a water molecule (Wat 470, not shown) forms H-bond with an oxygen atom of the γ-phosphate. On the other hand, an O-atom of the β-phosphate is liganded (through Wat 535) to the side-chain of Asp 218 at the N-terminus of helix 8. The position of Mg^{2+} with its characteristic interaction distances [87] is well defined in the structure: it is complexed with O-atoms of both β- and γ-phosphates, but does not interact directly with the protein. The α-phosphate is involved here in an interaction with the side-chain of Lys 219, known from previous structures with MgADP [6] MnAMP-PNP [69] MgAMP-PNP [73] and with MgAMP-PCP [72].

An interesting feature of both structures is that the side-chain of the conserved Lys 215 exhibits a considerable flexibility indicated by the low electron density in this region. The distances of its last resolved atom (CG) in the MgATP-bound structure from an O-atom of the γ-phosphate and of the α-phosphate are 5.8 Å and 4.2 Å, respectively (Figure 4C). As mentioned above, a similar flexibility of Lys 215 has been observed in the structure, which binds MgAMP-PCP.

It is also notable that the average crystallographic B-factor values of the nucleotide phosphates are 55.3 and 62.0 $Å^2$, respectively, in the MgATP- and ATP-bound structures, not very different from the value of 53.1 $Å^2$, found previously in case of the bound MgAMP-PCP [72], but different from the lower values (46.9 and 32.6 $Å^2$, respectively) found in the cases of MnAMP-PNP bound to pig muscle [69] and MgAMPPNP bound to Tm [73] PGKs. Considering the full occupation of the nucleotides (proved by the low B-factors of the adenosine moieties), this provides an argument in favour of the increased mobility of the nucleotide phosphate-chain in these structures. In addition, similar to MgAMP-PCP-bound structure, the mobility of MgATP phosphates correlates well with the increased flexibility of helix 13 (characterized by the B-factors of 55.9 and 38.9 $Å^2$, respectively), to which their interactions are practically absent (cf. also Table 1).

Thus, the conformation and orientation of MgATP phosphates, observed in the crystal structure, are different from those characteristics of either the metal complexes of MgAMP-PNP or MgAMP-PCP, but resembles better to the latter one. Superimposition of the nucleotides revealed an intermediate position of MgATP phosphates between the two analogues, which is more close to that of MgAMP-PCP phosphates (Figure 5A). These different conformations of the nucleotides cannot be due to different crystallization conditions (low or high salt concentrations), since the complexes containing bound Mg(Mn)AMP-PNP and MgAMP-PCP have been crystallized under similar conditions (at low ionic strength from PEG), while the complexes containing bound MgATP and MgAMP-PCP with more similar conformations have been crystallized under different conditions (at high ionic strength from citrate and at low ionic strength from PEG, respectively).

Indeed, various conformations of the phosphate chains of the analogues AMP-PNP and AMP-PCP have been observed in crystal structures of their complexes not only with PGK, but with other proteins, too [72]. Similar variation of the phosphate conformations can be collected from the literature for the natural nucleotide, ATP itself as it binds to various proteins in their crystal structures (Figure 5B). These observations suggest a general mobility of the bound nucleotide phosphates, which may allow a considerable variation of the conformation of the bound nucleotide triphosphate chain, as actually depicted from the various crystal structures of PGK (Figure 5A).

Thus, we may assume that these phosphates can fluctuate between two alternative positions, which was originally observed in separate crystal structures with MnAMP-PNP [69] and MgAMP-PCP[72]. Since the crystal always represents a rigid structure, one can usually observe only one of the alternative phosphate conformations in one of the alternative position. Further arguments for the assumed fluctuation have been provided by solution experiments of kinetic, stability and thiol-reactivity studies [72] (cf. Chapters II. 2.5 and 2.8).

It is also notable that conformation and interactions of MgATP phosphates (Figure 4C) greatly differ from those of the other nucleotide substrate, MgADP (Figures 4A and 4B), while the contacts and conformation of MgADP are surprisingly more similar to those of the ATP-analogue MgAMP-PNP (Figures 4E, 4F and 4G). The differences in the crystallographic B-factor values of the nucleotide phosphates and of the backbone atoms of helix 13 correlate well with each other and with the different binding modes of the nucleotide substrates (Table 1). Besides the crystal structures, different binding modes of MgATP and MgADP are also supported by solution phase experiments. The two nucleotide substrates have markedly different effects on the calorimetric stability of PGK (cf. Chapter II. 2.8.) as well as on the conformation of helix 13, reflected by the reactivity of the thiols located on this helix (cf. Chapter II. 2.5.). Furthermore, Mg^{2+}-ion strengthens the binding of ADP, while has no effect on ATP binding (cf. Chapter VIII.). Most probably, complexation with the metal ion affects differently the phosphate chain flexibility of MgADP and MgATP. In contrast to the rigidly bound MgADP (whose phosphates are linked to Asp 374 (belonging to helix 13) through well defined interactions of Mg^{2+}, in case of MgATP the metal ion does not interact with any protein atoms. Therefore, complex formation with Mg^{2+}-ion does not restrict the mobility of MgATP phosphates. The relevance of phosphate chain flexibility of MgATP and its fluctuation between two alternative sites (in contrast to the rigidly bound MgADP) in respect of the catalysis are discussed separately (cf. Chapter VIII.).

1.5. Difficulties in Identification of 3-Phosphoglycerate (PG) Site and the Hinge-Bending Hypothesis

At the time of identification of the nucleotide-binding site the location of the site for the other substrate 3-phospho-D-glycerate (PG) proved to be difficult and experimental progress became very slow. Attempts to generate a binary complex between PGK and PG were met with initial failure with both the yeast and horse-muscle enzymes. The substrate 1,3-bisphospho-D-glycerate (bPG) was judged too unstable in aqueous solution to use, so attention became focused on PG. However, trials of soaking crystals of the enzyme in solutions of the soluble and stable 3PG produced either no observable binding to both horse muscle PGK in the presence of sulphate [57, 58] and yeast PGK in the presence of selenate [5] or caused extensive conformational changes in the horse muscle enzyme in the presence of tartrate, that were too large to be analysed at the given resolution [4], thus the binding site of PG remained obscured.

In the (temporary) absence of experimental information on the total extent of the active site of PGK, it was possible to speculate about it and to consider further the implications of the binding of the nucleotide substrates. It was already known, that all nucleotide substrates in both the horse and yeast enzymes bound at the same site on the C-domain in an extended conformation with the phosphate chain, containing the transferred phospho-group, situated in the neck region pointing towards the N-domain. Given this mode of binding it seemed highly unlikely that the phosphoglycerate substrate could be itself bound to the C-domain if the most probable direct transfer of the phospho-group (cf. Chapter V. 5.1.) is to occur. The most obvious locations for the phosphoglycerate binding site would in fact be in the neck region or the nearby region N-domain near to the nucleotide's polyphosphate chain. Examination of the model with horse muscle PGK, however, seemed to rule out the neck region because its component elements, the loops between β-strand K and helix 13, and β-strand L and helix 14, each composed of triple glycine sequences (residues 371, 372, 373 and 394, 395, 396 according to the mammalian PGK numbering). The complete lack of functional side-chains in this region ruled out the neck region as having the possibility of a binding site for PG close to the nucleotide substrate, or any conceivable catalytic site for phospho-transfer. In contrast, the inner face of the N-domain opposite the nucleotide binding site and closest to the polyphosphate chain included many functional side-chains which could act both in phosphoglycerate binding and in catalysis. There was in particular a region that became known as the "basic patch" which included six positively charged residues (Arg-s 38, 65, 122 and 170, as well as His-s 62 and 169) in addition to other functional side-chains. In the absence of clear experimental evidence for the location of the PG binding site, this concentrated patch of basic residues was considered to have the most appropriate characteristics for the binding of the phosphoglycerate substrates with their multiple negative charges.

However binding in this 'basic patch' would place the phosphoglycerate about 12-15 Å distant from the potential position of the γ-phosphate of ATP, too far for direct phospho-transfer to occur. The only solution to this structural dilemma seemed to involve a large-scale conformation change to the structure of PGK, such as to bring the proposed binding sites for the nucleotide and phosphoglycerate substrates closer together for catalysis to occur. Such a proposed conformational change might be facilitated by the marked two-domain structure of PGK, because a movement of the two domains towards one another would be achievable if the neck region acted as a kind of "hinge" about which the movement occurred. Independent experimental evidence of a large-scale conformation change in PGK on substrate binding was indeed obtained from X-ray solution scattering (cf. Chapter II. 2.9.) [102, 103].

While the failure of determination of PG-site of the horse muscle PGK has led to the "hinge-bending" hypothesis and to the assumption of this site at the inner surface of the N-domain, the same experience with the yeast enzyme was interpreted in quite different way. This has led to a marked divergence in the subsequent approaches of the groups working on the two enzyme species. In the case of the yeast enzyme it was assumed that no binding of PG had taken place because PG was already bound to the so-called "native" enzyme [5]. A density feature in the neck region was ascribed to bound PG and informed subsequent biophysical study of the yeast enzyme whether using X-ray or NMR analyses (for the latter cf. Chapter II. 2.1.). An important consequence of this decision was that because the "bound" PG was close to the position of the γ-phosphate of bound ATP, there seemed to be no need for any large scale hinge-bending conformational change as proposed for the horse-muscle enzyme. Given the high degree of structural and sequential similarity between the yeast and horse PGK, it is very unlikely that the two enzymes would differ in a basic feature of their catalytic activities. In fact the binding interaction between the "bound PG" and the enzyme was very tenuous and amounted too little more than a single hydrogen bond. This, and the lack of any significant functional side-chains that could act as catalytic groups in the region between the proposed bound substrates, prevented any further advance in the understanding of the activity of yeast PGK to be made. It now seems almost certain that the identification of the additional density in the neck region was incorrect, as of course were the deductions made from it.

As discussed in detail in the following chapters the early hypothesis about the "hinge-bending" conformational change, which has been put forward by Banks et al. [4], even in the absence of definitive knowledge of about the PG site, has been approved later by several lines of experimental evidences.

1.6. Binding Details of 3-Phosphoglycerate (PG)

In the case of horse-muscle PGK, the inability to soak PG into the crystals was ascribed to the possible competitive inhibition of the binding site by sulphate ions from the mother liquor of crystallisation. In fact both horse and yeast PGK's were crystallised from high concentrations of ammonium sulphate, and sulphate ions can readily substitute for phosphate groups in protein binding sites [104, 105]. Therefore the further study of the active site in horse PGK needed to find a crystallisation medium lacking high concentrations of sulphate ions, or probably any anions, and this proved time consuming. The solution to this problem was achieved by changing the horse-muscle PGK for the homologous pig-muscle enzyme. The new enzyme could be crystallised in the presence of 3-phosphoglycerate from the non-ionic precipitant, polyethylene glycol-8000. The new crystal structure was solved by molecular replacement using the known horse-muscle PGK structure [15]. The amino acid sequence of pig muscle PGK at that time was not known, but as expected from the known 96 % amino-acid homology between horse, human and mouse PGK's [24], the X-ray map indicated very few amino-acid substitutions (e.g., at residues 148 and 387), compared with horse PGK. The presently known sequence of pig muscle PGK (cf. Figure 2) is entirely consistent with this finding. Subsequent refinement of the pig PGK crystal structure revealed a large additional electron density feature near the 'basic patch' on the N-domain, presumably marking the position of the bound PG moiety. This density was identified as the phosphate group of PG. Further refinement including this phosphate atom revealed the remainder of the PG molecule, which could be fitted in its entirety into the electron density. This showed that the bound PG is located in a shallow cavity in the inner face of the N-domain, i.e., exactly at the site suggested by Banks et al. [4], as illustrated in Figure 6A.

The details of PG binding to pig muscle PGK is depicted in Figure 7A. Its 3-phosphate group made a double hydrogen-bond interaction with each of the guanidine groups of Arg-s 65, 122 and 170, which follow β-strands B and D as well as α-helix 5, respectively. The bridge oxygen makes a further hydrogen-bond to the imidazole of His 62 located on the loop that follows β-strands B. The 2-hydroxyl of PG, which defines the D-isomer, is hydrogen-bonded to Asp 23 and Asn 25 that follow β-strand A. Finally, the 1-carboxyl group of PG interacts with a water molecule (Wat 423) that is hydrogen-bonded at the N-terminal end of helix 14 by the main-chain N atoms of Gly-s 395, 396 and 397. The carboxyl-group, although is not within H-bonding distance, also involved in electrostatic interactions with the guanidine group of Arg 38. The 3-phosphoglycerate (PG) substrate appears to be rigidly bound by these multiple interactions in an orientation that places its 1-carboxyl group, which is directly involved in the phospho-transfer, nearest the position of the γ-phosphate of the bound MgATP. This binding site does not coincide with the density feature identified as the bound PG in yeast PGK, and no electron density was found in the map of pig muscle PGK complexed with PG at the binding site proposed for PG in the yeast enzyme. Later X-ray studies with the yeast enzyme [70, 106] have approved the site of PG as determined with pig muscle PGK.

Up to now there are several PGK structures containing bound PG. All of them are ternary complexes containing also bound nucleotide, Mg(Mn)AMP-PNP [69, 73], MgADP [80, 84] or MgAMP-PCP [72]. There are also two recent PG·MgADP ternary complexes containing also the transition state analogues, MgF_3^- or AlF_4^- published at the time of writing this book [76] and are mainly discussed in Chapter IX. The binding mode of PG in all complexes is largely similar and apparently independent of the enzyme source, of the protein conformational state and of crystallization conditions (cf. Figures 6A, 6B and 6C). There are, however, small, but significant changes in the interaction of PG that accompany the transformation of the binary PG complex into the ternary ones. The details of PG binding in these complexes are compared in Figures 7A-7F). An interesting change e.g., that Arg 65 undergoes a change in orientation of its guanidine-group compared with the binary complex [69] so as to be prepared for formation of an important salt bridge with Asp 218 (not shown) that stabilizes the active conformation and discussed later [76]. The major change seen in all ternary complexes concerns the conserved Arg 38/Y38/Tb39/Tm36 which moves from the site it occupied in the binary complex to a new position where its guanidine-group can form a double H-bond to the 1-carboxyl group of PG. Thereby the weak interaction existing in the binary complex between these groups (Figure 7A) is largely strengthened in the ternary ones

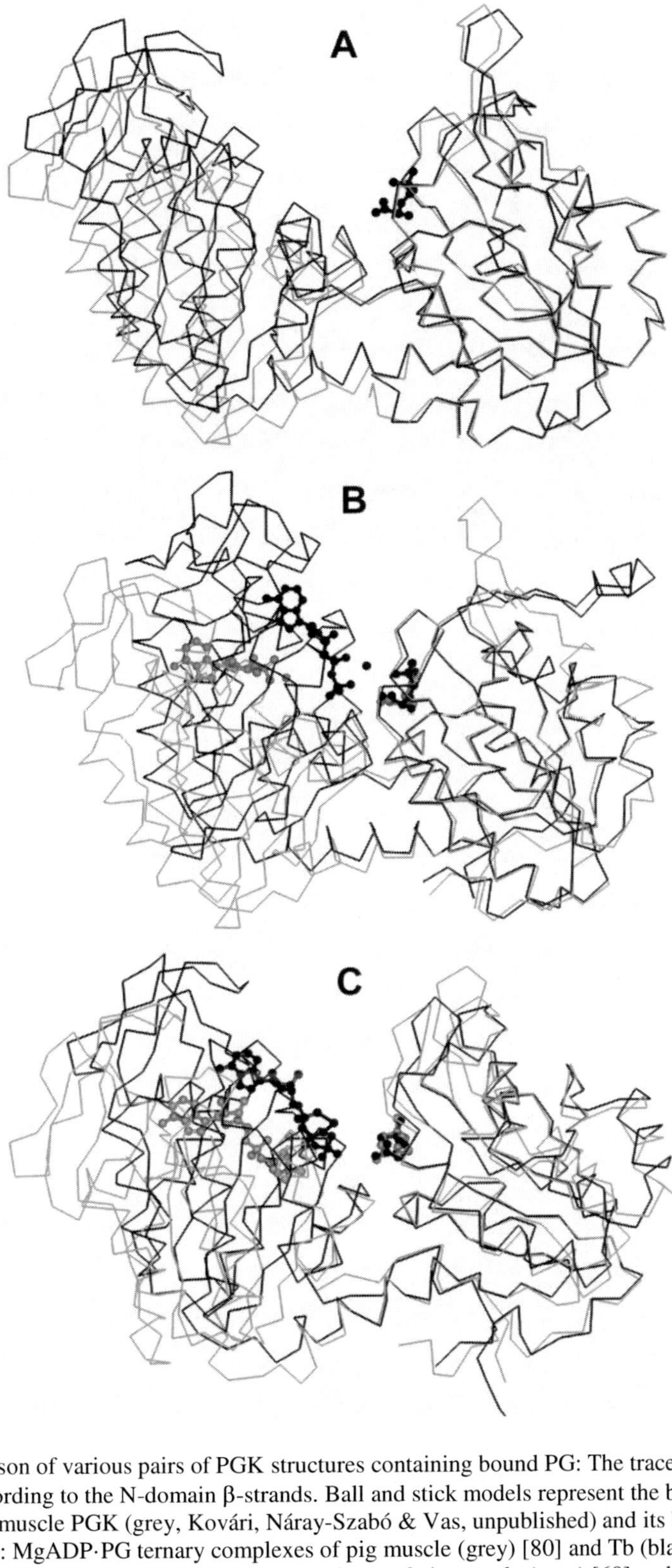

Figure 6. Comparison of various pairs of PGK structures containing bound PG: The traces of Cα atoms are superimposed according to the N-domain β-strands. Ball and stick models represent the bound ligands. A: Substrate-free pig muscle PGK (grey, Kovári, Náray-Szabó & Vas, unpublished) and its binary complex with PG (black) [15], B: MgADP·PG ternary complexes of pig muscle (grey) [80] and Tb (black) [84] PGK-s, respectively, C: Mg(Mn)AMP-PNP·PG ternary complexes of pig muscle (grey) [69] and Tm (black) [73] PGKs, respectively.

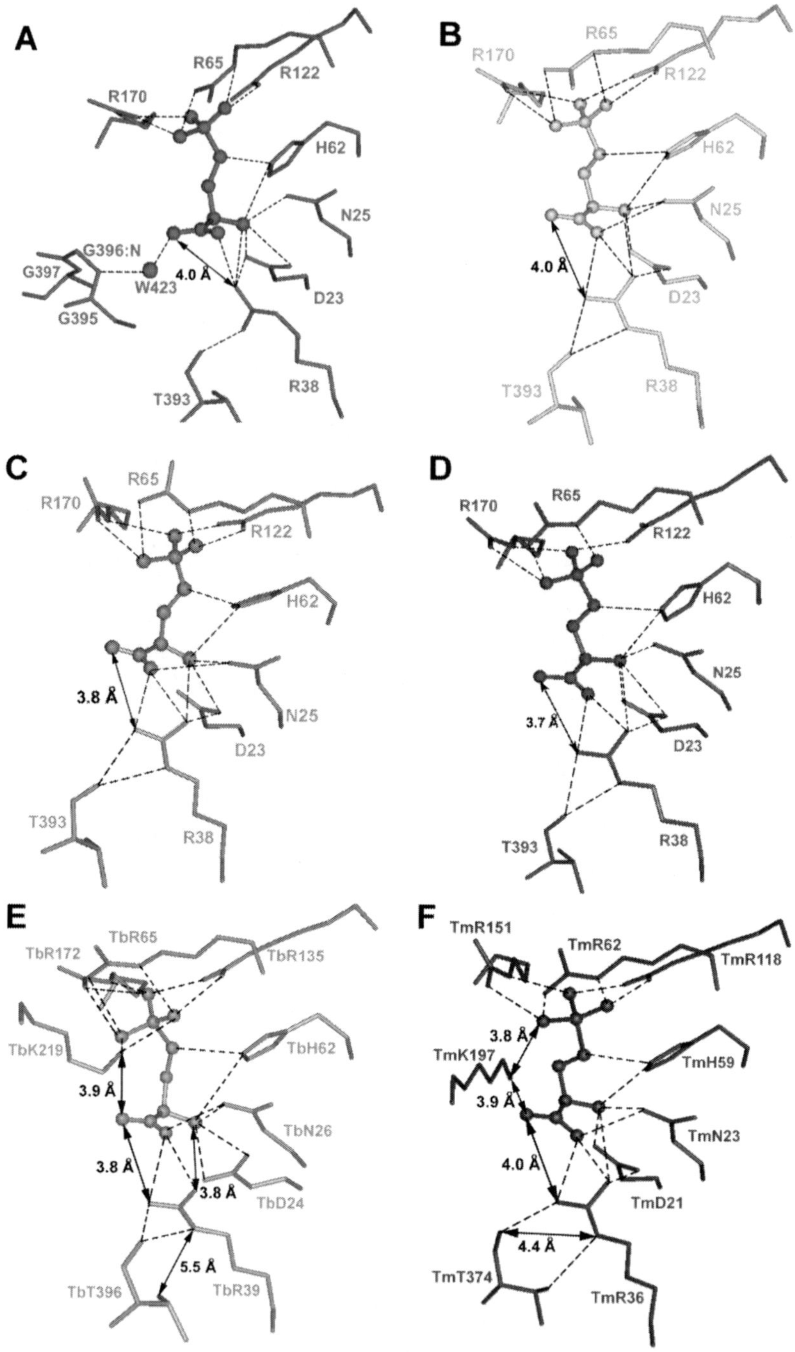

Figure 7. Binding details of PG as bound to various crystal structures: Ball and stick models represent the bound PG-s, while stick models label the interacting side-chains. Dashed lines connect the interacting atoms (the limiting distances are 3.5 and 4.0 Å-s for H-bonds and electrostatic interactions, respectively).
A: PG binary complex of pig muscle [15], B: PG·MgAMP-PCP ternary complex of pig muscle [72], C: PG·MnAMP-PNP ternary complex of pig muscle [69], D: PG·MgADP ternary complex of pig muscle [80], E: PG·MgADP ternary complex of Tb [84], F: PG·MgAMP-PNP ternary complex of Tm [73] PGKs.

(Figures 7B-7F). At the same time in the ternary complexes of Tm (Figure 7F) and Tb (Figure 7E) PGKs the H-bond interaction of the guanidine-group of Arg 38/Y38/Tb39/Tm36 with the OH-group of PG is weakened or disappears entirely. These changes may help to orient the PG carboxylic group for the nucleophylic attack on the γ-phosphorous atom of MgATP during catalysis. It is remarkable that only these two ternary complexes exhibit a domain-closed conformation [73, 84]. Some other changes, e.g., weakening of the H-bond of Asn 25/Y25/Tb26/Tm23 (ND2) with the OH-group of PG and the disruption of the H-bond of Arg 38/Y38/Tb39/Tm36 (NH1) with Asp 23/Y23/Tb24/Tm21 (OD2) (i.e., with the residue which permanently fixes the OH-group of PG in all four structures) occur only in the closed structure of TbPGK (Figure 7E), but have not yet taken place in the partially closed structure of TmPGK (Figure 7F). As pointed out later (cf. Chapter II. 2.7.) Arg 38 is an essential catalytic residue and therefore it is important that, as seen in Figures 7A-7F, in addition to its interaction with the carboxylate of PG, it also forms interaction(s) with the main chain carbonyl-group of Thr 393. Through these interactions Arg 38, which interacts with PG, also connects the two domains. These interactions of Arg 38 are multiplied and strengthened in the ternary complexes, in which they play an important role in promoting communication between the two domains (for detailed discussion cf. Chapter IX.).

There is another remarkable interaction of PG, namely the interaction with the conserved Lys 215/Y213/Tb219/Tm197, which exist exclusively in the ternary complexes of Tb [84] (Figure 7E) and Tm [73] (Figure 7F) PGK-s as well as in the TSA-complex of HsPGK [76] (Figures 77A and 77B). In these structures the ε-amino group of Lys 215/Y213/Tb219/Tm197 interacts with the 3-phosphate group and less strongly with the 1-carboxyl group of PG. These interactions are apparently related to the conformational state of the enzyme molecule, since these complexes contain the two domains in closed positions. In all of the other structures with open conformation this lysine residue is far away from the bound PG. This difference is illustrated by superimposition of PG and its surroundings in two pairs of open and closed structures (Figures 8A and 8B). The importance of this conserved lysyl-side-chain in the relative domain movement and in catalysis will be discussed later (cf. Chapters II. 2.9., V. 5.6. and IX.).

In summary, the binding interactions of either substrates (PG or the nucleotide) characteristic of their binary complexes in the main features are largely remained the same in the presence of the second substrate in the respective ternary complexes. As shown above, there are however, some significant differences, which are related to the conformational state (i.e., the relative domain positions) of the enzyme.

1.7. Conformational Changes Caused by PG

While in the binary complexes of *B. stearothermophilus* PGK with MgADP [6] and of pig muscle PGK with MgATP [7] the relative position of the two domains is the same as in the substrate free horse [4] or pig muscle [72] enzyme, there is a significant conformational change of the relative domain position in the pig enzyme with bound PG [15] as compared to

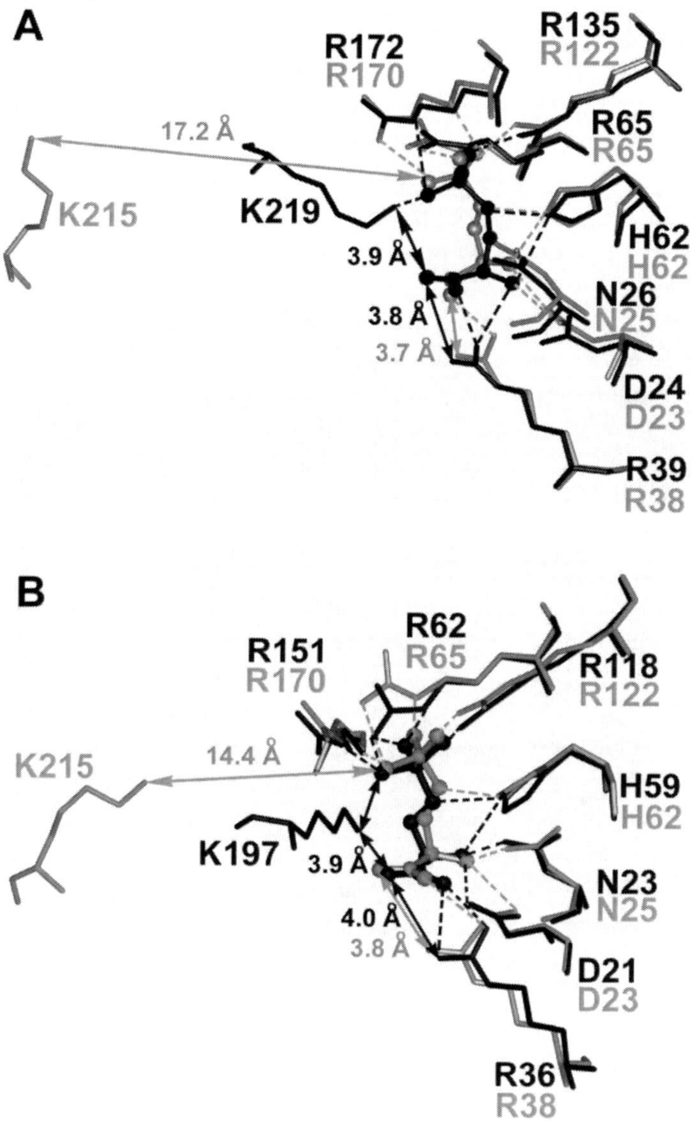

Figure 8. Participation of Lys 215/Tb219/Tm197 in PG binding depends on domain closure: The bound PG molecules are superimposed in (A) PG·MgADP ternary complexes of pig muscle (grey) [80] and Tb (black) [84] PGK-s, respectively; and in (B) PG·Mg(Mn)AMP-PNP ternary complexes of pig muscle (grey) [69] and Tm (black) [73] PGK-s, respectively. The surroundings of PG are depicted in the way as shown in Figure 7.

the horse enzyme, whether in its native or nucleotide-bound form [4]. Given the very high degree of homology between the pig muscle and horse muscle enzymes this difference must be associated with the binding of the PG substrate. Such effect is not induced by either of the nucleotide substrates in their binary complexes. The conformational change was best modelled as a rigid body rotation of the N-domain relative to the C-domain of 7.7° in a direction so as to close the gap between them. This conformational change is illustrated by Figure 6A, in which the PG bound structure is superimposed on the structure of the substrate-free PGK according to the backbone atoms of the inner β-strands of the N-terminal domains. This conformational change would push the bound PG considerably closer the position of the bound nucleotide substrate than would be the case of the native enzyme. However, although the gap between the substrate binding sites was significantly reduced, it is not completely eliminated. As discussed below concerning the conformations of the ternary complexes, the elimination of the gap could only be observed much later in the crystal structures of two different ternary complexes from Tm [73] and Tb [84] PGKs, exhibiting closed domain conformations (cf. Figures 6B and 6C, respectively). These structures at the same time provided the first direct evidence in favour of the hinge-bending hypothesis that was put forward almost two decades earlier by Banks et al. [4]. The crystal structures of the much later determined TSA-complexes of HsPGK are also entirely consistent with the occurrence of domain closure [76], details of which is outlined in Chapter IX. At the time of publication of the first closed structures, however, the highly sophisticated mechanism of domain closure was not understandable.

A further significant conformational change in the PG·PGK binary complex from pig muscle involves helix 13 (residues 371-382) in the enzyme's active site. In the native horse or pig muscle PGK-s this helix is mostly disordered (the crystallographic B-factor values are around 50 $Å^2$), especially, at its C-terminus. In contrast, the electron-density map of pig PG·PGK binary complex shows helix 13 to be ordered and characterized by a much lower B-factor (22.3 $Å^2$). Thus, it is reasonable to ascribe the ordering of the helix to the presence of the bound PG substrate, similar to the above described positioning of the nucleotide phosphates (cf. Chapter I. 1.4.). As with the nucleotides, the relationship between ordering of helix 13 and the restriction of the reactivity of the two fast reacting thiols in this helix also holds with PG [92, 93] (cf. Chapters II. 2.5 and II. 2.8). Thus, the reduced reactivity of these groups in the presence of PG confirms the observation that the binding of PG causes the ordering of helix 13. Exactly how the ordering of helix 13 is affected by the binding of PG at a site more than 10 Å from the helix was unclear at the time of solving the structure of the PG binary complex.

Another remarkable conformational change, occurring upon the binding of PG, is the movement of the whole helix 14 with respect of the remaining part of the N-domain, especially, with respect of helix 1. This "piston-like" motion of helix 14 was recognised later by Bernstein and Hol [74] by comparing the X-ray structures of PG- and MgADP-bound binary complexes of pig muscle [15] and Bs [6] PGKs, respectively. The occurrence of this motion was suggested as a consequence of the movement of the PG binding loop. A more detailed molecular basis of the movement was proposed much later [107-109] and will be discussed in Chapter IX.

1.8. Conformation of the PGK Ternary Complexes: Problem of the Crystallographic Observation of Domain Closure

Structural studies of the binary substrate complexes have shown that the nucleotide and PG substrates are bound at sites on C- and N-terminal domains, respectively, separated by about 10-15 Å-s. In spite of the fact that PG binding brings the N-domain closer to the C-domain by a domain movement of 7.7° rotation, the remaining distance (more than 9 Å, see Figure 9A) would be still too large for the direct in line transfer of ATP γ-phosphate to the carboxylate of PG without covalent intermediate, which is the most likely chemical mechanism for the PGK catalysed reaction suggested by other type of solution experiments (cf. Chapter V. 5.1.).

The previously proposed "hinge-bending" conformational change was still sufficient to resolve this difficulty. Namely, a "hinge-bending" type of relative domain movement would increase the probability of a closer contact of the two bound substrates, appropriate for the reaction. This proposal was in agreement with independent findings of a substantial conformational change of PGK molecule on the formation of a ternary complex with the two substrates: the radius of gyration of PGK was observed to decrease by 1.1-1.2Å±0.3Å [102, 103] (and also supported by a later study [109]), indicating a more compact shape for the ternary complex as suggested above (various studies on the substrate caused conformational changes are discussed separately, cf. Chapters II. 2.5. and II. 2.9.).

Thus, in the light of the open conformations of the PGK binary complexes in their crystal structures and of the substantial conformational changes in the presence of both substrates (or analogues) in solution, it was plausible to assume that domain closure occurs upon binding of both substrates, i.e., only in the ternary complex. On this basis the crystal structures of PGK ternary complexes are expected to exhibit domain-closed conformations. In contrast to this expectation, however, the first ternary complex determined by crystallography at 2.0 Å resolution, the complex of pig muscle PGK with MnAMP-PNP and PG [69] was obtained only in a slightly folded conformation (represented by orange traces in Figures 3B and 3C, similar to the PG binary complex (black trace in Figure 6A) [15]. Thus, the distance between the putative reactive groups of the two substrates remained around 9 Å (Figure 9A) in this essentially open conformation. This ternary complex was produced by diffusing MnAMP-PNP into the pre-grown crystal of the PG binary complex, thus, the crystal lattice forces was thought to prevent the large scale conformational changes that otherwise would have occur in solution of the ternary complex. In a 2.4 Å resolution structure of the ternary complex of an active mutant of yeast PGK, produced by co-crystallization with the same ligands, however, was also obtained in a largely open conformation (grey trace in Figure 3B) [70] again apparently inconsistent with the hinge-bending behaviour. Later on, co-crystallization in the presence of PG and MgADP and inorganic phosphate as precipitant has also yielded crystals of pig muscle PGK exhibiting open conformation (grey trace in Figure 6B) [80], that has been attributed to the mobility of the β-phosphate of MgADP which does not occupy a well-defined position in this 1.8 Å resolution structure (Figure 9B). A repeated appearance of the open conformation (2.5 Å resolution) of pig muscle PGK was observed when it was co-crystallized with MgAMP-PCP and PG (cyan trace in Figure 3C) ([72]. From complementary studies of single-crystal microspectrophotometry [110] it was proposed that the open conformation of this latter ternary complex probably does not represent the average protein conformation in solution. Rather, the source-specific protein-protein interactions operating in the crystal lattice may have selected and stabilized a particular protein conformation, as found also in other cases [111-113]. This problem of crystallographic studies is discussed below in detail in the light of analysis of crystal lattice forces operating in various PGK crystals.

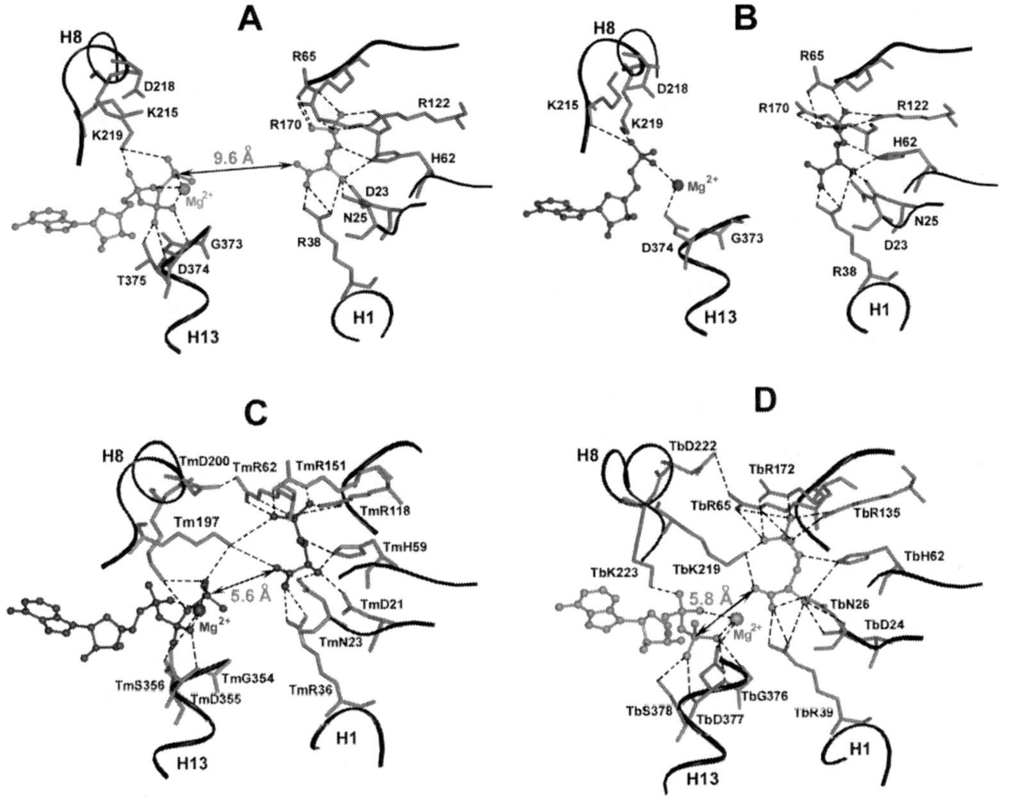

Figure 9. Relative positions of the two bound substrates (analogues) in various PGK ternary complexes: The bound ligands are shown as ball and stick models, the interacting side-chains are indicated as stick models. Some nearby secondary structural elements are illustrated as ribbons. Dashed lines similar to Figures 4, 5 and 7 connect the interacting atoms. A and B: Open structures of PG·MnAMP-PNP ternary complex of pig muscle [69], and PG·MgADP ternary complex of pig muscle [80], C and D: Closed structures of Tm ternary complex [73] and Tb [84] PGK-s, respectively.

Thus, in spite of considerable experimentation with PGK-s from horse, pig and yeast, variously liganded with substrates and substrate analogues, no crystals containing the putative folded (closed), catalytically competent form of the enzyme have been observed. Thus, the conformation assumed by PGK during catalysis remained unknown for long time. This unsatisfactory situation was resolved when two independent crystal structures were almost simultaneously determined and both suddenly showed the domain-closed conformation of PGK. The crystals were obtained of the abortive ternary complex with PG and MgADP (green trace in Figure 3A and black trace in Figure 6B) [74] using PGK from Tb, the causative agent of sleeping sickness, as well as of the ternary complex of the bacterial Tm PGK with PG and MgAMP-PNP (blue and black traces in Figures 3B and 6C, respectively) [73], crystallized from the precipitants, phosphate and PEG, respectively. In these crystals the apparently homologous PGK molecules appeared in a folded state with the N- and C-domains in contact with one another, with the two ligands "trapped" between them. Thereby direct evidences for the validity of the hinge-bending hypothesis and for the occurrence of domain closure have been provided. The hypothesis is even further evidenced much later by the completely closed crystal structures of the TSA-complexes of human PGK [76] (cf. Chapter IX.).

Table 2. Distances (Å) characteristics of the relative domain positions

	PGK complex	Conformation	§Arg65(Cα)-Lys215(Cα)	Between reacting atoms of substrates
Pig muscle	MgADP·PG ternary [80]	slightly closed	24.3	-
	MgAMP-PNP·PG ternary [69]	slightly closed	23.7	9.6[a]
	PG binary [15]	slightly closed	23.8	-
	substrate free (Kovári et al. unpublished)	open	25.5	-
	MgAMP-PCP·PG ternary [72]	slightly closed	23.9	14.4[a]
Horse muscle	substrate free [4]	open	26.0	-
B. stearo	MgADP binary [6]	open	25.1	-
T. brucei	MgADP·PG ternary [74]	completely closed	11.1	5.3[b]
T. maritima	MgAMP-PNP·PG ternary [73]	largely closed	13.8	5.6[a]
Human	MgADP·MgF_3^-·PG [76]	completely closed	11.2	2.12 and 2.17[c]

§The numbering refers to mammalian PGK sequence

[a]between an O-atom of PG carboxylate from the P-atom of the terminal phosphate of the analogue of MgATP

[b]between an O-atom of MgADP β-phosphate and the P-atom of 1-phosphate of the putative bPG (the latter one is modelled into this structure in place of PG)

[c]The distances of the O2 atom of PG and the O3B atom of MgADP from the Mg-atom of the transition state analogue MgF3, respectively

In the first two 'closed' forms of PGK the nucleotide and PG substrates are bound to the same sites as observed in the 'open' form, but are located 6Å closer together (Figures 9C and 9D) as a result of the conformational change. Thus, in the closed form the two substrates are almost in the appropriate distance (cf. Table 2) and alignment for catalytic interaction, showing, for the first time, that these crystals contain the PGK molecule in a catalytically competent form. In the open ternary complexes of pig muscle PGK, however, the distances between the two bound ligands are much larger (Table 2), than those in the closed ternary molecules. As discussed below, these open and closed structures together provided a framework to deduce valuable information about the mechanism of domain closure as well as about the architecture of the active site.

1.9. Protein Conformer Selection by Crystal Packing Contacts

In order to understand how the conformational transition of PGK from the open to the active closed state takes place it would have been advantageous to obtain crystal structures of both extreme forms from the same enzyme source. In spite of all kinds of efforts, however, it could not be achieved: open and closed structures with different relative domain positions have been determined of the enzyme from different sources. Thus, the difficulties arose from their sequence differences (up to 50%) had to be considered in the pair wise comparisons.

It was rather intriguing and requires a special issue that pig muscle PGK ternary complex has never been crystallized in the closed conformation, although the same ternary complexes were crystallized under the same conditions as in cases of the two closed structures of Tb and Tm PGK-s, available at that time. Namely, open structures of the pig muscle enzyme have been published for the complexes of PG and MnAMP-PNP from PEG [69], of PG and MgAMP-PCP from PEG [72] as well as of PG and MgADP from phosphate [80]. On the other hand, no open structures of Tb and Tm PGKs have been reported so far.

One may argue that the sequence differences *per* se may influence the state of equilibrium between the open and closed states, by stabilising one conformation over the other. This argument, however, cannot be hold since there are several lines of experimental evidences [102, 103, 114, 115] in favour of existence of a more compact conformation (compatible with the closed form) in solution of the ternary enzyme-substrate complexes of various PGK species, including yeast and pig muscle PGKs. It is reasonable to assume that during the catalytic cycle all PGK species temporarily adopt both open and closed conformations, irrespective of the differences in their sequence. Since the specific catalytic activities of various PGKs are similar to each other, the conformational transition, required by the function should occur with the same probability.

Thus, it seems to be very probable that the sequential differences may influence the appearance of one or the other extreme conformation in the crystalline state, simply by stabilizing one conformation over the other, through formation of different protein-protein contacts in the crystal lattice. In order to get evidence in favour or against this assumption, the sequence differences were surveyed and analysed in the crystal contact surfaces. To this end, two independent pairs of PGK crystal structures were compared bearing the molecules with open (O) or closed (C) conformation, namely pig muscle PGK·PG·MgAMP-PCP (O1) [72] and Tm PGK·PG·MgAMP-PNP (C1) [73] ternary complexes, both of them crystallized from PEG as well as pig muscle PGK·PG·MgADP (O2) [80] and Tb PGK·PG·MgADP (C2) [84] ternary complexes, both of them crystallized from high concentration phosphate solutions. On this basis, crystal contact forming residues (CFR) were identified and the conserved or non-conserved character of their side-chains, their sequential position and their chemical nature were investigated [116].Inspecting the contacts in the four investigated PGK crystals has revealed their source-dependent nature. The statistical analysis of CFR-s is given in Table 3. In each case almost all CFR-s belong to the non-conserved, sequentially variant residues. Participation of a conserved residue in formation of a crystal contact appears to be a rare exception. From the total number of CFR-s (38) present at different sequential positions in the two open conformations together (O1 and O2) as much as 26 CFR-s possess side-chains

Table 3. Distribution of the crystal contact forming residues (CFR) of the molecules present in some open and closed PGK structures (adapted from [116])

Conformation	Crystalline complex of PGK	Total no. of CFR-s	Conserved CFR-s	CFR-s with different sequential position			CFR-s with identical sequential position	
				Total number[§]	Side-chain differences in the same position of the other conformation[x]	Side-chain similarities in the same position of the other conformation[x]	In pairs of similar conformations	In pairs of different conformations
OPEN	MgAMP-PCP·3PG (**O1**)	**23**	**1** (Lys 191)	O1+O2: **38**	**26** (from these 11 CFR)	**12** (from these 3 CFR)	O1-O2: **16** (from these 7 CFR in C1+C2)	O1-C1: **7**
	MgADP·3PG (**O2**)	**31**	**none**					
CLOSED	MgAMP-PNP·3PG (**C1**)	**32**	**2** (Pro 42, Lys 173)	C1+C2: **67**	**52** (from these 13 CFR)	**15** (from these 2 CFR)	C1-C2: **10** (from these 3 CFR in O1+O2)	O2-C2: **5**
	MgADP·3PG (**C2**)	**45**	**none**					

The numbers represent the sum of CFR-s participating in side-chain-side-chain and side-chain-backbone interactions either directly or through a water molecule (CFR-s involved in backbone-backbone interactions are not counted).

[§]This number is less than the sum of CFR-s in O1+O2 and C1+C2, respectively, due to the part of CFR-s in identical sequential position.

[x]The residues of the other conformation are counted independently of their CFR-nature.

chemically different from those of the residues occupying the same sequential position (independent of their CFR-nature) of either of the two closed conformations (C1 or C2). Similarly, from the total number of CFR-s (67) present at different sequential positions in the two closed conformations together (C1 and C2) as much as 52 CFR-s possess different side-chains from those of the residues occupying the same sequential position of either of the two open conformations (O1 or O2). These figures may explain why pig muscle PGK cannot form crystal structures with the closed molecular conformation: it does not have those residues in 52 contact forming positions which are present in either the Tb or Tm enzyme. Furthermore, *vice versa*, Tb and Tm PGK-s cannot produce crystal structures with molecules of the open conformation, since they do not have those residues in 26 contact forming positions which are present in the pig muscle enzyme.

It is reasonable to assume that not all CFR-s are equally important in determining one extreme conformation or the other. One can assume that the crucial ones should possess identical sequential positions in pairs of similar conformations (e.g., O1 and O2 or C1 and C2). Naturally this analysis has led to smaller numbers: 16 or 10 CFR-s is in identical positions in the two open or in the two closed structures, respectively. The importance of these residues may be underlined by the fact that a much smaller number of CFR-s with identical sequential positions can be counted (7 or 5, respectively) if the comparison is made between the pairs of different conformations, i.e., between O1 and C1 or O2 and C2.

The above described CFR analysis of two open (O1 and O2) and two closed (C1 and C2) crystal structures has revealed great similarity between the same type of conformations not only at the level of CFR-s, but sometimes even better among the contact forming structural elements (such as helices, β-strands or loops) of the interacting molecules. Visualization of the intermolecular contact surfaces at three dimensions in the crystal forms (O1 and O2) of pig muscle PGK, in spite of the different nature of the ternary complexes and of different crystallization conditions (PEG or phosphate), shows a high extent of similarity. Figure 10 illustrates the spatial arrangement of the molecules within the crystal lattices and a selection of characteristic contacts in the structures of O1 and O2. Figures 10A and 10B show in both cases a face to back arrangement of the molecules. Although the definitive side-chain and backbone contacts are partly different, they are common in the sense that the *heterologous* contacting surfaces are formed. Furthermore, in both crystals the contact surfaces are formed from the same structural elements of the molecule (Figures 10C *vs.* 10D and 10E *vs.* 10F).

For comparison, Figure 11 shows the spatial arrangement and some details of selected contact surfaces of the enzyme molecules, possessing largely closed conformations, in two independent crystal forms of Tm (C1) and Tb (C2) PGK-s, respectively. In spite of the different sources and the different nature of the enzyme-ligand complexes, in both crystals a back-to-back arrangement of the molecules with *homologous* contacts exist (Figures 11A and 11B) details of which are depicted in Figures 11C, 11D, 11E and 11F). These characteristics are strikingly different from those of the open structures.

Thus, the analysis presented in this work clearly shows that the differences or similarities in the crystal lattice contacts exist at various levels: in the sequential position of CFR-s, in the chemical nature of CFR-s and in the molecular steric position of the various structural elements (helices, β-strand, and loops) constituting the interacting molecular interfaces. It is difficult to decide what the most crucial factor is in determining the protein-protein contacts within a particular crystal. The contacts, however, should be responsible for stabilization and the appearance of a certain conformational state of a protein molecule in the given crystal.

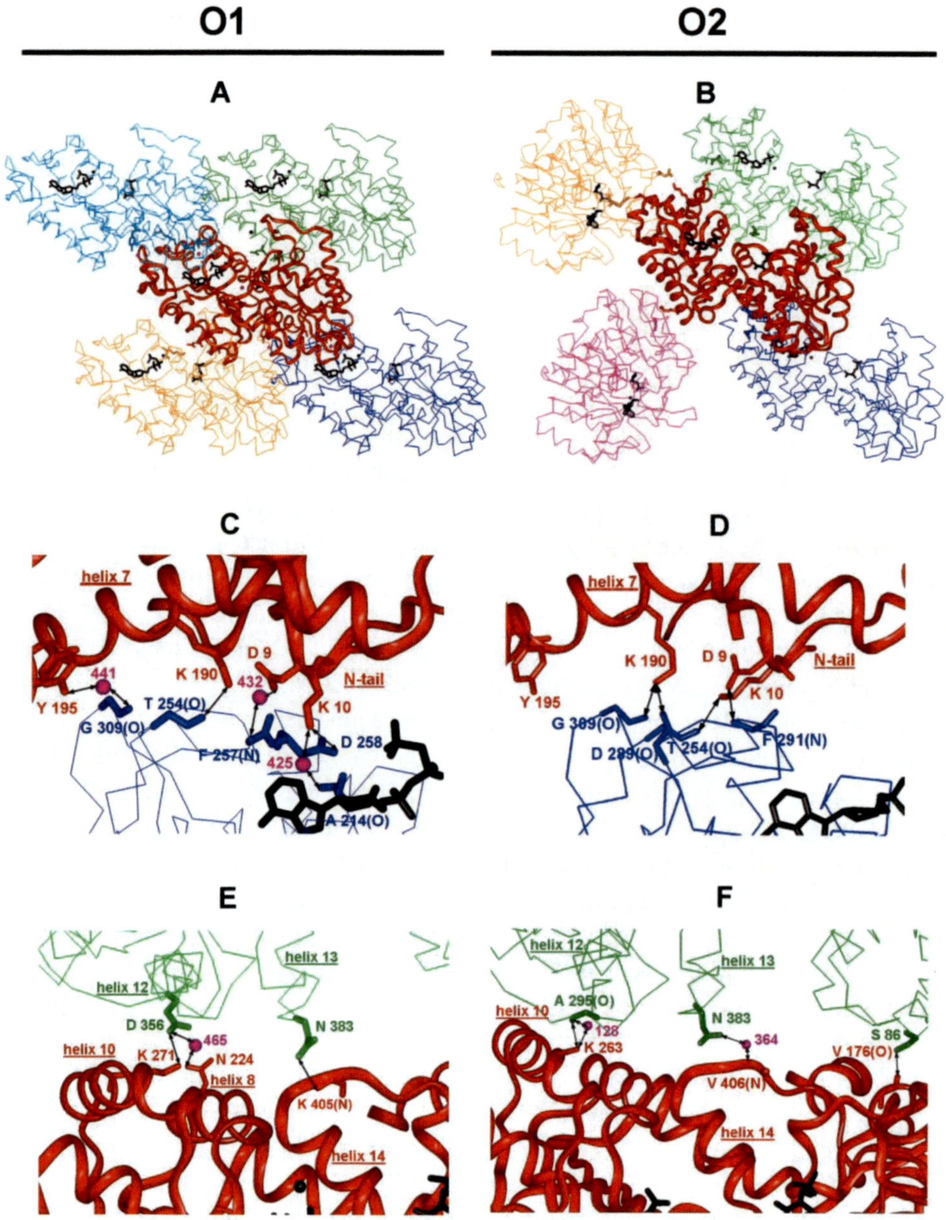

Figure 10. Spatial arrangement of protein molecules in the crystal forms of pig muscle PGK possessing open domain positions (adapted from [116]): Selected molecules as positioned in the crystals of PG·MgAMP-PCP (O1) [72], and PG·MgADP (O2) [80], ternary complexes of pig muscle PGK are depicted in panels (A) and (B), respectively. Details of characteristic contact surfaces of the molecules are shown in panels (C) and (E) for the structure O1 or in panels (D) and (F) for the structure O2. The interacting residues, backbone atoms and the bound ligands are marked as stick models. Water molecules involved in the interactions are shown as pink balls. Black arrows indicate the interacting distances (H-bonds or salt bridges). The colour codes of the symmetry equivalent molecules are as follows: Panels (A), (C) and (E): red (x,y,z), blue (x-1,y,z), orange (x,y,z-1), cyan (x,y,z+1), green (x+1,y,z); and panels (B), (D) and (F): red (x,y,z), orange (-x,0.5+y,-z), green (x,y,z+1), blue (x-1,y,z), purple (-x-1,0.5+y,-z-1).

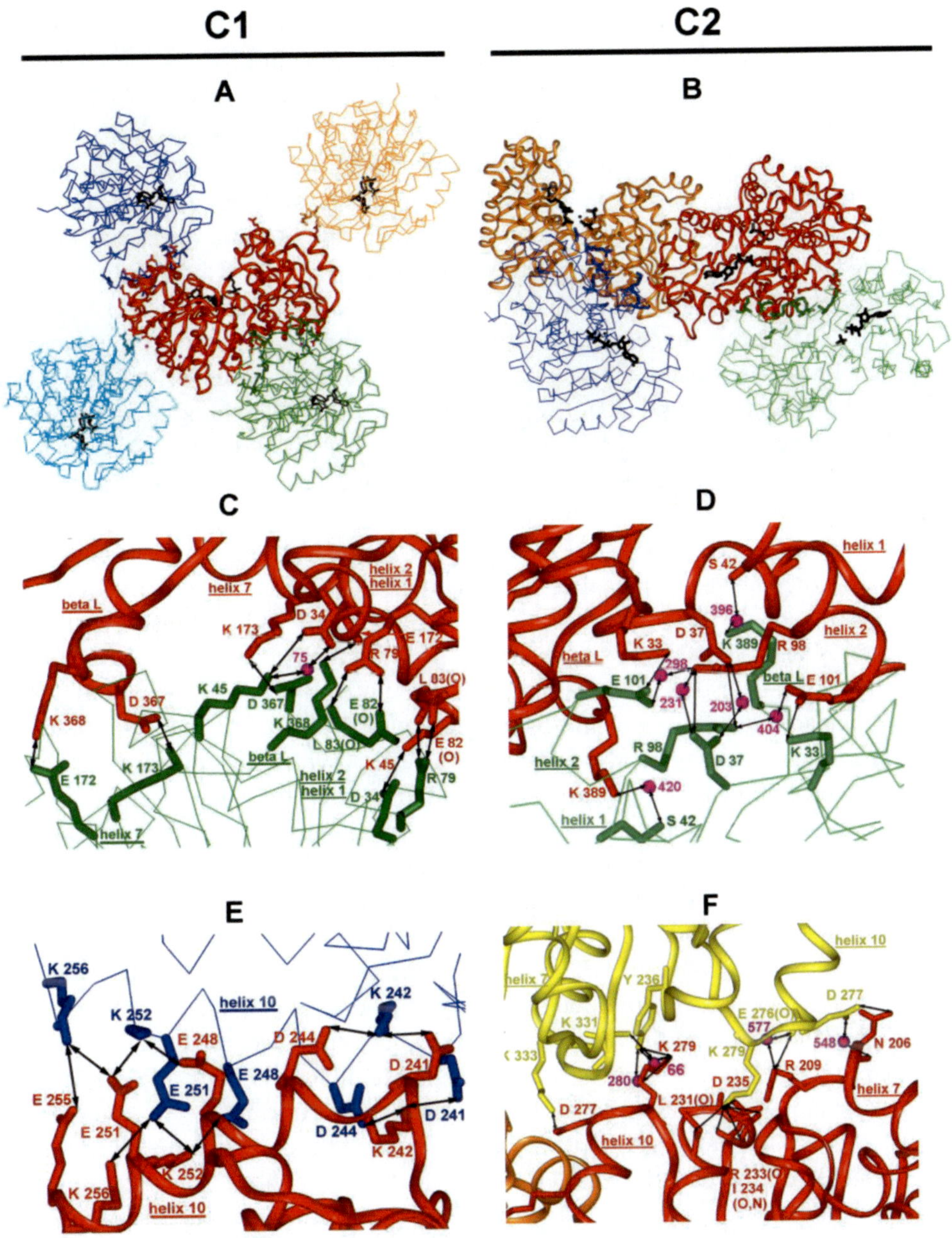

Figure 11. Spatial arrangement of protein molecules in the crystal forms of Tm and Tb PGKs possessing closed domain positions (adapted from [116]): Selected molecules as positioned in the crystals of PG·MgAMP-PNP complex of TmPGK (C1) [73] and PG·MgADP complex of TbPGK (C2) [84] ternary complexes are shown in panels (A) and (B), respectively. Details of characteristic contact surfaces of the molecules are shown in panels (C) and (E) for the structure of C1 or in panels (D) and (F) for the structure of C2. The interactions are illustrated in the same way as for Figure 10. The colour code of the symmetry equivalent molecules of panels (A), (C) and (E) is as follows: red (x,y,z), orange (x-0.5,0.5-y,-z-1), green (-x,-y-1,z), blue (-x,-y,z), cyan (0.5+x,0.5-y,-z). In case of TbPGK the four molecules in the picture constitutes an asymmetric unit.

In order to understand why the protein conformation existing only in minority in solution (e.g., the open conformation of PGK in its ternary enzyme-substrate complex) can be stabilized in crystalline form (e.g., the O1 and O2 open structures of pig muscle PGK), one has to consider the crystallization process at the molecular level. In the initial solution, open and closed conformers exist in equilibrium. Once the crystallization begins, the solution phase open conformer will be gradually depleted and replenished from the remaining closed conformer, because protein-protein interactions in the crystal are energetically favoured between the molecules with the open conformation. To this end, the open conformer of pig muscle PGK will be crystallized despite to its minority in solution. This is entirely consistent with the view of the non-specific nature of the crystal packing contacts [112, 117-119].

Thus, in general, the lattice constraints can perturb the conformational state and thereby may alter the functional state of a protein, as noted in several instances (cf. reviews [111, 120] and also some recent papers [121, 122], including nucleotide binding to crystalline PGK [110]. Thus, it is not surprising that two significantly different, open and closed, conformational states of PGK are stabilized by different crystal packing forces, as it was actually found.

The above analysis of the crystal contacts with two independent pairs of open and closed PGK crystal structures have revealed the importance of source specificity in attaining a unique conformational state in the crystal. The results clearly indicate that the difference in the sequence is the main factor, which directs the selection of different conformers in the crystal lattice. To the best of our knowledge the present work has provided the first example of such a comparative analysis of the molecular contact surfaces of the same protein in its various crystal forms. It may be also notable, that PGK provides a special example where the generally accepted approach of the large-scale variation of crystallization conditions, in order to get crystals of different protein conformations, apparently does not work. Such strong relationship between the conformational state of the crystalline protein and the characteristic source-dependent nature of crystal contact forming residues may need to be considered for crystal growing of other proteins, too.

The inability of pig muscle PGK to close its domains as a result of lattice forces in the crystalline state, suggested from the above analysis of crystal contacts, however, is entirely consistent with the results of the independent SAXS measurements carried out with solutions of various enzyme substrate complexes of the highly homologous human PGK [109]. In this work conformational changes compatible with domain closure has been detected with the functional MgATP·PG ternary complex, while no such changes were observed with either PG- or MgATP binary complexes. This result, at the same time, supports the view that simultaneous binding of both substrates to PGK is required for complete domain closure (cf. Chapter II. 2.9.). On the other hand, it has to be noted that, in contrast to the appearance of the open form of the crystalline pig muscle PGK, in the crystal structure of the TSA-complex of HsPGK, determined much later [76] the protein exhibits a completely closed conformation, in spite of the conservative nature of the mammalian PGK sequences. The orthorhombic $P2_12_12_1$ space group of HsPGK crystal, however, is different from the monoclinic $P2_1$ space group of all types of pig muscle PGK crystals, but the same as the orthorhombic space group of TbPGK crystal. Thus, the different geometric arrangements of pig and human PGK molecules in their crystals may lead to different type of crystal contacts and therefore stabilize the greatly different open and closed conformations, respectively.

1.10. Initial Proposal for the Possible Catalytic Mechanism in Relation with the Domain Closure

Even a little after the time of publication of the hypothetical hinge-bending structural change its possible nature was explored by considering what and where the 'hinge' might be on the PGK molecule. The neck region of the PGK molecule is made up two α-helices: helix 7, residues 189-202, and helix 14, residues 396-404 (cf. Figure 1A). These are the last elements of secondary structure in the N- and C-domains, respectively. These two helices are in contact with one another and thus the simplest type of molecular hinge to envisage would correspond to the contact normal - perpendicular to the two helix axes (Figure 12). In these terms the operation of the hinge would correspond to a rotation of one helix relative to the other about their contact normal [123]. A calculation of this hinge bending motion was set up using the atomic co-ordinates of horse PGK. The polypeptide chain was broken at residue 203, i.e., at the C-terminus of helix 7, de-coupling the two domains, which were then treated as rigid bodies in the subsequent rotation. The goal of the rotation was to reduce the distance between the two substrates, the ATP in its known binding site, and the PG in its proposed binding site in the 'basic patch'. A penalty function for the rotation could be estimated from the increase in the number of bad contacts (≥ 3.9Å) between protein atoms as the rotation occurred. At a rotation of the hinge by 26° the distance between the two substrates had reduced from about 12 Å to 5.5 Å, with a penalty of only 6 unfavourable contacts. The inter-substrate distance could be further reduced by increasing the rotation beyond 26° but only at the cost of a steep increase in the number of unfavourable contacts. A rotation about the hinge of 26° was seen to be a reasonable compromise between achieving the goal without producing unfavourable inter-domain contacts. This rotation of 26° about the helix normal also reduced the radius of gyration of horse PGK from 23.4 Å to 21.6 Å. This reduction of 1.8 Å is somewhat greater than the 1.43 Å upper limit observed in the X-ray solution studies of the yeast PGK [102]. However, given that the real conformational change is certain to be much more complex than the simple 'mechanical' model using rigid protein domains described above, the agreement between the observed and modelled change in the radius of gyration is perhaps acceptable. In any event the underlying intention was to determine whether the observed change in radius of gyration of PGK on the formation of the ternary complex could correspond to the kind of hinge-bending conformational change required to bring both substrates together in a catalytic complex. The modelling shows that this is possible.

Later on, in the course of determination of the open structure of the crystalline pseudo-ternary complex of pig muscle PGK some further progress could be made concerning the understanding the "hinge-bending" conformational change. Although, this structure was found in a partially folded state similar to the PG binary complex, rather than the expected fully folded state, it has been possible to propose a plausible catalytic mechanism for the enzyme. The consequence of limiting the "hinge-bending" conformational change in the ternary complex to the same value observed in the PG binary complex results in the 1-carboxyl of the bound PG still being too distant from the γ-phosphate of the bound AMP-PNP for the known in-line direct transfer of the phospho group to occur. Based on the above considerations, it is possible that the packing in the crystal lattice restricts the full

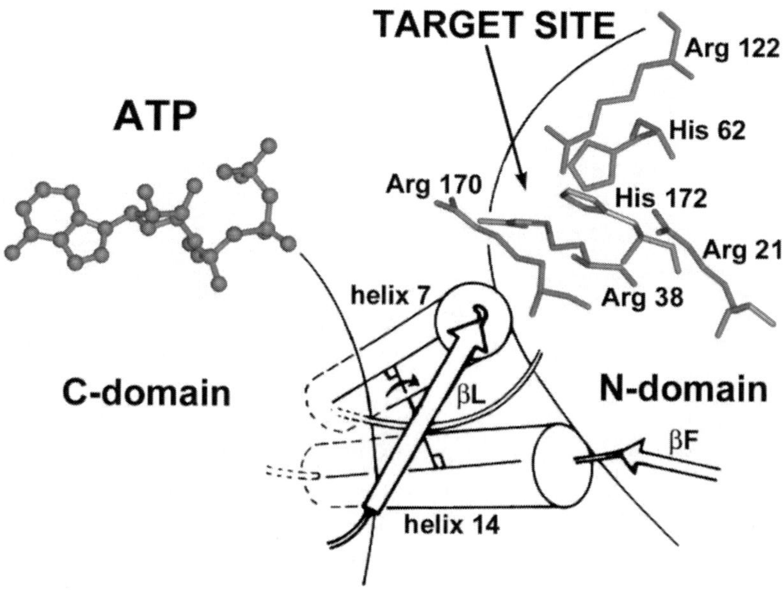

Figure 12. A "helix scissors" mechanism of domain closure: A schematic diagram illustrating the proposed "hinge" in PGK along with the relative positions of the ATP and the target site, based on [123].

conformational change. Examination of the electron density maps revealed a cleft in the protein structure lying between the nucleotide and PG binding sites. The cleft resulted from the lack of close packing of a small number of mostly hydrophobic side-chains from each domain. It was possible to model the closure of the cleft by rotating the two domains as rigid bodies to give a more close-packed domain interface. The radius of gyration (*R*g) of the ternary complex was calculated before and after this cleft closure. The *R*g of the ternary complex before closure was 23.36 Å, the same as for the PG binary complex. After closure of the cleft the *R*g reduced to 21.99 Å. This value represented a reduction from the *R*g for the native horse enzyme of 1.55 Å, which is in reasonable agreement with the reduction of 1.1-1.2 Å ± 0.3 Å reported from X-ray solution scattering with yeast [102], pig muscle [103] and much later with human [109] PGK-s to accompany the formation of the MgATP·PG ternary complex. The result of this modelling was to bring the γ-phosphate of AMP-PNP close to the 1-carboxyl of PG, and close to the N-terminus of helix 14. Helices 13 and 14 move closer together to form, with the side-chain of Arg 38, a tight structure surrounding the reaction groups of the two substrates.

The structure around the γ-phosphate of the nucleotide and the 1-carboxyl group of PG, at the catalytic center of PGK also suggested a possible catalytic mechanism for the enzymatic reaction (Figure 13), the main features of which was approved much later (Chapter IX.). The key protein groups were seen to be the guanidine side-chain of Arg 38, and the helix dipoles [83, 124] of helices 13 and 14, which have important binding roles for the negatively charged reactants. It was noted that the N-termini of both helices 13 and 14 have a highly conserved Gly-Gly-Gly sequence which is characteristic of phosphate binding helices [105], and that Arg 38 was identified at this time as an essential catalytic residue of PGK by site-directed mutagenesis [125]. The metal ion associated with the nucleotide substrates also seems to play an important role in the catalysis.

Initially (Figure 13A), the relative positions of the bound ATP and PG are as interpreted in the pseudo-ternary complex, with one exception. Apart from its orientating effect, the co-ordination of the metal ion to the γ-phosphate, and the hydrogen-bonding of the amido group of Asn 336 to the oxygen bridging the β- and γ-phosphates, may also increase the partial positive charge on the phosphorus atom of the γ-phosphate. This enhances the conditions for nucleophilic attack by the bound PG.

The closure of the cleft described above would reduce the inter-substrate distance to about 5Å as required by the associative transfer, and at the same time bring the γ-phosphate closer to Arg 38 and the N-terminus of helix 14 (Figure 13B). Movement of the γ-phosphate to a position between the N-terminus of helix 14 on the one hand and the guanidine group of Arg 38 on the other may constitute "capture" of the γ-phosphate by these two strongly interacting groups. This capture may fix the "activated" γ-phosphate in a location for nucleophilic attack by the 1-carboxyl group of PG, leading to the formation of a transient penta-coordinate transition state of the γ-phosphorus. These rearrangements may disturb the previous coordination of the metal so that its position linking the α- β- and γ-phosphates is changed to that observed with ADP where it links α- and β-phosphates and Asp 374. The change of the metal coordination is followed by collapse of the transition state to form a metal-ADP and bPG complex bound to the enzyme, which then reverses the conformational changes and releases the products (Figure 13C).

This initial proposal of the catalytic mechanism, even if it is derived from the crystal structure of an unproductive open conformation, is essentially supported and completed by later, more direct observations. As anticipated the closed structure of PG·MgADP ternary complex of TbPGK (Figure 14) provided more clear evidence of the enzyme's mechanism of catalysis. Modelling the substrate bPG (black in Fig 14B) from the bound PG (grey in Figure 14A) and the bound phosphate ion (dark grey in Figure 14A), as observed on separate PGK molecules within the same crystal structure [84], places the phosphorus atom of the transferable 1-phosphate group of bPG just about 4 Å from the β-phosphate oxygen nucleophile of the bound MgADP. This short distance and the proximity of the magnesium ion strongly support the earlier evidence that phospho-transfer occurs through an associative mechanism [46] in which a charged transition state must be stabilised [38]. This transition state was modelled into the closed structure of TbPGK as a penta-coordinated transferring phospho-group, with its characteristic planar configuration, equidistant from the carbonyl oxygen of the bound PG and the oxygen of the β-phosphate of the bound MgADP (Figure 14C). At first instance only three specific stabilising interactions of this negatively charged phospho-group are assigned [74]:

1. The positively charged magnesium ion which, in the presence of bPG, switches from co-ordinating the α- and β-phosphates of ADP to bridging the β-phosphate of ADP and the 1-phosphate of bPG.
2. The guanidine group of Arg Tb39/38/Tm36, which also binds to the PG carboxylate group.
3. The main-chain peptide group of Gly Tb399/396/Tm377 in the first turn of helix 14.

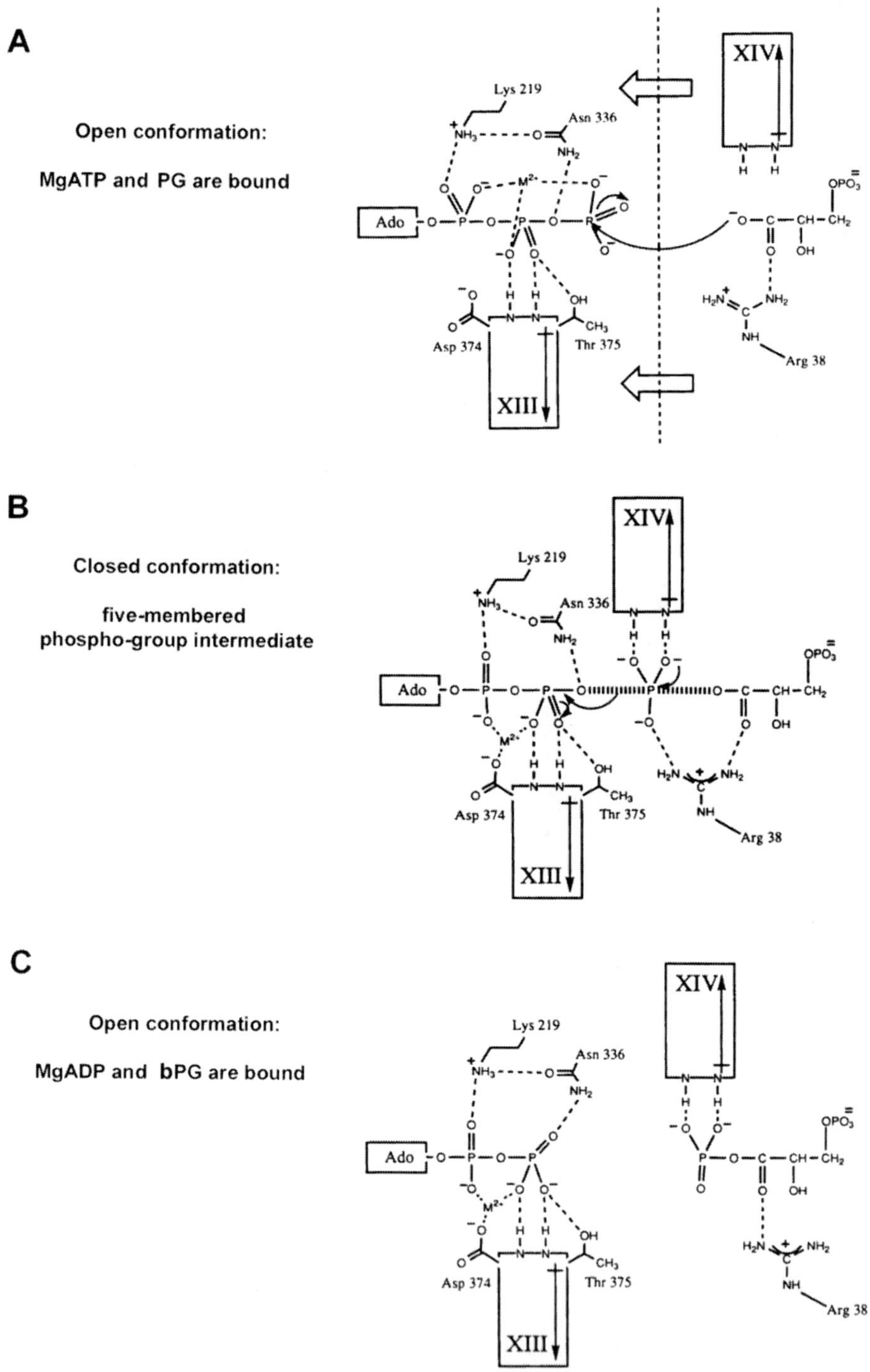

Figure 13. Schematic representation of the proposed catalytic mechanism (adapted from [69]): Helices XIII and XIV are represented by boxes in which the helix dipoles are shown in the conventional manner. Broken lines denote H-bonds. The large arrows in (A) represent the direction of the proposed conformational change, and the heavy broken lines in (B) represent the axial bonds of the proposed penta-coordinated transition state complex.

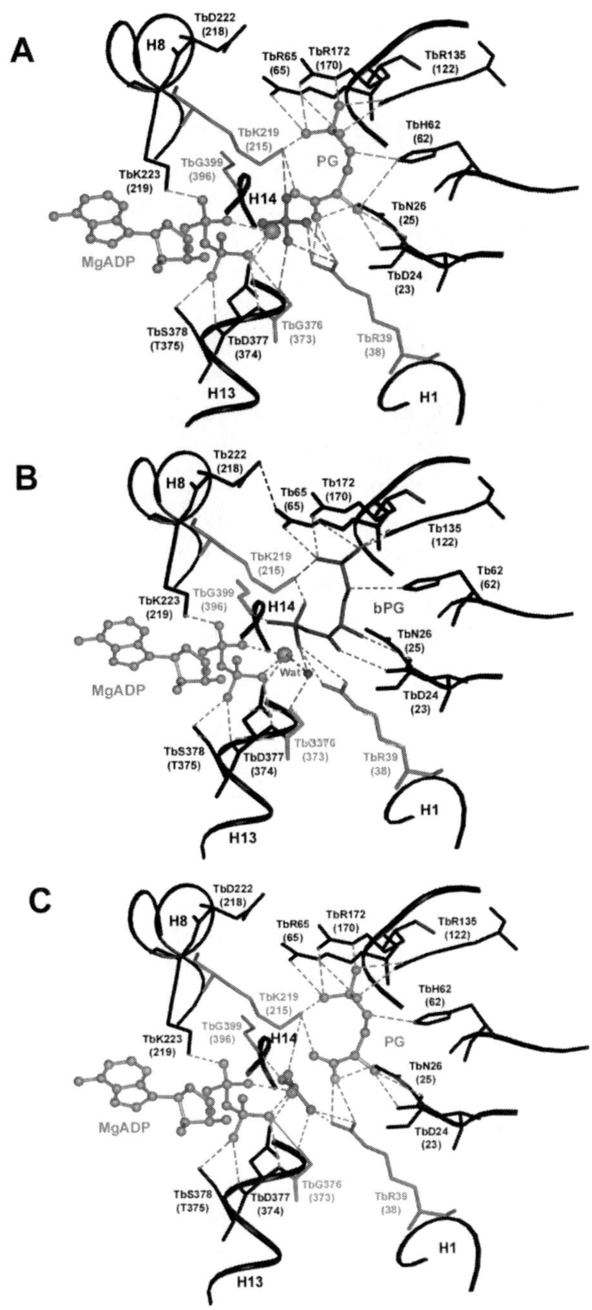

Figure 14. The positions of the catalytic residues in the active site of PGK after domain closure: The surrounding helices (black ribbon) and the interacting residues (black stick models, except of the grey coloured Arg Tb39/38, Lys Tb219/215, Gly Tb376/373 and Gly Tb399/396) around the bound substrates (grey ball and stick models) are shown as determined from the closed crystal structures of TbPGK) [84]. In (A) the dark grey phosphate ion is the one that observed in this crystal structure of TbPGK ternary complex with MgADP and PG crystallised from phosphate [84]. In (B) the possible binding mode of bPG (black stick model) is illustrated as determined by docking into this structure in place of PG. In (C) the trigonal transferring phospho-group is shown (dark grey ball and stick model) as modelled Bernstein and Hol into the space between the bound PG and MgADP [84]. The dashed lines represent H-bonding (up to 3.5 Å) and ionic interactions (up to 4.0 Å).

Later on, from the other closed structure of TmPGK [73] Lys Tm197/215/Tb219 was also suggested, as another positively charged residue, to be involved in the stabilisation of the transition state. On this basis Bernstein et al. [84] further refined their model and included Lys Tb219/215/Tm197 among the residues coordinating the transferring phospho-group, although the electron density of this side-chain could not be localised in the structure of TbPGK, but it was possible to place it in its most favourable conformation (as shown in Figure 14B). The direct role of this lysyl side-chain in the catalysis was approved much later by site-directed mutagenesis [107], by transient kinetic studies of the mutant [126] as well as by the structural studies with HsPGK [76]. Finally, the positive ends of the dipoles of helices 13 and 14 point directly towards the negatively charged phospho-group, which further stabilises the transition state.

The main features of this proposal [84] for the catalytic mechanism of PGK from its abortive ternary complex are very similar to those suggested previously from the PG·MnAMP-PNP pseudo-ternary complex of pig-muscle PGK [69]. It now seems likely that even without determination of the catalytically competent closed structure; the broad outline of the catalytic mechanism of PGK has been described, although the role of the metal ion is still rather speculative. Later on, the model was further refined on the basis of mutational and structural studies carried out with HsPGK and summarised in Chapters II. 2.7. and IX.

1.11. Main Features of Domain Closure: Identification of the Hinges

Perhaps the most interesting revelation from the structure of the folded form of PGK concerns how the large conformational change that is necessary for catalysis is induced by the substrates. Initially, the main important structural elements involved in domain closure have been identified at the level of the secondary structure. This has led to identification of the possible hinge-points in the molecule. Structural comparison of the folded form of the abortive ternary complex of TbPGK [74] with the open forms of pig PGK, and particularly its PG binary complex [15], has yielded not only hitherto unknown structural detail on the catalytic cycle of the enzyme, but important observation of synergistic effects in the substrate-induced conformational changes has been also made. This is based on taking the observed structures of horse [4], porcine [15], Bs [18], and TbPGK [74], as representative, respectively, of the unliganded, PG binary-, MgADP binary-, and ternary-complexes of PGK. On this basis, the structures of the binary complex of MgADP are similar to that of the unliganded form, showing that the binding of nucleotides alone generates no large-scale conformational change. This observation is approved later for the case of MgATP binary complex, too [7]. In contrast, the structure of the PG binary complex does exhibit a conformational change: one in which the N- and C-domains move together by an 8° rotation about a hinge near the N-terminus of the domain-linking helix 7. This rotation would cause the two substrates to move together, but not sufficiently to completely close the gap between them. However in the PG·MgADP ternary complex of TbPGK a rotation of the two domains of 32° is observed which is sufficiently large to move the two substrates close enough together for catalysis to occur. In this case the largest part of the rotation occurs at the C-terminus of the domain-linking helix 7. This is called C-hinge. The N- and C-hinges of PGK are illustrated in Figures 15A and 15B, respectively.

In the light of these observations Bernstein et al. [74] propose that the complete closure of the active site cleft of PGK is brought about by the synergistic combination of the conformational changes induced by PG and MgADP substrates. They suggest that PG binding to the arginine-rich 'basic patch' induces a reorientation of its binding loop which results in a 'piston-like' motion of helix 14, located in the neck region between the two domains, to generate the first 8° bend in the 'hinge'. Helix 14 is then in an appropriate position to make key interactions with the C-domain once MgADP is also bound. The binding of MgADP induces rearrangement of residues at the N-terminus of helix 13, also in the neck region, which couple with changes induced by PG binding to extend the β-sheet interactions between β-strands K and L, preceding helix 13 and helix 14, respectively (for labelling of the secondary structural elements see Figure 1A). These interactions affect the conformational change and stabilise the active form of the enzyme. These changes only occur in the presence of both substrates and act to prevent the futile hydrolysis of bPG or MgATP when bound singly.

This description provided a phenomenological explanation of the main features of domain closure, but it was still far from understanding the operation of the defined molecular hinges at the level of interactions of side-chain and backbone atoms. Furthermore, an extensive comparison of two independent pairs of open and closed ternary complex structures has allowed definition of further hinges; among them the most important one is at the β-strand L [80]. Each pair, chosen for comparison consisted themselves the same nature of the ternary complexes, crystallised under similar conditions. Namely, the structures are compared in the following pairs:

Pair 1, open and closed structures of pig muscle [80] and Tb [84] PGK-s, respectively, both of them complexed with MgADP and PG and crystallised in the presence of phosphate; Pair 2, open and closed structures of pig muscle [69] and Tm [73] PGK-s, both of them complexed with Mg(Mn)AMP-PNP and PG and crystallised in the presence of PEG.

The basis of comparison of remote species is the fact that all PGK species should temporarily adopt both open and closed conformations in solution during the catalytic cycle. The hinges, which are operational during the domain movement, should therefore be present in all the species, irrespective of the differences in sequence. Thus, comparison of the open structures of two different pig muscle ternary complexes with the corresponding closed structures from remote species was carried out with the aims *i)* to locate all the hinges involved in the transition between the open and the closed states, *ii)* to identify differences between the interactions of the substrates with the enzyme in the two states and *iii)* if possible, to describe the effect of the substrate interactions on the conformational state of the hinge(s).

New features of domain closure have emerged from these comparisons. A superimposition of all four PGK ternary complexes optimising the alignment of the backbone atoms of the core β-strands in the N- and C-terminal domains, respectively, is shown in Figures 15A and 15B. The conformation of each individual domain, except for a few source-specific insertions, is similar in all structures even when the relative domain positions are widely different. Thus, the domains move almost as rigid bodies relative to each other.

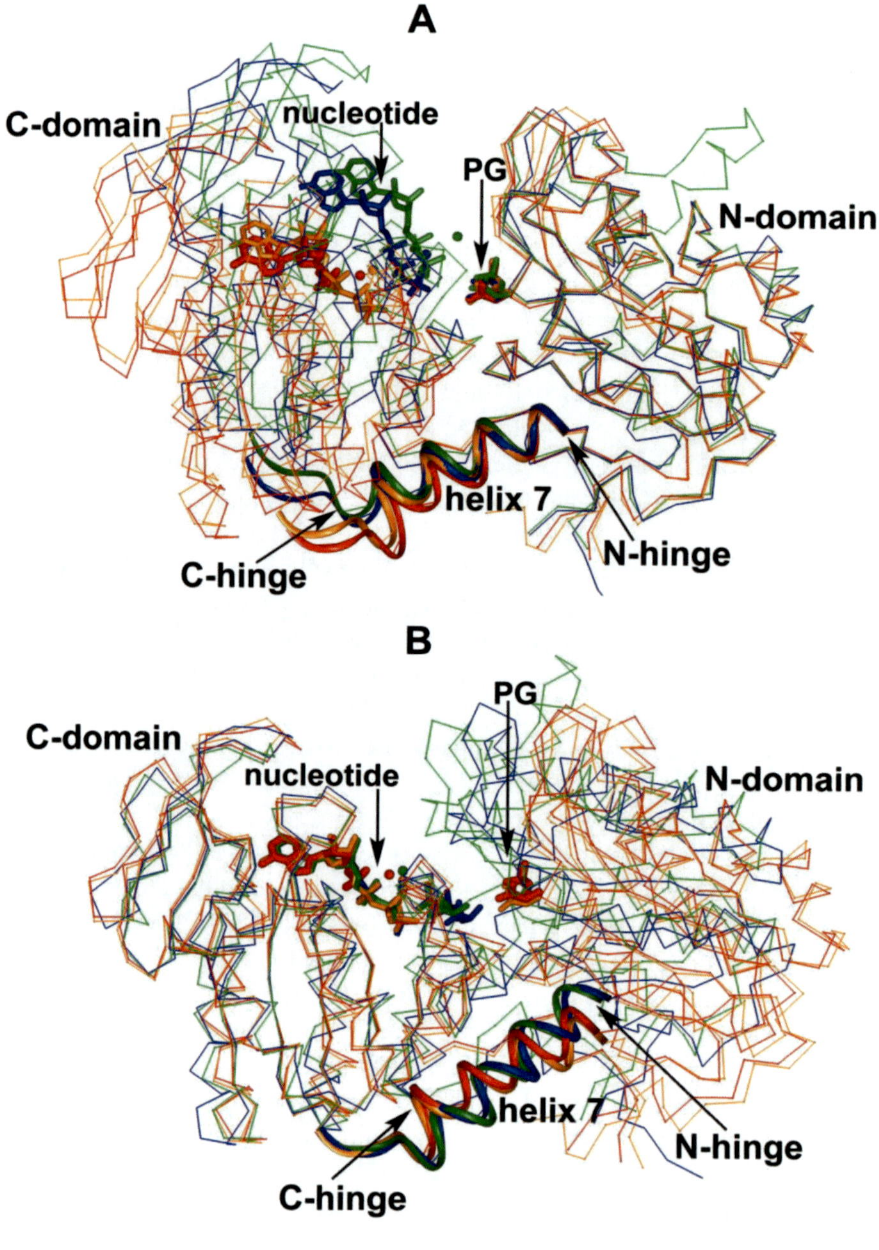

Figure 15. Comparison of the domain positions in the PGK ternary complexes (adapted from [80]): The Cα traces of four PGK ternary complexes: those of PG and MgADP with pig muscle (red) [80] and of the Tb (green) [84], as well as those of PG and Mn/MgAMP-PNP with pig muscle (orange) [69] and of Tm (blue) [73] PGK-s are shown superposed according to the backbone atoms of the core β-strands of the N-domain (A) and of the C-domain (B), respectively. The bound ligands are shown as stick models. In both figures the interdomain helix 7 is highlighted by ribbons of the respective colours.

Of the two known hinge points at the ends of helix 7 suggested previously [74], only the conformation of the C-hinge is significantly different in the open and the corresponding closed ternary structures (Figure 15A). This is confirmed by the rms deviation data of the superimposed helix 7 plus C-domain core β-strands which are much higher than the respective data obtained by separate superimposition of these structural elements (Table 4). Much smaller differences were observed in the data when helix 7 and the N-domain core β-strands were chosen for superimposition, which indicated that in all four ternary complexes the N-hinge is similarly closed, possibly due to the presence of PG bound uniformly to all four structures (Figure 15A).

In addition to these hinges at each ends of helix 7, the presence of another important hinge operating in β-strand L was observed as illustrated in Figure 16A. This strand connects helices 13 and 14, and is situated between the two domains. Helix 13 constitutes an integral part of the C-terminal domain, while helix 14 moves together with the N-terminal domain. These features are supported by the rms deviation data obtained from superposition of the respective structural elements (Table 4) which show that the relative movement of helices 13 and 14 is as large as the relative movement of the C- and N-terminal domains observed when their core β-strands were superimposed. A smaller movement of helices 13 and 14 relative to the C-terminal and N-terminal domains, respectively, can be also deduced from the rms data in Table 4, in agreement with previous observations [84].

Table 4. RMS deviation of various pairs of superimposed elements of the compared open (pig muscle) and closed (Tb and Tm) PGK structures (adapted from [80])

Superimposed part	Sequence numbering (muscle PGK)	Number of superimposed atoms	RMS-deviation (Å/Å²) in pairs of PGK structures: MgADP·PG ternary complex pig - *T. brucei*	RMS-deviation (Å/Å²) in pairs of PGK structures: MgAMP-PNP·PG ternary complex pig - *T. maritima*
helix 7	189-197	72	0.243	0.393
C-domain (core β strands)	207-212(G) 231-236(H) 277-282(I)	144	0.379	0.429
helix 7+C-domain		216	**1.167**	**1.18**
N-domain (core β strands)	17-22(A) 57-61(B) 91-96(C)	136	0.414	0.390
helix 7+N-domain		208	0.578	0.428
C-domain+N-domain		280	**2.583**	**1.905**
helix 13	374-379	48	0.218	0.288
helix 14	399-403	40	0.721	0.350
helix 13+C-domain		192	0.543	0.523
helix 14+N-domain		176	0.659	0.486
helix 13+helix 14			**1.946**	**1.386**
β-strand L	388-397	88	**0.832**	**0.898**
helix 8	219-228*	80	1.279	0.959
	222-228	56	0.821	0.344
helix 8+C-domain		200	0.787	0.686
helix 8+helix 13		104	1.203	0.727
helix 8+helix 14		96	1.978	1.537

*not used in further comparisons

The relative movement of helices 13 and 14 is depicted in Figures 16B and 16C, which also show the domain rotations by movements of the respective core β-strands. β-Strand L, which connects these two helices, must therefore undergo a large conformational change. The rms deviation data (Table 4) support the existence of a significant hinge at this location.

In addition, modelling studies (Szilágyi and Vas, personal communication) showed that the particular backbone conformation of the inter-domain region which is characteristic to the closed PGK molecules, and consists of part of helix 13 and the β-strand L, can position the two domains closer together. This modelling has also demonstrated quite clearly that without operation of the hinges at helix 7, the polypeptide chain would break. Therefore the hinge region at β-strand L and the hinges at helix 7 must act in concert, and both regions possibly have important roles in the positioning of the two domains.

Superimposition of all four PGK structures to optimise the alignment of various structural elements of the two domains and of the inter-domain region has also led to other new observations. It is notable that all the relevant parts, highlighted by the ribbons, are well separated into pairs according to the basic conformational state of the molecule.

The operation of the new putative hinge point in β-strand L results not only in movement of helix 14, but also of the remaining part of the C-terminal end of the chain (helix 15 and the C-tail). This is clear in Figure 16B where the molecules are superimposed according to helix 13. The C-terminal end of the polypeptide chain (residues 398-416) including helices 14 and 15 constitutes an integral part of the N-domain. This feature of the PGK molecule has been known for a long time (as illustrated in Figure 1B), but the role it plays in the folding process [127-129] or in the function [23, 128] has not yet been properly understood.

Figure 16B also shows that helix 13 undergoes a conformational change at its N-terminus, which is located close to the bound nucleotide phosphates. This is one aspect of the conformational changes (ordering) of helix 13 due to nucleotide binding, as discussed in Chapter I. 1.4. At the same time the whole helix 13 is shifted relative to helix 14 when the molecules are superimposed according to helix 14 (Figure 16C).

Helix 8 moves considerably relative to the whole C-domain. Superimposition, optimising the overlap of the core β-strands of the C-domain (Figure 17A), clearly indicates that helix 8 moves more or less independently of this domain, although it constitutes an integral part of the C-domain (cf. also rms deviation data in Table 4). As a consequence, helix 8 also moves relative to helix 13 (not shown); the latter one is being moved more or less together with the remaining part of the C-domain. The relative movement of helices 8 and 13 is important in respect of formation of the active site (cf. Chapter IX.). The pronounced movement of helix 8 occurs at a hinge between its N-terminus and the C-terminus of β-strand G, which connects helices 7 and 8. The operation of this hinge is directly mediated through an interaction with the nucleotide as discussed later (Chapter X.). The C-hinge of helix 7 is situated at the N-terminus of β-strand G and thus may be indirectly affected.

It is also evident from Figures 17A and 17B that helix 8 undergoes substantial conformational changes, especially in its N-terminal part located close to the bound nucleotide. The superimposition optimised for the C-terminal part of helix 8, shows clearly that the first turn of this helix is better formed in the closed structures, but that the rest of it is more ordered in the open structures. The H-bonds along this helix are significantly weaker in both closed structures, resulting in elongation of the helix as a whole, in spite of the fact that its first turn has become more ordered.

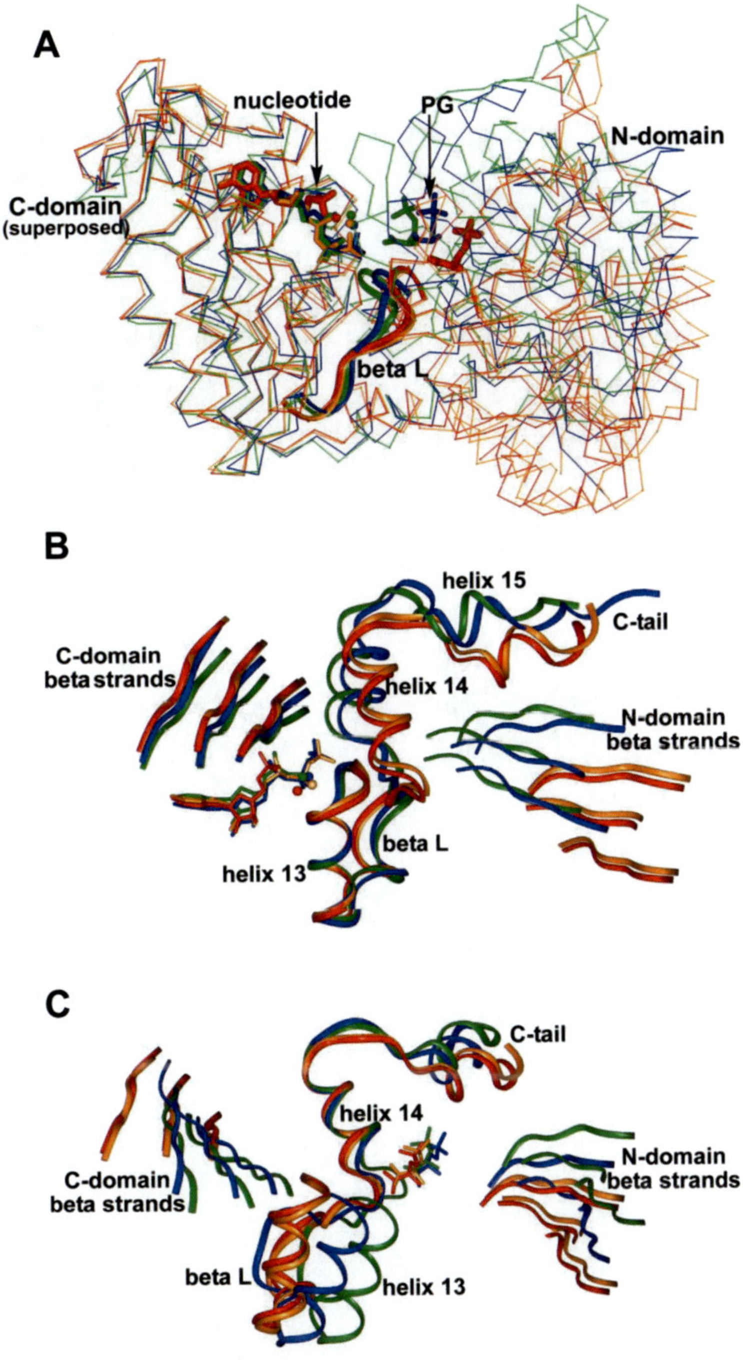

Figure 16. Operation of the main hinge at β-strand L (adapted from [80]): In (A) the Cα traces of four PGK ternary complexes: those of PG and MgADP with pig muscle (red) [80] and of the Tb (green) [84], as well as those of PG and Mn/MgAMP-PNP with pig muscle (orange) [69] and of Tm (blue) [73] PGK-s are shown superposed according to the backbone atoms of the core β-strands of the C-domain. The β-strand L is highlighted as ribbons of the respective colours. In (B) and (C) the four molecules are superposed according to the backbone atoms of helix 13 and helix 14, respectively, and for clarity the Cα traces are not shown, only the position of the N- and C-domain β-strands are illustrated by ribbons.

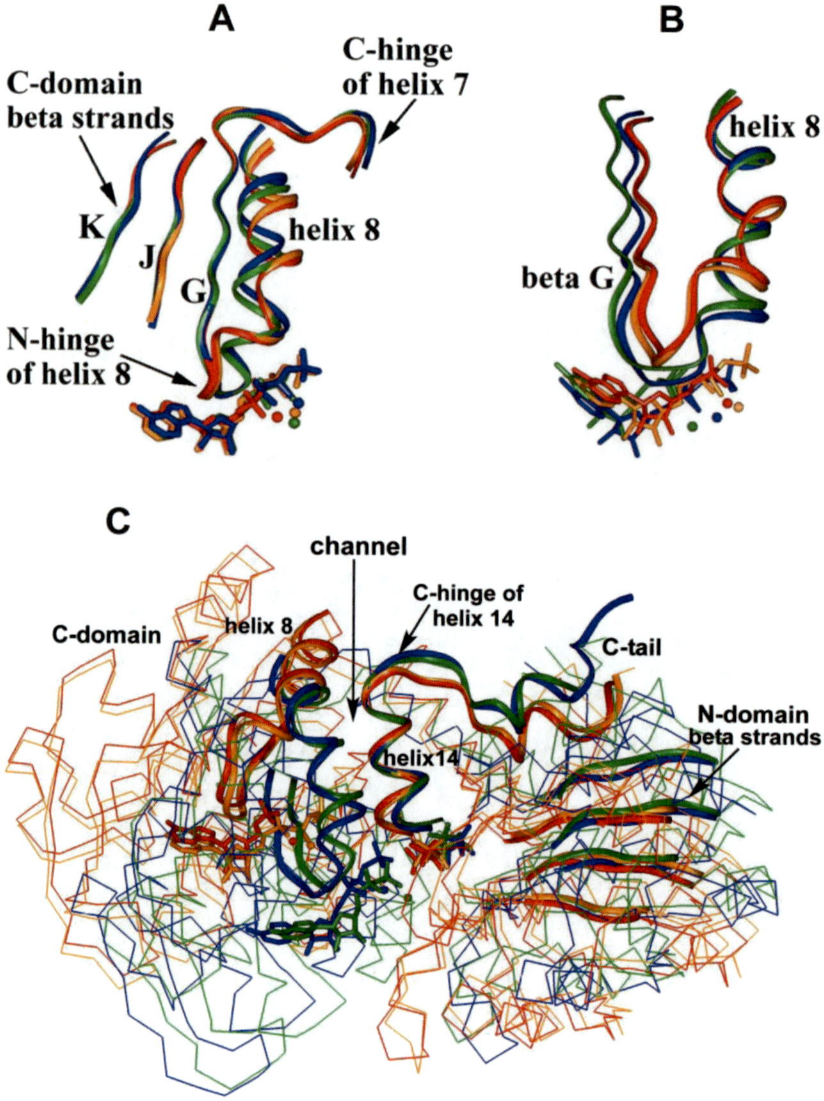

Figure 17. Important changes in the secondary structure between the open and closed states (adapted from [80]): Four PGK ternary complexes: those of PG and MgADP with pig muscle (red) [80] and of the Tb (green) [84], as well as those of PG and Mn/MgAMP-PNP with pig muscle (orange) [69] and of Tm (blue) [73] PGK-s are superimposed according to the backbone atoms of (A) the core β-strands of the C-domain, (B) the C-terminal part of helix 8 and (C) the whole helix 14. In the latter case the PGK molecules are also represented by the Cα traces. Ribbons of the same colour code highlight the investigated secondary structural elements and the bound ligands are also shown in stick models.

Helix 8 also moves relative to helix 14 (Figure 17C). This motion brings helix 8 of the C-domain into a position more parallel to helix 14 (which belongs to the N-domain) and thus the open channel between the two domains can be closed by complementary side-chain interactions, as it has been already proposed [69]. From this superimposition, which optimises the overlap of helix 14, an additional hinge has been observed between this helix and the rest of the C-terminal part. The C-tail including helix 15 moves together with the core β-strands of the N-domain, while helix 14 does not completely follow this motion (Figure 17C).

Thus, in addition to the hinge point at β-strand L, which is mainly responsible for directing the relative domain positions, two hinge points have been described: one associated with the loop between β-strand G and helix 8 and the other located between helix 14 and the short helix 15 at the C-terminus of the molecule. These two hinge points play an important role in closing the channel between the domains.

It is also notable that location of the above-described hinges has been approved much later when the X-ray structure of the closed conformation of HsPGK has been determined [76]. As it is discussed in Chapter IX., HsPGK crystallised in a closed conformation highly similar to that of TbPGK. Thus, the more straightforward direct structural comparison of the sequentially highly similar mammalian PGKs (pig muscle and human) gave very similar results to the above analysis.

This identification of PGK hinges from X-ray structural data, together with various enzymological, site-directed mutagenesis and other structural (NMR, X-ray scattering, fluorescence, etc.) studies of the solubilised enzyme, as discussed in the following Chapters, have finally led to a detailed description of the differences between the interactions of the substrates with the enzyme in the two states and allowed to describe the effect of the substrate interactions on the conformational state of the hinges. Thus, the molecular mechanism of PGK domain closure and the accompanying events at the catalytic site, at the level of atomic interactions, will be outlined below.

In: Phosphoglycerate Kinase: A Hinge-Bending Enzyme ISBN: 978-1-62808-836-6
Editor: Mária Vas

Chapter II

Structural Information as Derived from Studies with Solubilized PGK

Mária Vas
Institute of Enzymology, Research Centre for Natural Sciences,
Hungarian Academy of Sciences, Budapest, Hungary

Although in solution it is not yet possible to obtain as detailed structural picture about proteins with a size of PGK as obtained by crystallography, any solution data concerning the structure are extremely important. PGK, similar to other proteins, functions in solution, i.e., in a physical state, essentially different from the crystal. While in the crystal the inherent flexibility of protein molecules may be inhibited or distorted by the lattice forces, in solution the functionally relevant conformational changes, e.g., the large-scale domain motions are allowed to take place. Thus, in contrast to the static picture given by crystallography, solution studies can report about the functioning conformational states of protein molecules.

2.1. NMR Spectroscopy

Among the various biophysical methods NMR is the most powerful one in determining the three-dimensional structure of proteins in solution. Although, even now (at the beginning of the 21st century) there is still a size-limit for macromolecules, it is able to provide sufficiently detailed structural information about solubilised proteins. Furthermore, it is an important complementary method to crystallography, not only for basic structural point of view, since it can also report about dynamic properties of proteins and can determine the mobility of various local regions of the molecule.

The first NMR studies were undertaken with yeast PGK and provided useful structural information much before the X-ray structure was available. Even the primary sequence was not known when Tanswell and coworkers [130], using proton NMR spectroscopy, were able to identify three His residues. Furthermore, a pH-titration enabled assignment of C-2 and C-4 imidazole protons of each of three histidines and the determination of their p*K* values as 8.20 (one of them) and 7.05 (two of them), respectively. The former histidine was deduced to lie buried within the polypeptide chain while the latter pair lies in practically identical environments on the surface of the enzyme exposed to solvent. At that time the assignment of each His was not possible, but introduction of the paramagnetic lanthanide cations (Gd, Pr or Eu) in place of Mg^{2+} in MgATP or MgADP enabled to collect geometrical data and to map the active-site region around the bound lanthanide. A line width titration with Gd^{3+} allowed an estimation of the relative distances of the nucleotide protons and of the three His residues from the paramagnetic probe. The metal-histidine distances were calculated from the relaxation data and found that the histidines are close to the phosphate chain of ATP. Two of the three His resonances were also broadened by binding of PG. On this basis it was postulated that at least part of His residues are located at or close to the active site. In a following work resonances of the two His residues were assigned as resonances from two well conserved histidines, namely His Y62/62 and His Y167/169 [131], that were shown later to be the constituents of basic patch on the inner surface of the N-domain [132], where the PG binding site, indeed, was later localised by crystallography [15].

There is also another remarkable observation that was made already in this early NMR work [130]. The distances of His residues from the MgATP site were found to be considerably shorter (6-8 Å) than those (at least 12-15 Å) seen by X-ray structure determination [5, 60]. As pointed out by Joao and Williams [36], this difference may be due to the inherent differences of the two physical methods applied, but can be equally due to the different behaviours of the solubilised (more flexible) and of the crystalline (more rigid) protein.

In subsequent NMR studies [131-134] procedures were adopted in order to enhance the resolution of the observed resonances and extend the assignment to the various protein side-chains.

In a one dimensional proton NMR spectroscopic measurements the photochemically induced dynamic nuclear polarization (CIDNP) technique was used with flavin dyes to identify His, Trp and Tyr resonances in case of both yeast and horse muscle PGK-s [133]. Surprisingly, the resonances of the conserved residues of the two different PGKs could not be similarly detected. A remarkable observation, however, has made with the horse muscle PGK. The addition of PG and MgATP together resulted in the appearance of two additional resonances in the CIDNP spectrum due to a His residue (tentatively assigned to His Y388/390) that is inaccessible to flavin in both the enzyme alone or in its binary complex with PG. Since this His residue is located in the interdomain region [4] this observation indicates the occurrence of a conformational change in this region, upon the simultaneous action of both substrates, that may be important in the domain movement.

A better resolved assignment of NMR resonances are greatly helped by protonation of titratable groups and by chemical modification [131] or by genetically engineered mutation of certain side-chains [132]. For example, the pH-dependence of His resonances (combined with crystallographic data) has been used in making assignment. This was further supported by NMR investigation of site-specific mutants of the 'basic pach' residues of the N-domain and let to firm sequence-specific resonance assignment to His Y62/62, His Y167/169 and His Y170/172. Typical proton NMR spectra of yeast PGK are illustrated in Figures 18A and B and a list of the assigned resonances in the aromatic region is shown in Table 5. In general, the most resolvable peaks in the spectra are assigned to resonances of the very exposed histidines or other (tyrosine and tryptophane) residues on the surface, but especially in the substrate-binding cleft and in the interdomain hinge region of the molecule. From relaxation data it was also deduced that the peaks of the NMR spectrum assigned above represent residues with relatively long relaxation times and therefore relatively high mobility. Comments similar to that given for the aromatic region of the spectrum also apply to the aliphatic region.

Table 5. Characterisation of the observable aromatic region of the NMR spectrum of yeast PGK (based on the work by Wilson and coworkers [131])

Assigned residue	Labelling of the NMR peaks[a]	Structural position of the assigned residues
His Y52/Asp 52 (exposed)	2a, 2b	N-domain (not conserved)
His Y53/Gly 53 (exposed)	9,.10, 11	N-domain (not conserved)
His Y123/124 (exposed)	1, 5	N-domain (not conserved)
His Y149/Ala 151 (exposed)		N-domain (not conserved)
His Y388/390 (exposed)	not seen	hinge region (β sheet L, conserved)
His Y62/62 (surface)		near 3-PG site (loop, conserved)
His Y167/169 (surface)	3, 4, 5a, 12	near 3-PG site (loop, not conserved)
His Y170/172 (surface)		near 3-PG site (loop, not conserved)
Tyr Y193/195 (hinge)	14a, 8a	hinge region (helix, not conserved)
Tyr Y380/Trp 382 (buried)	not seen	C-domain (helix, not conserved)
Trp Y308/310 (surface)	7b, 8c	C-domain (adenine site, not conserved)
Trp Y333/335 (buried)	not seen	C-domain (β sheet, conserved)
Phe Y163/165 (near hinge)	15-18 and 7-14	N-domain, close to 3-PG site (loop, conserved)
Phe Y185/187 (hinge)		hinge region (helix, not conserved)
Phe Y194/196 (hinge)		hinge region (helix, not conserved)
Phe Y340/342 (buried)	not seen	near adenine site (loop, conserved)

[a]numbers refer to the labelling seen in Figure 18A.

These higher resolution spectra also permitted to study the enzyme-substrate interaction in greater detail. Table 6 summarises the observations concerning PG and the nucleotides (analogues). For PG its influence on the resonances of three His-s in the 'basic patch' as well as those of Tyr Y193 and Phe-s Y163, Y185 and Y194 in the interdomain region are demonstrated. This result suggested binding of PG to the 'basic patch' of the N-domain even before it is evidenced by crystallography. For the nucleotide substrates (binding of which to the C-domain was known at that time from crystallography [4, 5]), surprisingly, influence on the same resonances as for PG was noted. Since the effect was the highest for the tetraphosphate analogue of MgATP and the smallest for MgADP, it was plausible to assume that the phosphates of the nucleotides is not as far from the 'basic patch' of the N-domain as observed in the crystal and, at least part of the time, may be able to interact with it. – Experiments were also done with ^{31}P-labelled substrates [131]. It was found that the paramagnetic Gd^{3+} (in place of Mg^{2+}) added to PGK in the presence of AMP-PNP and PG broadens the ^{31}P-resonances of both ligands. The most obvious interpretation of this interesting result is that, for some of the time the metal Gd^{3+} is very close to the phosphates of both ligands. The significance of these findings will be discussed later (Chapter VIII.) in the context of the function.

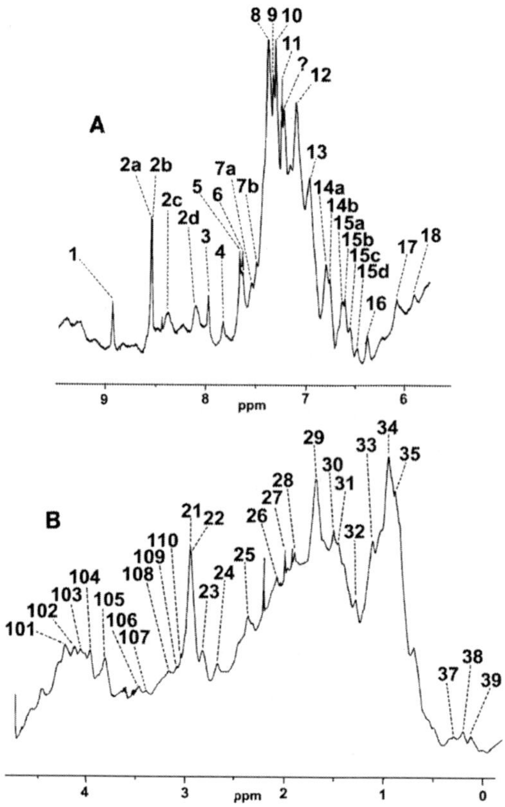

Figure 18. The 500-MHz proton-NMR spectra of yeast PGK: The upper figure (A) shows the aromatic resonances and the lower figure (B) shows the aliphatic part of the spectrum, based on the work by Wilson and coworkers [131]. Assignment of part of the resonances is listed in Table 5.

Table 6. Proton NMR resonances of yeast PGK affected by binding of various ligands

Ligand[a]	Peaks[b] shifted (s) or considerably broadened (b)
PG	1, 3, (4), 5, 12, 14a
Adenosine	8(s)(b), 12(s)(b), 13(s)(b), 15(s)(b)
MgADP	same as for adenosine plus 3(s), 5a(s), 12(s), 14a(b), 7a(b)
MgATP	same as for adenosine plus 3(s), 5a(s), 12(s), 4(b), 7a(b), 14a(b), 15b(s), 16(s), 17(b)
MgAMP-PNP	same as for MgATP
Mg-PPP_i	3(s), 4(s), 5a(s), 12(s), 7a(B), 14a(B), 15b(b)
Adenosine + Mg-PPP_i	additive effects of the two ligands
$MgAP_4$	same as for ATP but stronger effects

[a]binding were studied over a wide range of concentrations
[b]numbers refer to the labelling seen in Figure 18A

In the followings an extension of the proton NMR study of yeast PGK towards the two-dimensional (2D) methods and selective deuteration of Phe residues allowed not only an improved resolution of the aromatic region of the spectrum, but also further insight into the effects of MgATP and PG on the spectrum and thereby on the enzyme conformation [134]. Especially, interesting changes have been observed upon PG binding in chemical shift and dipolar couplings that can be interpreted as being due to a change in the mutual orientations of Phe Y185/187, Tyr Y193/195, Tyr Y48/Phe 48, His Y52/Asn52, and His Y53/Gly 53. These residues are located in helices 1 and 7 (this latter interdomain helix is numbered as 5 in yeast PGK structure [5]), thus, these helices are reoriented as a result of PG binding. The reorientation of these two helices has been indeed observed and confirmed by later X-ray crystallographic studies [15, 74]. A further interesting observation was that NMR resonances in the interdomain region are perturbed less by nucleotide binding than by PG binding. This indicated that the conformational changes in the interdomain region of the enzyme induced by PG binding are different and probably larger in magnitude than those by MgATP binding.

Detailed Studies on Interaction with the Nucleotides

Among the substrates most of the NMR studies are dealing with the binding of the nucleotide substrates to PGK, i.e., of the ligands possessing great variety of stereochemical conformations, at least in their free form in solution. Free ADP and ATP are floppy molecules with two main internal mobilities with large amplitudes and low activation energies, namely, the glycosidic bond rotation of the adenine with respect of ribose, and the reorientation of the phosphate chain with respect to the adenosine moiety, in addition to the low-amplitude ribose puckering. The configuration complexity is further increased due to the various possible ways of interaction of the nucleotide phosphates with the metal ion (Mg^{2+}). For an enzyme bound nucleotide only a single (or a greatly restricted number) specific conformation(s) is (are) reasonable to assume. In case of PGK a further complication is the observed differences in MgADP and MgATP binding to the enzyme (cf. below), as well as to the uncertainties in the number of the nucleotide sites of PGK.

Even in an early NMR study [135] it was observed that the metal-ion differently affects the binding modes of ADP and ATP to PGK. Namely, using the paramagnetic Mn^{2+} (that can effectively replace Mg^{2+} [3]) proton NMR titration curves indicated that the metal complexes of various nucleotide diphosphates (ADP, IDP, GDP) binds approximately two orders of magnitude more tightly to PGK, compared to the free nucleotide. This synergism only operates in binding of the Mn-nucleotide *di*phosphates, and not for the *tri*phosphates [135]. Furthermore, ^{31}P NMR measurements indicated that sulfate ion (a known crystal growing agent in protein crystallography) interferes with Mg^{2+} chelation to the enzyme bound ATP and ADP. The effect on enzyme complexes is stronger for the E·MgATP complex than for the E·MgADP complex [136].

Another relevant observation made in the latter work, which is however similar for all substrates, that ^{31}P chemical shifts of enzyme bound MgADP, MgATP and PG (either bound separately in the binary complexes or bound together in the ternary complexes) do not show any discernible variation in the pH range 6.2-9.0. Since all these substrates contain titratable phosphate groups in this pH range in free solution, this result suggests that in the enzyme bound substrates these phosphate moieties are well sequestered from bulk solution. However, such a conclusion is based on the implicit but questionable assumption that pH dependence of the chemical shifts of substrate nuclei in the enzyme-bound complexes are determined by the same factors as those in free solution [137].

In the knowledge of detailed resonance assignments in the proton NMR spectrum of yeast PGK, careful titrations were carried out with both ATP [138] and ADP [139], both in the absence and presence of Mg^{2+}. While slight differences have been observed in interactions of MgADP and MgATP phosphates with the residues of Asp Y372/374 and Thr Y373/375 at the N-terminus of helix 13 (corresponding to helix 12 according to the numbering of yeast PGK structure), binding of the adenine moiety of both nucleotides is found to be very much similar in the C-domain. The only aromatic residue in the C-domain was Phe Y342/Trp 344, which have been assigned and its resonances are seen to be affected by nucleotide binding at the hydrophobic site.

The most remarkable findings of these titrations [138, 139], however, are the similar behaviours of ATP and ADP. From the change in the chemical shift of the basic patch histidine resonancies (His Y62/62, Y167/169 and Y170/172) and the nucleotide C8H, C2H and C1'H resonancies, it became apparent that there may be at least two binding sites for each nucleotide. Downfield shifts observed for the above histidine resonancies at low nucleotide/enzyme ratio indicates that the primary binding site involves electrostatic interactions between the nucleotide phosphate chain and the basic patch region of the N-terminal domain. The secondary binding site is shown to involve predominantly hydrophobic interactions between the adenosine moiety and the protein. Evidence from the other two-dimensional NMR experiments [140] suggests that the secondary site is equivalent to the crystallographically observed catalytic site. Furthermore, the affinity of the catalytic site is increased relative to the primary electrostatic site with increasing Mg^{2+} concentration, which is valid for both ADP and ATP. Thus, the primary binding site of both ADP and ATP, in the absence of Mg^{2+}, involves strong electrostatic interaction between the di-or triphosphate chain of the nucleotide substrate and the PG-binding site in the N-domain. This binding mode is most probably an unproductive binding mode of the nucleotides and may be similar to binding of other inhibitory anions to this site (see below). Only in the presence of excess Mg^{2+} relative to either ADP or ATP assures a primary binding to a predominantly hydrophobic site, which is characterised by the specific interaction of the adenosine moiety with the catalytic binding site in the C-domain. This picture, however, different from the results obtained by Ray et al., who also detected two binding sites for both ATP and MgATP [141], as well as for ADP and MgADP [142], but in contrast to Fairbrother et al. [138], suggested that the interaction with the catalytic site is much stronger than with the secondary site.

There are several other works, which mainly deal with the conformation of the bound nucleotide itself. Pappu and Serpersu prepared perdeuterated yeast PGK and by this way the proton signals of the bound substrates were visible in the ^{1}H NMR spectrum even at low substrate to enzyme ratio [143]. This method greatly improved the two-dimensional correlated spectroscopy: it eliminated the overlap between the enzyme and substrate nuclear Overhauser effect cross-peaks and also permitted observation of the weak transfer NOE cross-peaks between the substrate protons that are greater than 4 Å apart. The intensity of NOE cross-peaks was used to determine the interproton distances of the enzyme-bound MgATP. These distances were then used in model building studies to determine the conformation of the bound MgATP [144]. The average conformation is *anti* with O4'/C2' *endo* ribose pucker and *trans, gauche* about the C4'-C5' bond. This result is apparently compatible with the high resolution crystal structures of PGK nucleotide complexes, where both C2' *endo* [4] and O4' *endo* [6, 69] ribose ring conformations have been observed. Since the two conformations are energetically not very different, it is very likely that in solution the enzyme-bound nucleotide may not have a single rigid conformation at the active site, instead, it may be flexible to vary its conformation within a range representing an ensemble of conformations.

In order to determine the location of the cation with respect to the phosphate chain of the nucleotides in case of ATP-utilizing enzymes, the natural Mg^{2+} is often substituted by the paramagnetic cations Mn^{2+} or Co^{2+} *in vitro.* This substitution allows the determination of the structure of the enzyme-bound substrates through measurement of distance-dependent enhancement of the spin-lattice relaxation rates of substrate nuclei due to dipolar interactions with the paramagnetic cation [145]. ^{31}P relaxation measurement of this kind have been published for the enzyme-bound nucleotide complexes of yeast PGK and determined the distances of Co^{2+} from the phosphates of both ADP and ATP [141]. It was found that the cation is directly coordinated to all the phosphate groups of the nucleotide in both the CoADP and the CoATP complexes. The metal-nucleus distances were all the range of 2.7-4.1 Å, almost appropriate for direct coordination. The coordination of the metal to both α- and β-phosphates of ADP is in good agreement with all the crystallographic data obtained with various PGK-s [5, 6, 84]. However, for ATP bound to PGK, the conclusion of direct coordination of the cation to all three phosphate groups does not agree with the X-ray results that place the Mg^{2+} coordinated only to the γ-phosphate [4] or to β- and γ-phosphates [7] of ATP. Coordination to all three phosphates have been observed, however, in the crystal structures of Tm [73], yeast [70] and pig muscle [69] PGK-s where the analogue AMP-PNP substituted the bound ATP. It is difficult to decide whether these observations indicate real differences between crystal and solution or reflect a kind of mobility of the metal with respect of the nucleotide phosphates. This problem will be discussed in Chapter VIII. in the context of role of metal in the catalysed phospho-transfer.

While ^{31}P relaxation measurements allowed only to determine the location of the metal cation with respect to the phosphate chain, the availability of ADP and ATP uniformly labelled with ^{13}C made it possible to determine a sufficient number of distances between the cation (which is chelated to the phosphate chain) and the nuclei in the adenosine moiety [146]. In this work a complete characterisation of the conformations of MnADP and MnATP bound to the active site of yeast PGK is presented. The relatively long lifetime of the paramagnetic complexes with Mn^{2+} as the cation allowed determination of the distances in the expected range of 5-12 Å. The ^{13}C relaxation measurements were performed on the enzyme-bound complexes E·Mn(^{13}C)ADP and E·Mn(^{13}C)ATP at three distinct ^{13}C NMR frequencies. The frequency dependence of the relaxation data were analysed in order to evaluate distances for all ten ^{13}C nuclei in the adenosine moieties of E·MnADP and E·MnATP. These distance data, taken together with the previously determined cation-^{31}P distances [141], have been used as constraints in molecular modelling, in which molecular dynamics simulations and energy minimalisation have been performed and led to the conformations of the bound nucleotides, compatible with the distance data shown in Table 7. The atomic numbering and the nomenclature of the torsion angles characteristic of the nucleotides are illustrated in Figure 19. The model assumed a single enzyme-bound conformation for each of the nucleotides. The details of the enzyme-bound MnADP and MnATP conformations, characterised by the various torsion angles (listed in Table 8) are well distinguishable from each other, indicating that structural alterations may occur in the enzyme-bound reaction complex as the enzyme turns over. For example, when the adenosine moieties in the bound structures of MnADP and MnATP are superposed (Figure 20A), the cation is found to be displaced by about 2.4 Å-s, suggesting that the structural changes may involve movement of the metal-ion. A similar, but somewhat smaller (about 1-1.5 Å) change in the position of the metal ion was noted by comparison of the crystal structures of the bound MgADP and the ATP-analogue MgAMP-PNP (Figure 5A). Apart from this similarity, however, the NMR-determined structures of the enzyme-bound nucleotides are significantly different from those in published X-ray crystal structures of the enzyme nucleotide complexes (Tables 7 and 8 as well as Figure 21).

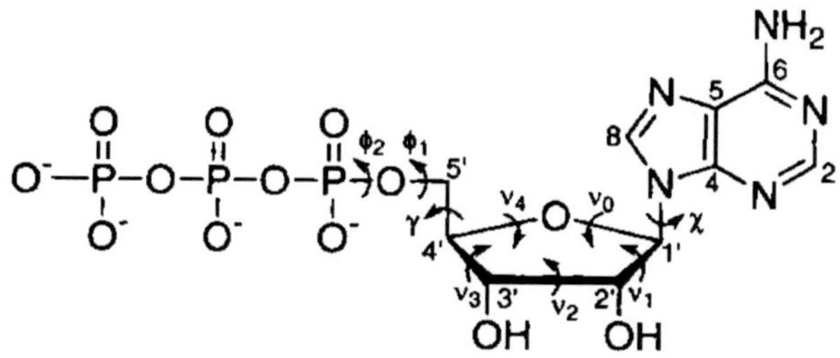

Figure 19. Numbering of the carbon atoms and designation of the torsion angles in ATP molecule as defined in [146] (adapted with permission from [146]; Copyright (1999) American Chemical Society): Curved arrows show dihedral angles that characterise the nucleotide conformation. The cation bound to the phosphates is not shown.

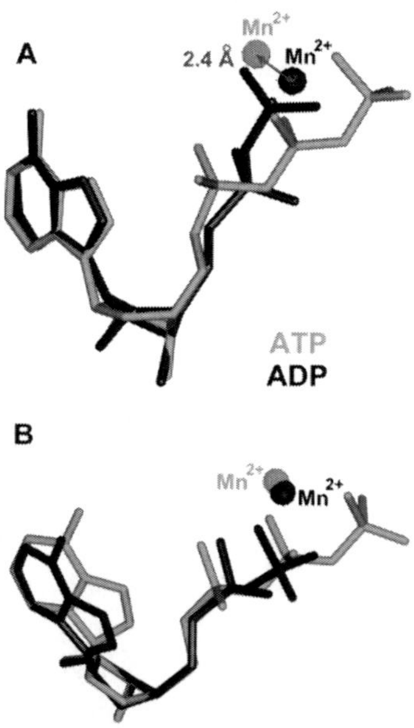

Figure 20. Conformations of PGK bound MnADP and MnATP bound as determined by NMR: The NMR models of MnADP (black) and MnATP (gray) obtained by Ragunathan and coworkers [146] (the coordinates were kindly provided by the authors) are superimposed according to all heavy atoms of adenosine group (A) or according to all heavy atoms of the ribose ring (B).

Table 7. Mn^{2+}-^{13}C distances for the labeled carbons in the nucleotide calculated from NMR relaxation measurements and their comparison with the X-ray structural data (adapted with permission from [146]; Copyright 1999 American Chemical Society)

Carbon atom	MnADP[a] (NMR model)	MgADP[b] (X-ray structure as bound to Bs PGK [18])	MnATP[a] (NMR model)	MgATP[c] (X-ray structure as bound to yeast PGK [5])
C2	9.1	11.5	9.0	12.7
C4	8.6	9.9	8.1	11.0
C5	8.1	11.0	7.3	11.9
C6	8.1	12.3	7.7	13.1
C8	7.6	9.4	6.7	10.2
C1’	8.7	7.5	9.0	8.6
C2’	8.6	7.4	8.6	8.0
C3’	8.8	6.0	8.6	7.2
C4’	8.2	5.3	8.2	6.7
C5’	6.6	4.9	6.7	7.0

[a]Distances for the model were derived from the final minimized structures and were within ± 0.2 Å
[b]PDB file name 1PHP
[c]PDB file name 3PGK

Table 8. Various torsion angles characterizing the NMR-determined enzyme-bound nucleotide structures compared with those from published X-ray structures (adapted with permission from [146]; Copyright (1999) American Chemical Society)

Designation	Torsion	MnADP[a] (NMR model)	MgADP[b] (X-ray structure as bound to Bs PGK [18])	MnATP[a] (NMR model)	MgATP[c] (X-ray structure as bound to yeast PGK [5])
χ	O4'-C1'-N9-C8	33.0	66.9	35.3	-8.5
γ	O5'-C5'-C4'-C3'	50.0	175.4	47.2	-109.3
φ_1	C4'-C5'-O5'-Pα	-174.0	119.0	165.7	-83.2
φ_2	C5'-O5'-Pα-Oαβ	-70.0	54.8	59.5	-148.5
υ_0	C4'-O4'-C1'-C2'	32.3	-25.5	10.8	2.7
υ_1	O4'-C1'-C2'-C3'	-6.5	37.8	12.9	22.3
υ_2	C1'-C2'-C3'-C4'	-19.6	-34.2	-30.1	31.7
υ_3	C2'-C3'-C4'-O4'	39.4	20.1	36.3	-31.8
υ_4	C3'-C4'-O4'-C1'	-46.0	3.2	-29.9	18.6

[a]The values for the angles in the model were derived from the conformation search on energy-minimized structures.
[b]PDB file name 1PHP
[c]PDB file name 3PGK

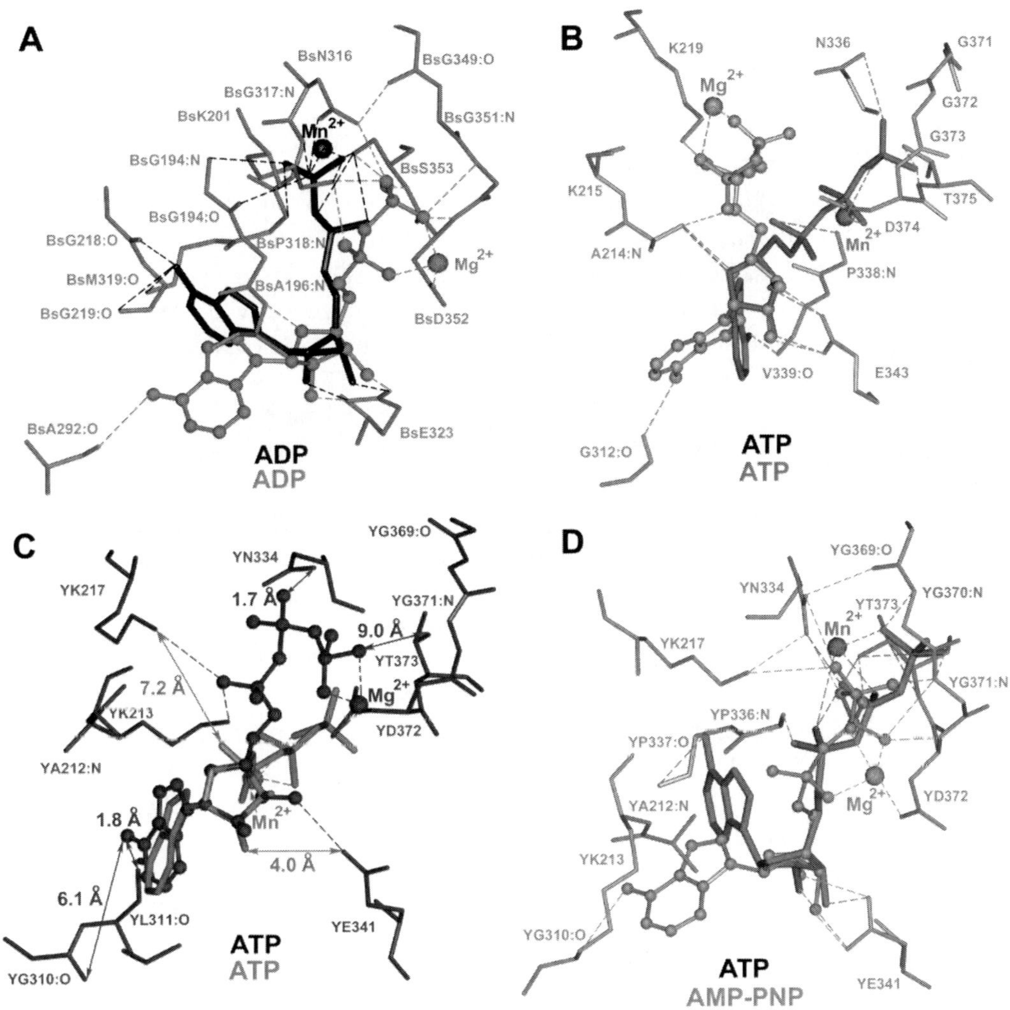

Figure 21. Comparison of the NMR-derived structures of PGK bound MnADP and MnATP with the crystallographically determined nucleotide conformations: The NMR model of MnADP (black sticks) is superposed onto MgADP (grey ball and sticks) that bound in the crystal structure of *Bs*PGK [6] (A). The NMR model of MnATP (black sticks) is superposed with MgATP (grey ball and sticks) as bound to pig muscle PGK [7] (B), and with MgATP (black ball and sticks) as bound to yeast PGK [5] (C), as well as with MgAMP-PNP (grey ball and sticks) as bound to yeast PGK [70] (D). The superimpositions were done according to the heavy atoms of the ribose rings. In all cases the NMR-modelled nucleotides are shown by stick models, while the crystallographically determined nucleotides are illustrated by ball and stick models and the surrounding part of the protein by stick models. The atomic interactions are indicated by dashed lines. In some cases specified distances are indicated by arrows.

The smallest deviations from the corresponding crystal structures were obtained when the nucleotides were superimposed according to the ribose ring atoms. However, as seen in Figures 21A, B, C and D, even in this case the relative position of the adenine and ribose rings (determined by the torsion angle χ) is spectacularly different in the structures derived from NMR-modelling and obtained by crystallographic refinement. Furthermore, the positions of the metal are largely different in the structures of two different types and no any possible contacts of Mn^{2+} with the protein are seen in case of NMR-determined structures (cf. Figure 21). Naturally, it cannot be excluded that these differences may be due to operation of the crystal lattice forces. It is notable, however, that the conformations of MnADP and MnATP, as determined by NMR, fit more or less well to the nucleotide site and able to form (although sometimes only few) atomic contacts which would stabilise the interactions of PGK even with these different nucleotide conformations.

As for the nucleotide phosphates, in case of MnADP simultaneous interactions with Lys Bs201/219/Y217 (from helix 8), Asn Bs316/336/Y334 (from β-strand J) and Ser Bs353/Thr 375/Thr Y373 (from helix 13) looks very probable, in agreement with the X-ray structure (Figure 21A). In case of MnATP only the interaction with Asn 336/Y334 and Thr 375/Y373 are detectable when it is superposed with the crystal structure of MgATP binary complex of pig muscle PGK [7] (Figure 21B). Thus, the interaction of MgATP phosphates with Lys 219/Y217 at helix 8 (seen in the crystal structure) apparently cannot be formed in case of the NMR-modelled conformation of MnATP. Moreover, no any interaction with the protein can be deduced if the superposition is done into a much less refined structure of MgATP binary complex of yeast PGK [5] (Figure 21C). The latter one has been considered for comparison by Ragunathan et al. [146], since this is the only ATP-bound structure of yeast PGK. Here the comparison is also presented with the better refined AMP-PNP-bound structure of yeast PGK [70] (Figure 21D). Similar to Figure 21B, this shows the possibility of well-defined interactions of MnATP phosphates with Thr Y373/375, Gly Y370/372:N and Gly Y371/373:N at the N-terminus of helix 13. Taken together, these pictures may be in agreement with the assumption of fluctuation of ATP-phosphates between two alternative sites at helices 8 and 13 (described in Chapter I. 1.4.).

The most interesting feature of Figs 21A-D is the illustration of accommodation of adenine ring by the protein in a position quite different from the known well resolved crystal structures. This is in agreement with the low degree of specificity of the hydrophobic adenine binding pocket, and also in line with the enzyme activity of PGK towards GTP and ITP [12]. Furthermore, the above comparison of NMR and X-ray structures of PGK-bound nucleotides draw our attention to the fact that the well-established hydrophobic pocket of PGK in the C-domain may also be able to adopt the nucleotide rings in positions different from those determined by crystallography (cf. Chapter I. 1.4.).

Relationship between the Sites of PG and General Anions

The binding of various anionic inhibitors to yeast PGK has been studied by monitoring the changes in its ^{1}H-NMR spectrum following addition of the anions. The use of anionic diamagnetic $[Co(CN)_6]^{3-}$as well as paramagnetic $[Cr(CN)_6]^{3-}$ and $[Fe(CN)_6]^{3-}$ probes has enabled the location of the primary anion binding site in the 'basic-patch' region of the N-terminal domain [131, 140]. These inhibitors and a variety of other large negatively charged inhibitors [147, 148] were found to affect the resonances of the 'basic patch' histidine residues (His Y62/62, His Y167/169 and His Y170/172), i.e., the same resonances as does the substrate PG itself (see above). Indeed, it was shown that binding of anions is directly competitive with the binding of PG. Placing of anions into the 'basic patch' in the crystal structure of yeast PGK [5] by modelling, has revealed Arg Y65/65 and Arg Y168/170 as further important interacting residues.

Besides the strong primary anionic site, NMR evidence was also presented for a secondary (much looser) anion binding site involving certain other surface histidines on the N-domain. It was therefore concluded that there was an extended area on the N-domain over which an anion could bind [131]. Later on, surface lysine residues were also detected as constituents of the secondary anionic site. Since there are 42 Lys residues in the molecule of yeast PGK, it was not possible to locate the site(s) of secondary binding. It was however speculated that this site may involve the totally conserved lysines Y213/215 and Y217/219, which interact with the phosphate chain of the bound nucleotide substrates.

Anion binding to PGK has a special interest since anions at low concentrations activate, while at high concentrations inhibit the enzyme reaction (cf. Chapter VI.) At the time of detection of these primary and secondary anionic sites, it was suggested that the primary site is responsible for the observed activation by promoting the two-step dissociation of the tightly bound product, bPG [149], while the secondary site leads to inhibition through simple competition with the nucleotide substrate [140]. This model, however, could not satisfactorily explain, even qualitatively, the competitive inhibition by anions with respect of PG.

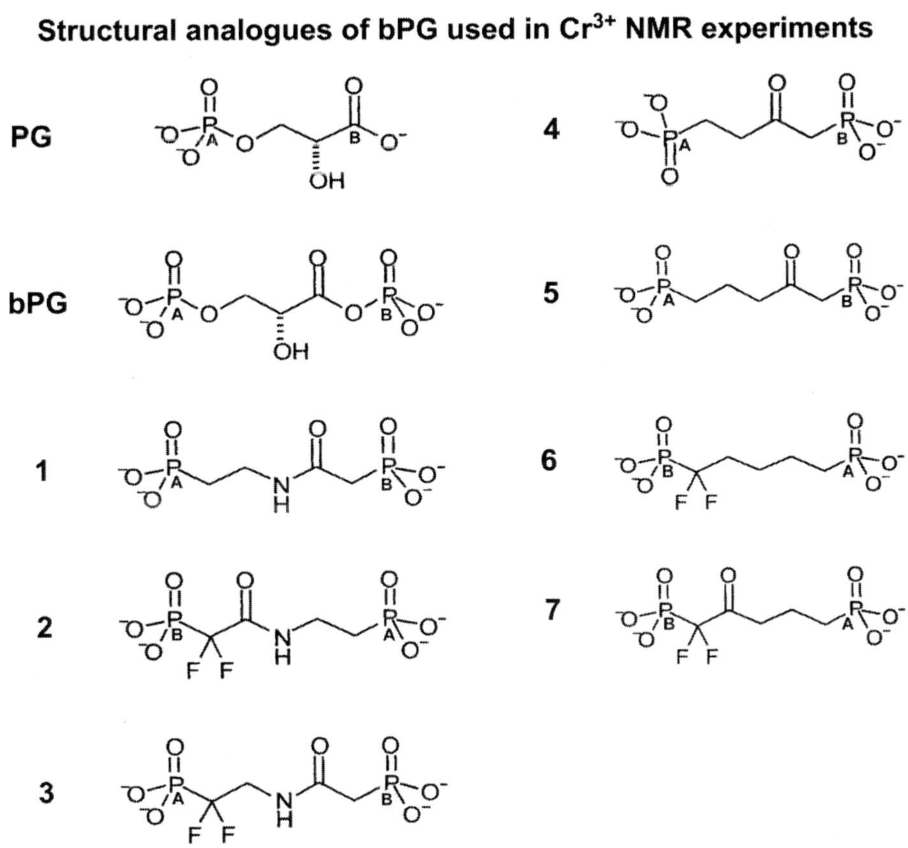

Figure 22. Structures of the analogues of bPG used in Cr^{3+} NMR experiments carried out by Jakeman et al. [150]: The two ends of the molecules are labelled by A and B. In PG and bPG molecules these denote the non-transferred 3-P and the transferred 1-P end, respectively. In other compounds (labelled from 1 to 7) the labels are somewhat arbitrary but have been placed to best match the orientation of PG.

Mimicking bPG Binding by using Its Analogues

The substrate bPG itself is a very unstable compound due to the acyl-phosphate character of its 1-P. Therefore, PGK could not be crystallized with bound bPG. Instead, a range of bisphosphonate analogues of bPG were synthesised [151, 152]. Nonscissile methanephosphonic acids replace the two phosphate monoesters of bPG and multiple fluorine substitution for hydrogen in the alpha-methylene groups of the phosphonic acid lead to several stable, tight binding analogues (listed in Figure 22) [150]. The ligands were designed to have similar electronic distribution to bPG and PG in having either a simple ketone or an amide group in analogues positions to the substrates. Their binding modes to PGK have been determined in the presence of CrADP, the substitution inert paramagnetic analogue of MgADP, by using NMR. In order to distinguish the orientation of the enzyme-bound analogues an elegant approach was developed and applied [150]. The principle of the method is illustrated in Figure 23.

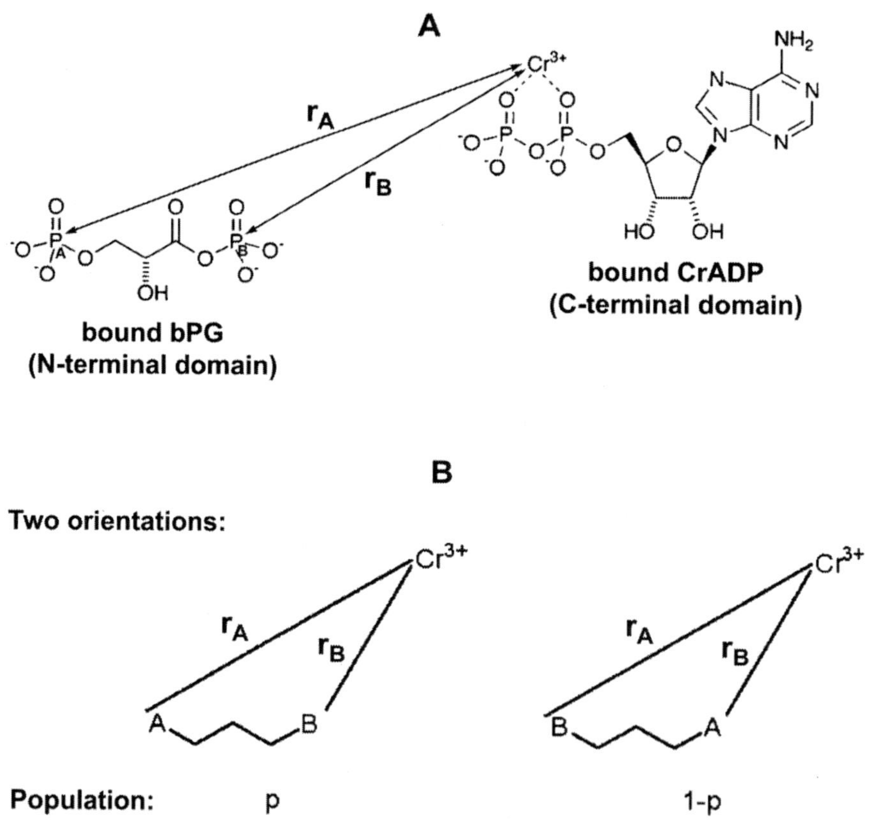

Figure 23. The principle of determination of orientation of bPG by Cr^{3+} NMR experiments as suggested by Jakeman et al. [150]: In (A) a schematic picture of the binding sites of PGK for bPG and CrADP is shown. In the active conformation the two domains come closer together such that the β-phosphorous is close to the terminal phosphate of ADP. In (B) the two possible orientations of the bound bPG or analogue (the compound A-----B) with respect to Cr^{3+} are illustrated and represented by the populations p and $(1-p)$. Note that r_A is always larger than r_B.

In the experiments the effect of Cr^{3+} on the relaxation times of the two ^{31}P nuclei in the ligand (in Figure 23 denoted as A and B), was measured and used the relative broadening to determine the distance to each nucleus and hence the orientation of the ligand. Thus, the apparent ^{31}P-chromium distance ratios were obtained from measuring the change in line width of the ^{31}P signal of the analogue with the addition of CrADP. Since the ^{1}H NMR spectra acquired on titration of the ligands with PGK, do not indicate occurring of large conformational changes, it is assumed that the enzyme remained in the open conformation upon titration and the following equation was derived for estimation of the conformational preference of a ligand:

$$r_A / r_B = \sqrt[6]{\left\{\frac{34p + (1-p)}{p + 34(1-p)}\right\}} \qquad \text{Eq. 1}$$

Where the ratio r_A/r_B is fixed and can be estimated from the open crystal structure as 1.8 and p and $(1\text{-}p)$ are the probabilities of binding in orientations I and II, respectively. This function is depicted in Figure 24, which shows that relatively small portion of binding in the alternative conformation drastically reduce the apparent distance ratio. The results are summarised in Table 9. None of the distance ratios measured approaches the theoretical limit of 1.8, the largest one is being 1.49. This discrepancy cannot be due to domain closure upon ligand binding, as the ratio of r_A/r_B should be greater and not smaller, due to the closure. The most likely explanation for the reduced ratio that the bPG analogues can bind in both possible orientations. The conformational preference range for the investigated ligands, however, varies in a wide range between 29 and 94 % (cf. Table 9).

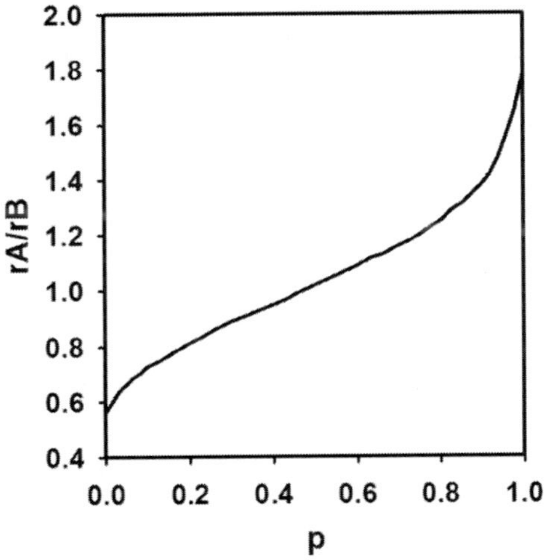

Figure 24. Dependence of r_A/ r_B on the population of the first orientation of bPG analogue: The curve represents the plot of Eq. 1, based on Jakeman et al. [150].

Due to the similar electronic distribution of the ligands to the substrates bPG and PG, it was expected that the non-symmetrical ligands would bind such that the carbonyl oxygen would occupy a location similar to that of carboxylate of PG. The results presented for the non-fluorinated analogues **1,4** and **5** (Table 9) show that they do indeed bind preferentially in the expected orientation, but the conformational preference is not too strong.

The most striking result is that almost all of the α,α-difluoromethanephosphonate analogues bind with high affinities [152] and moreover, with the more highly charged difluoromethanephosphonate group at the 3-phosphate end. This was true whether the fluorines were placed at what was expected to be the 1-phosphate end (**2**) or the 3-phosphate end (**3**). Figure 22 indicates that in all cases the difluoromethanephosphonate group is positioned at the 3-phosphate end (one of the few exceptions is (**7**)). This finding makes sense of the affinities, characterised by the values of the dissociation constants, listed in Table 9. It can be assumed that the greater charge density induced by the difluoro-substitution is the main driving force for the conformational preference seen. The results do agree with the long standing findings that the 3-phosphate should be double ionised to bind effectively [153] and that the binding of carboxylate of PG is less important [154].

Table 9. Relative Cr^{3+} to ^{31}P distances measured using paramagnetic titrations
(based on the work done by Jakeman and coworkers [150])

Substrate and its analouges	r_A/r_B	Preference for 3-PG orientation[a] (%)	Dissociation constant[b] (μM)
PG	1.80[c]	Unknown	110
1	1.05±0.1	58	675
2	0.87	29	6
3	1.49±0.2	94	2
4	1.12	67	2
5	1.02	53	19
6	0.89	32	4
7	large uncertainty	<50	3

Formulas of the compounds numbered from 1 to 7 are shown in Figure 22.
[a]Calculated from Eq. 1
[b]Taken from Jakeman and coworkers [152]
[c]Determined from the crystal structure [5]

Thus, a difluoromethanephosphonate group at the 3-phosphate end is always favourable and makes the ligand bind in this orientation even if this means disrupting other substrate-like interactions elsewhere in the molecule. This effect can be seen from the comparison of compounds **2** and **3**, both of which bind with the fluorines at the 3-phosphate end, despite putting the carbonyl group of **2** in the "incorrect" location. However, such "incorrectness" only reduces the affinity from 2 μM (**2**) to 6 μM (**3**). This conclusion is also in line with the results of a previous study, which showed that in the bPG molecule substitution by a single difluoromethanephosphonate group always improved the affinity, but that substitution of difluoromethanephosphonates at both ends of the ligands conferred no additional affinity and often reduced the affinity [152].

These results point out the importance of the basic patch as a region of high positive charge to locate the 3-phosphate of the substrate [155]. It may also explain why mutations (cf. Chapter II. 2.7.) in the basic patch have relatively little effect on the kinetics of the reaction [36, 156].

Studies on the Arrangement of the two Substrates in the PGK Ternary Complexes

In the NMR work with the paramagnetic cations and yeast PGK [141], the frequency dependence of the ^{31}P relaxation times of PG was used to determine a Mn^{2+}-^{31}P distance (i.e., the distance between the ADP bound metal from the phosphate of PG) and found to be 11.1±0.3 Å, suggesting that PGK·MgADP·PG abortive ternary complex is in the open conformation. Similar conclusion was reached from Mn^{2+} EPR results with yeast PGK [157] (see below). This finding is in sharp contrast to the crystallographically observed closed conformation of the same type of ternary complex of TbPGK [74].

Another NMR study was also devoted to determination of the arrangement of the two substrates at the PGK active site. In this work β,γ-bidentate CrATP, a stable exhange inert metal-analogue of MgATP was used as a probe of the conformation of the substrates at the active site of yeast PGK [158]. Previous studies showed that CrATP is not a substrate for PGK [159], it acts as a competitive inhibitor with respect of MgATP [160] and inactivates PGK by forming a stoichiometric coordination complex at the enzyme active site [161]. CrATP was also used in determination of the stereoselectivity of PGK, i.e., the conformation of ATP preferred by PGK (cf. Chapter II. 2.3.). The paramagnetic effects of Cr^{3+} on the longitudal relaxation rates of the ^{1}H, ^{31}P, and ^{13}C nuclei of PG were examined to determine the distances between the enzyme bound CrATP and the nuclei of PG in the ternary PGK·CrATP·PG complex [158]. This ternary complex, in contrast to the above abortive complex, may mimic better the properties of the functional PGK·MgATP·PG complex. The results show the close proximity of the substrates in the enzyme active site, which is consistent with the hinge-bending mechanism of the enzyme for bringing the two substrates together. The models constructed from the determined distances and found 4.5 Å distance of the nucleophile O-atom of PG to the P-atom of the ATP γ-phosphate, entirely compatible with the proposal of the direct transfer of the phospho-group between the bound substrates.

2.2. EPR Spectroscopy

PGK as all ATP-dependent enzymes uses MgATP and MgADP chelates as substrates, where Mg^{2+}-ion is coordinated to the oxygens of the phosphate groups of the nucleotides. In principle, it can form mono-, bi- and tridentate chelates. Additionally, binding to the α- and the β-phosphate groups introduces new chiral centers and thus leads to the occurrence of various diastereomers [162]. A detailed knowledge of the structure of the metal-nucleotide substrate or product complex, as it is bound to the enzyme active site, is required for understanding the mechanism of any enzyme action utilising ATP. There are several approaches to probe the coordination of the metal to the phosphate groups (cf. also Chapter II. 2.3.). This chapter is devoted to the method based on ^{17}O-containing phosphate-groups of the nucleotide and their interaction with the paramagnetic Mn^{2+} as monitored by electron paramagnetic resonance spectroscopy (EPR), e.g., [163].

In case of yeast PGK the structure of the MnADP complex bound at the active site has been investigated by EPR [157]. Inhomogeneous broadening in the EPR signals for Mn^{2+} resulting from unresolved superhyperfine coupling to ^{17}O regiospecifically incorporated to

ADP shows that Mn^{2+} is coordinated to the α- and β-phosphate groups of ADP. The EPR pattern for the enzyme·MnADP complex is characteristic of a predominantly axially symmetric zero-field splitting tensor. The symmetry and magnitude of the zero-field splitting interaction suggest that there is an additional negatively charged oxygen ligand in the coordination sphere of Mn^{2+}. A carboxylate from a residue of the protein is the most probably ligand because the magnitude and the symmetry of the zero-field splitting tensor in the MnADP complex with PGK are similar to those for Mn^{2+} in the formate stabilised dead-end complex of creatine kinase [164]. This conclusion is in good agreement with the crystallographic observation of ligation of both phosphates of MgADP to Asp Y372/374 through Mg^{2+} ion [6] (cf. Chapter I. 1.4.). In addition to this side-chain, there are also two or three water molecules in the coordination sphere of the metal-ion, as indicated by EPR measurements for solutions of PGK·MnADP complex in ^{17}O-enriched water.

In the work of Moore and Reed [157] effects of ^{17}O on the EPR signals from the stereospecifically labelled nucleotides, (R_P)-[α-^{17}O]ADP or (S_P)-[α-^{17}O]ADP, were used to determine the stereochemical configuration of MnADP at the active site. EPR spectra for Mn^{2+} in the enzymic complex with the (R_P)-isomer[3] show an inhomogeneous broadening due to superhyperfine coupling with ^{17}O, whereas spectra with the (S_P)-isomer are indistinguishable from those for matched samples with unlabeled ADP. These results show that PGK selectively binds the Λ configuration[4] of the (R_P)-isomer of the α,β-chelate of MnADP. The configurations of the diastereomers are illustrated in Figure 25.

The EPR spectrum for the PGK·MnATP complex differs from that of the corresponding ADP complex [135] and this observation indicates that there are differences in the coordination sphere of Mn^{2+} in the two complexes, in agreement with the crystallographic data (Chapter I. 1.4.). Signals in the EPR spectrum for the complex with ATP are, however, was found to be too broad to permit detailed investigation with ^{17}O labeled forms of ATP.

In an independent EPR study it was shown that the paramagnetic Mn^{2+} itself is able to interact with yeast PGK [166] with a K_d value of about 0.1 mM, however, the interaction is tightened (K_d is about 0.01 mM) in the presence of PG. From this study a proposal has been made that one of the functions of the bivalent metal cation is the coordination and neutralization of PG. This proposal, however, has not been confirmed by other structural studies.

Returning back to the EPR study with the ^{17}O labelled ADP, the dead-end ternary complex (PGK·MnADP·PG) was also investigated and found that the presence of PG does not alter the EPR spectrum [157]. The addition of vanadate, a transition state analogue (TSA) to this complex, however, causes marked changes in the spectral parameters. The experiments indicated that ADP remains bound as an α,β-bidentate ligand to Mn^{2+} in the dead-end complex with vanadate. A stable dead-end complex of PGK·MgADP·vanadate·PG (or with arsenate in place of vanadate) has also been detected in an independent study by ^{31}P NMR [142].

---

[3] The system, called (*R-S*), or CIP code, introduced by Cahn, Ingold and Prelog in 1966, is widely used as an extension of the IUPAC nomenclature rules. It defines the stereochemical configuration of the chiral P-atom: the arrangement of the four different coordinating atoms is either right-handed (*R, rectus*) or left-handed (*S, sinister*) around the Patom (labelled by subscript), according to the order of their atomic numbers.

[4] The screw sense (Λ and Δ) nomenclature is introduced by Cornelius and Cleland [165] and it is defined without any reference to the identity of the individual atoms (cf. Figure 25).

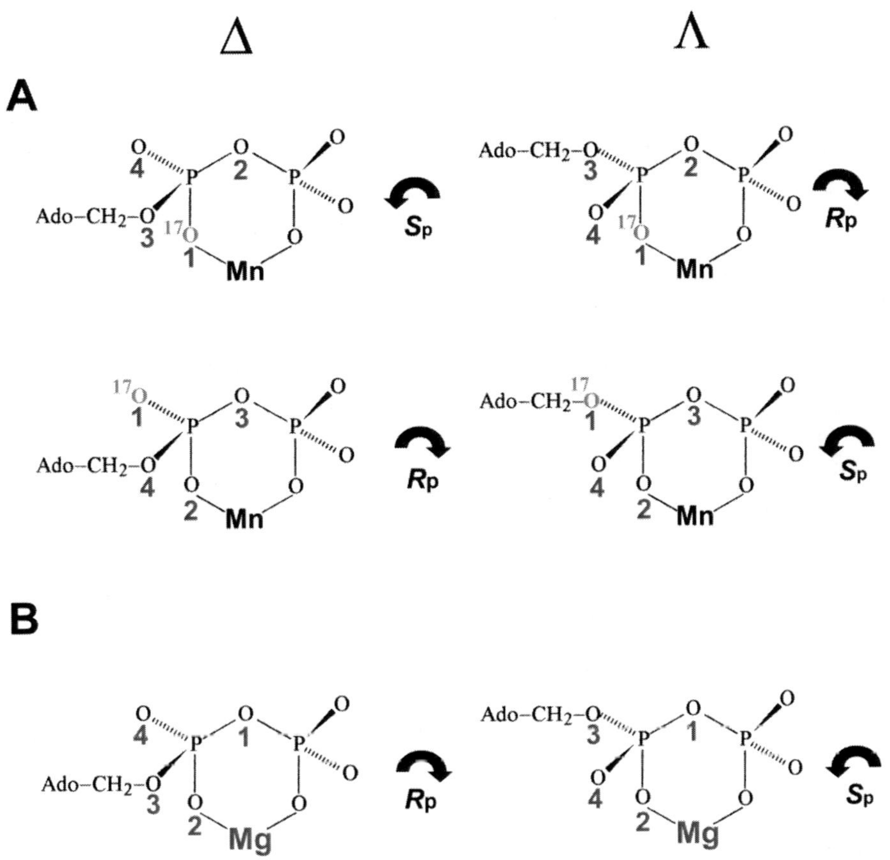

Figure 25. Stereoisomers of α,β-bidentate Mn[α-^{17}O]ADP (A) as well as MgADP (B): The classical R_P and S_P nomenclature is defined by the order of the atomic numbers of the individual atoms coordinated to the phosphorus, where the latter one represents itself as a chiral center. In case of coordinating atoms of the same type, the next neighbouring atom will be the determinant. The screw sense (Λ and Δ) nomenclature is that of Cornelius and Cleland [165]) and is defined without any reference to the identity of the individual atoms. In this system, the reference axis is a line through the metal perpendicular to the chelate ring, and the bound from the chelate ring to the rest of the molecule is the skew line defining either a left-hand (Λ) or a right-hand (Δ) helix. Note, that isomers of the same screw sense stereochemistry (i.e., with identical three-dimensional arrangement of atoms within the molecule) can be classified either as *R* or *S*.

The absence of any measurable spectral response upon binding of 3-phosphoglycerate to form the dead-end complex indicates that the missing phospho-group may be essential in bringing about ‘hinge-bending’ for enclosure of the active site. The hydrodynamic measurements on the dead-end combination of substrates support this concept [167] (cf. Chapter II. 2.9.). Experiments with anions, such as vanadate or arsenate were initiated in an attempt to find a substitute for the missing phospho-group in the dead-end complex. Vanadate is an inhibitor of several enzymes that catalyse phospho-transfer reactions [168], among them, of PGK [169]. The inhibitory action of vanadate is attributed to the ability of the pentavalent vanadium to adopt a stable trigonal-bipyramidal (TBP) geometry that is analogous to the transition state in phospho-transfer reactions [170] (cf. Chapter IX.). The EPR spectra for the dead-end complex of PGK with vanadate exhibit inhomogenous brodening both from α- and β-^{17}O-labeled forms of ADP. These observations demonstrate that Mn^{2+} remains bound to both phosphate groups of ADP in this complex. Although it is likely that changes in the EPR spectra for Mn^{2+} upon binding of vanadate are due to direct coordination between Mn^{2+} and an oxygen of vanadate, it was not possible to test this hypothesis by EPR experiments with ^{17}O-labeled vanadate because vanadate oxygens equilibrate rapidly with oxygens of water.

From the observation that the ^{31}P NMR signals, characteristic of the stable vanadate or arsenate-containing dead-end complex, could not be reversed upon removal of Mg^{2+} by the addition of excess EDTA, a covalent complex formation has been assumed [142]. ^{31}P NMR spectra revealed an especially strong interaction of the anions with PG. It was suggested that the enzyme catalyses the vanadylation at carbons 1 and/or 2 of PG. Studies with vanadium oxyanion using ^{51}V NMR show, indeed, evidence for non-enzymic formation of vanadium esters with glucose and ethanol [171].

Thus, EPR and NMR observations on the complex with vanadate are consistent with the notion that the enzyme undergoes a conformational change in this inhibitory complex, although the exact mode of binding of this anion has not been established.

2.3. Binding and Kinetic Data with Stereospecifically Labelled and Exchange-Inert Nucleotide-Metal Complexes

Nucleotide-metal complexes with well-defined stereospecificity are widely used for kinases to test the specific conformation of the nucleotide substrates that can interact productively with the enzymes. Thus, in addition to the above NMR and EPR structural studies enzyme kinetic and binding approaches provided important information for PGK, too.

The use of phosphorothioate analogues of ATP or ADP rests on the observation made by ^{31}P NMR spectroscopy that Mg^{2+} as hard metal ion coordinates preferentially to oxygen, and Cd^{2+} as soft metal ion coordinates to sulfur [159, 172]. Other cations, such as Co^{2+}, Mn^{2+} and Zn^{2+} are intermediate. The stability constants of Mg^{2+} and Cd^{2+} complexes of various phosphorothioate analogues of ATP and ADP have been determined and used to calculate the preference for O and S coordination [173]. The preference for sulfur over oxygen coordination for Cd^{2+} is estimated to be 60 regardless of the position of the phosphorothioate group. Preference for oxygen over sulfur coordination for Mg^{2+} is calculated to be 31,000 for MgATPβS. The different coordination of Mg^{2+} and Cd^{2+} has the consequence that the S_P-isomer of MgATPαS (or MgATPβS) has the same screw sense configuration (as defined by Cornelius and Cleland [165]) as the R_P-isomer of CdATPαS (or CdATPβS). This s illustrated by Figures 26A and 26B. Thus, in case of an enzyme reaction reversal of stereospecificity of the nucleotide is expected if the metal is coordinated to the phosphorotioate group under study during the rate-limiting step of the reaction. This approach has been applied to various enzyme systems and has been discussed in several reviews [38, 162].

In case of PGK catalysed reaction the stereospecificity and kinetic parameters of the diastereomeric αS and βS phosphorothioate analogues of ADP and ATP and their metal ion dependence were investigated by Jaffe and coworkers [159]. It was found that the S_P-isomer of MgATPαS is the preferred substrate for PGK: the $V_{max}(S_P)/V_{max}(R_P)$ is 235 and the S_P-isomer exhibits a K_m approximately 10-fold smaller than the R_P-isomer. The stereospecificity for the S_P diastereomer was also confirmed in the reverse reaction using the substrate MgADPαS. When Mg^{2+} is replaced by either Co^{2+} or Zn^{2+} this stereoselectivity disappeared and the ratio of $V_{max}(S_P)/V_{max}(R_P)$ became close to unity in both directions of the enzyme reaction. From this change of stereoselectivity with soft *versus* hard metal ion activators it was concluded that the metal ion is chelated to the α-phosphate of the nucleotide during the rate-limiting step of the reaction. Concerning coordination to the β-phosphate, however, the results are equivocal. Although the preference of the S_P-isomer of MgATPβS was also observed, even higher preference was obtained for the S_P-isomer of ZnATPβS: the $V_{max}(S_P)/V_{max}(R_P)$ are 1000 and 3000, respectively. The lack of metal-dependent stereospecificity is most unusual in kinase reactions and raises the question whether the metal is not chelated to the β-phosphate in the rate-limiting step of the reaction. However, the use of the reversal or relaxation of stereospecificity between hard and soft metals as a criterion of chelate structure leads to an unequivocal assignment of a particular phosphate as a metal ligand only if reversal or relaxation occurs.

Although the preference of soft metals for S-ligands and hard metals for O ligands is usually the dominant contribution to their stereoselectivity pattern for the chiral thiophosphates, other contributions such as the size of metal ion and interaction of substrate oxygen by H-bonding at the binding site may also play a role. Consequently, the magnitude of the change in stereoselectivity ratios with a series of metal ions differs from one enzyme to another. All possible patterns of reversal or relaxation of stereoselectivity with soft *versus* hard metals have been observed. For example in the hexokinase reaction no reversal for ATPαS, while reversal for ATPβS was observed [174]. Adenylate kinase also follows this pattern [175]. Another pattern of reversal occurring both with αS and βS includes the enzyme pyruvate [176], arginine [177] and creatine [178] kinases. The pattern of reversal for ATPαS, while no reversal for ATPβS, observed with PGK seems to be a unique one. It has to be noted, however, that, when there is no reversal or relaxation of stereospecificity, either the thiophosphate group involved is not a metal ligand or the enzyme interaction forces the metal to bind to the same atom, either oxygen or sulfur, for all metal ions. Thus, for the ATPβS complex of PGK, the simplest but by no means unique explanation is that the metal is not chelated to the β-phosphate of ATP. Unfortunately, other attempts to remove the ambiguity by investigating the β,γ-bidentate exchange inert Cr(III)ATP was unsuccessful since this compound was not a substrate for PGK (cf. below).

Using the thio-analogues of ATP, however, a further interesting observation has been made. The maximum velocities of PGK reaction decrease in the order of ATP, ATPαS, ATPβS and ATPγS: the V_{max} values relative to ATP as 100, are 60, 3 and 1. Thus, upon substitution of O with S in ATP, the rate decreases progressively as the substitution approaches the site of bond cleavage, indicating the importance of very precise atomic contacts at the enzyme active site.

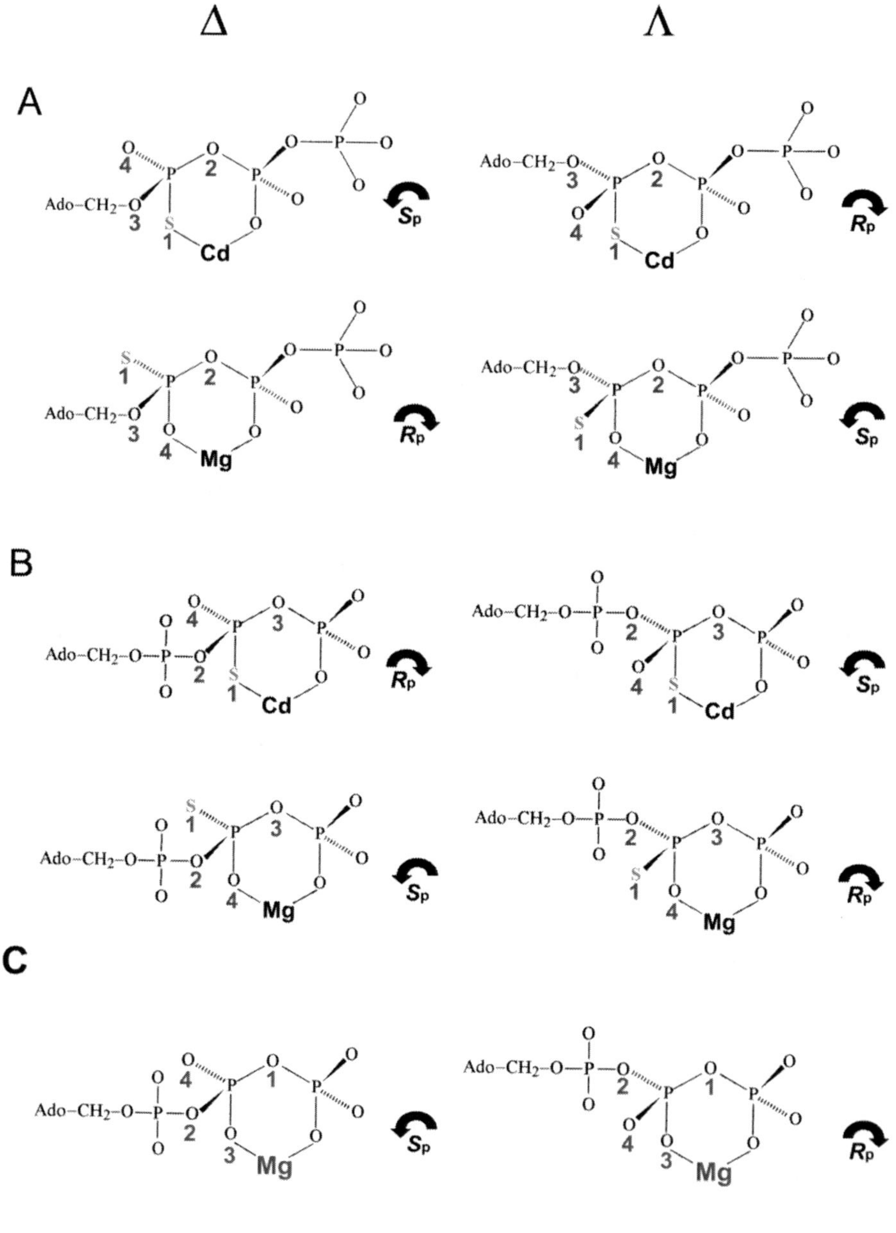

Figure 26. Stereoisomers of ATP(αS) and ATP(βS) complexed with Mg^{2+} and Cd^{2+}: The R_P and S_P nomenclature of diastereomers of ATP(αS) and ATP(βS) has been described by Eckstein and Goody [162]. The screw sense (Λ and Δ) nomenclature has been described by Cornelius and Cleland [165], cf. also legends to Figure 25. In (A) and (B) ATP(αS) and ATP(βS) are illustrated, respectively. For comparison, the stereoisomers of β,γ-bidentate complex of the natural MgATP is shown in (C).

The so-called exchange inert complexes of ATP with metals such as Cr(III) or Co(III) provide also advantageous experimental systems. The physiological substrate for kinases is primarily the Mg^{2+} chelate of ATP that exists in solution as a rapidly equilibrating mixture of various isomers, only one of which presumably possesses the proper steric structure and stereochemistry for the reaction with a specific enzyme. Through the knowledge of the structure and stereochemistry of the Mg^{2+} substrate and product complexes, one should be able to trace the movement of the metal during the reaction and thus gain a great deal of information about the mechanism of and the catalytic role of the metal in the enzyme-catalysed reaction. For this purpose a number of slowly exchanging, i.e., practically exchange-inert Co(III) and Cr(III) nucleotide complexes have been prepared [179-183].

With PGK the first observation was that the enzyme cannot handle either the bidentate or tridentate CrATP as substrates [160], which was confirmed later [159]. Both compounds, however are good competitive inhibitors and the tridentate analogue is somewhat stronger inhibitor than the bidentate one [160], which raises the possibility of coordination of the metal with all three phosphates of ATP in the Michaelis complex of the PGK-catalysed reaction.

When the metal-complex of ATP is β,γ-bidentate, the chiral β-phosphorus gives rise to Λ and Δ screw sense isomers (cf. also Fig 26.) The screw sense nomenclature is that of Cornelius and Cleland [165]. In this system, the reference axis is a line through which the metal is perpendicular to the chelate ring, and the bond from the chelate ring to the rest of the molecule is the skew line defining either the left-hand (Λ) or the right hand (Δ) helix. It was found that Δ isomer of Cr(III)ATP is more specific for PGK: it yielded a K_I value of 7.0 μM which is about an order of magnitude smaller than that of the Λ diastereomer (62.6 μM) [161]. The Δ isomer of Cr(III)ATP corresponds to the S_P isomer of β,γ-MgATP (cf. Figure 26). However, as pointed out by Cornelius and Cleland [165], the usage of *R,S* nomenclature has the difficulty that it does not always correspond to similar conformation (this definition is based on the numbering of the atoms coordinating the chiral phosphorus). Thus, only the screw sense nomenclature determines unequivocally the characteristic spatial conformation of diastereomers.

The application of the exchange inert bidentate CrATP isomers has further advantageous. The four free coordinating places of Cr(III) can be filled by either H_2O or NH_3 ligands. Thus, the tetraaquochromium-, tetraamminechromium- or the mixed $Cr(H_2O)_x(NH_3)_yATP$ complexes can be prepared and used for mapping of PGK active site due to their specific steric and chemical properties (cf. Figure 27). This is important, since it is not known which of the water molecules in the Mg^{2+} coordination sphere are in contact with the surface of the enzyme and therefore possibly involved in binding interactions.

In accordance with the hypothesis that Cr^{3+} requires a ligand from the enzyme in the enzyme-$Cr(H_2O)_4ATP$ complex, it was found that yeast PGK was inactivated by a time-dependent manner upon incubation with $Cr(H_2O)_4ATP$. Moreover, the inactivation rate has decreased systematically as amino-groups replace water molecules for coordination to Cr^{3+} in $Cr(H_2O)_x(NH_3)_yATP$ complexes. This supports the assumed coordination with PGK, since displacement of NH_3 from the coordination sphere of Cr^{3+} is more difficult than the displacement of H_2O. The presence of MgATP protected the enzyme against inactivation, i.e., against the formation of the coordination complex at the enzyme active site [161]. In this work it was also found that the other substrate PG, when present, reduced the observed inactivation rates. This reduction was proportional to the number of NH_3 ligands present in the coordination sphere of Cr^{3+} in the Cr-ATP complexes. This suggests that in the ternary CrATP·PG complex PG may be coordinated to the metal ion.

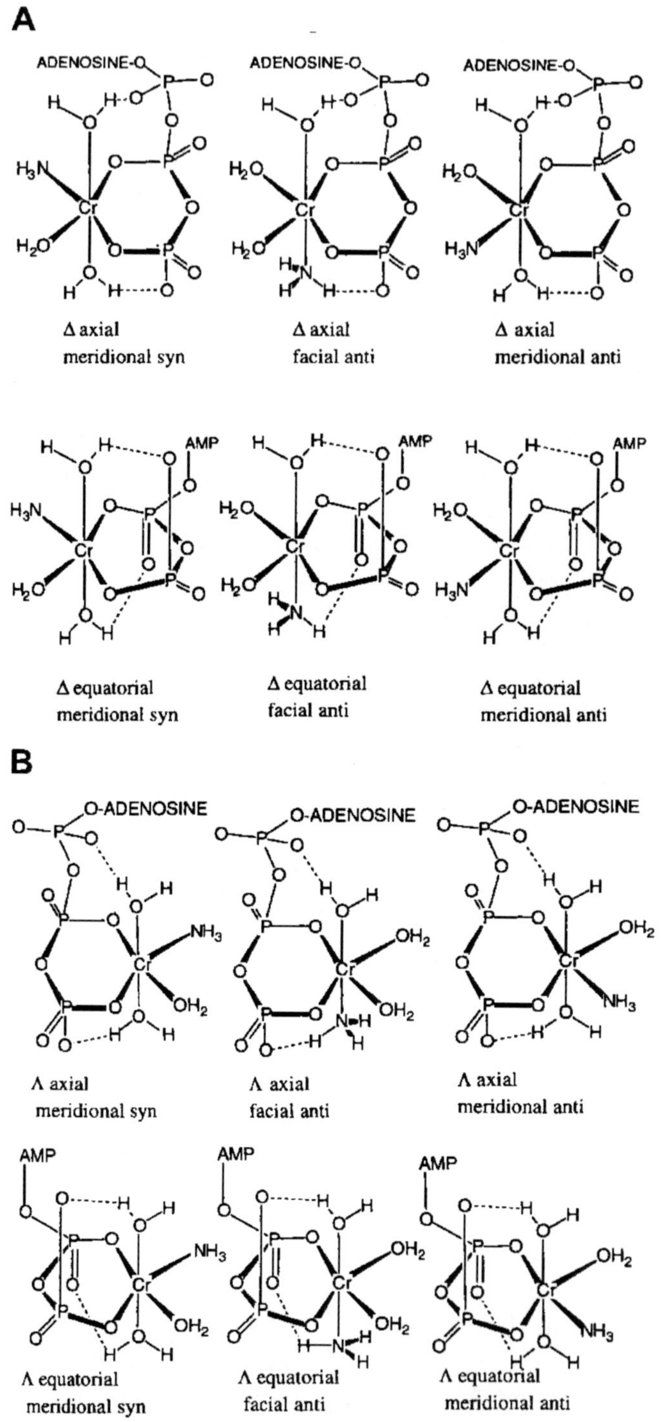

Figure 27. Stereoisomers of monoammine chromium(III)ATP (adapted with permission from Lester et al., [184]; Copyright 1998 American Chemical Society): In addition to the screw sense (Λ and Δ) isomers, the ammonia and the two coordinated phosphates can be either *facial* or *meridional*, and the ammonia can be either *syn* or *anti* with respect to the AMP. Finally, ring-puchering isomers with AMP *axial* or *equatorial* can occur when water molecule is *syn* to the AMP, allowing strong H-bonds with the α-phosphate.

A conclusion similar to the latter one was reached form an independent work using another exchange-inert complex with $Rh(H_2O)_4ATP$ [185]. Upon addition of this complex, the enzyme was inactivated, suggesting that a long-lived complex may be formed by donation of a ligand from the enzyme to the coordination sphere of Rh^{3+} in the enzyme-$Rh(H_2O)_nATP$ complex. Thus, the metal ion is possibly coordinated to a protein side-chain. In the presence of PG a one-turnover substrate activity of yeast PGK was detected by the aid of ^{31}P NMR. During this process transfer of the phospho-group between ATP and PG occurs with both substrates in the coordination sphere of the metal ion. Because of the slow ligand exchange rates of Rh^{3+}, the reaction product bPG remained coordinated with the metal ion. The reaction products $Rh(H_2O)_nADP$ and bPG remained bound to the enzyme even after removal of excess substrate. Thus, further turnovers are prevented. The main conclusions by Pappu and coworkers [185] from their findings that the metal ion may not only act as an electron sink to activate the electrophile, but it may also help to optimally align both substrates for phospho-transfer by coordination to both substrates. It is therefore likely that entry of PG into the coordination sphere of metal of a metal-ATP complex may start the proposed hinge-bending motion of PGK to form a closed active site between the two separate substrate binding domains. No direct coordination of PG to the metals has been observed, however, from the crystallographic data.

In order to localise the enzyme site where the metal ion is coordinated an analytical work has been carried out with the RhATP inactivated yeast PGK [186]. Cleavage of the inactivated enzyme with pepsin and subsequent separation of peptides by reverse-phase high-performance liquid chromatography gave two Rh-nucleotide bound peptides. One of the peptides corresponds to the C-terminal residues of helix 13 (equivalent to helix 14 of the muscle PGK) and the other to a part of helix 5 (equivalent to helix 7 of the muscle PGK), respectively. Neither of the metal-binding sites confirms the crystallographically determined [6, 70] Asp Y372/Bs352/374 to which the metal ion is suggested to be coordinated. Instead, metal binding to conserved glutamates, Glu Y398/400, Glu Y401/403 (in helix 13/14) and Glu Y190/192 (in helix 5/7) have been detected. Among them the most probable candidate is Glu Y398/400 located in the C-terminal peptide that is structurally closely associated with the N-terminal domain (cf. Chapter I. 1.) Mutation of this residue into either Ala or Lys caused activity lost by at least one order of magnitude. Furthermore, chemical modification of Glu Y398/400 slowed down significantly the RhATP caused inactivation. Thus, Glu Y398/400 may be a likely metal-coordination site of PGK. Therefore, the importance of the C-terminal peptide in PGK catalysis, shown previously by Ritco-Vonsovici and coworkers [187] as well as by Mas and Resplandor [188] may be due, at least in part, to the coordination of metal ion of the metal-ATP substrate. The metal coordination may align the C-terminal peptide to extend toward the N-terminal domain and form the closed active site.

In another attempt to localise the enzyme coordination site to the metal ion and at the same time to specify better the stereospecificity as well as the conformation of the nucleotide in the enzyme active site, a series of stereospecific isomers of bidentate β,γ-CrATP has been prepared [184]. In this work 12 isomers of monoammine chromium(III)ATP ($Cr(H_2O)_3(NH_3)ATP$) has been prepared (shown by Figure 27) and employed to evaluate the importance of each of the coordinated waters in the binding of MgATP to PGK. From the studies discussed above the enzyme was known to be specific for the Δ screw sense isomer of

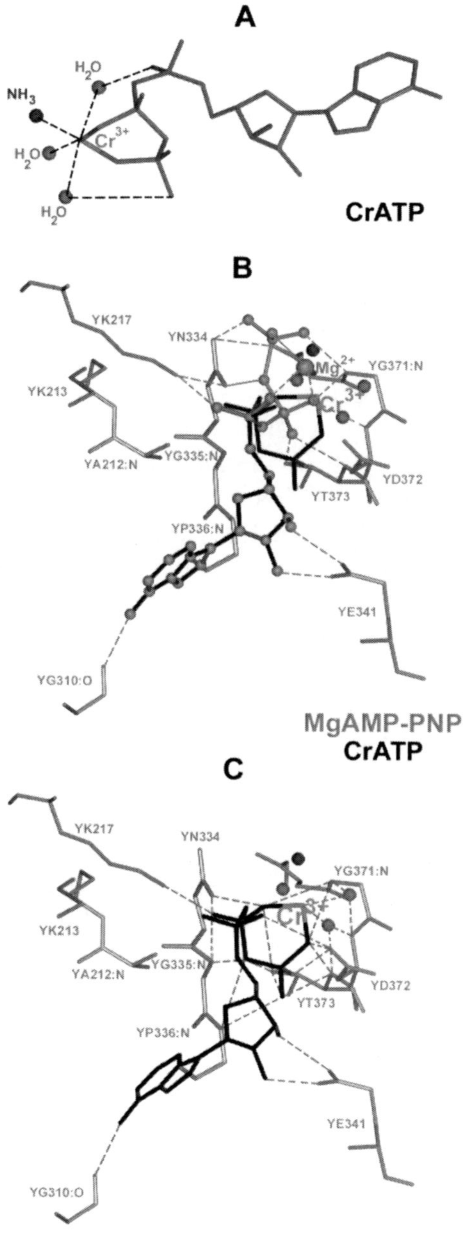

Figure 28. Modelling of an inhibiting isomer of monoammine chromium(III)ATP into the active site of yeast PGK (based on the work by Lester et al., [184]): The coordinates of Δ *axial-meridional-syn* isomer of monoammine chromium ATP, shown by red stick model in (A), have been provided by the authors. The chromium and nitrogen (from NH_3) atoms are also indicated. In (B) and (C) the isomer is superimposed with MgAMP-PNP (according to the heavy atoms of their adenosine rings), as bound in the yeast PGK crystal structure [70]. The surrounding side-chains (stick models) of yeast PGK as well as MgAMP-PNP (ball and stick model) are coloured as grey. The atomic interactions indicated by dashed lines: the interactions of MgAMP-PNP and of the isomer of monoammine chromium ATP are shown in (B) and (C), respectively. For clarity, in (C) the molecule of MgAMP-PNP is not shown, furthermore the interactions of chromium ion with only those waters are indicated which form connections with the peptide N-atoms of Gly-s Y371, Y372 and Y373.

MgATP. The present inhibition studies confirmed this and revealed a 4-fold preference for the Δ axial isomer over the Δ equatorial isomer. Two of the Δ axial isomers - axial-meridional-anti and axial-meridional-syn – binds more tightly than the third isomer, axial-facial-anti. The steric structure of the best inhibiting Δ axial-meridional-syn isomer is shown in Figure 28A. The axial preference is also found among the weekly bound Λ isomers. Besides the kinetic studies, the monoammine chromium(III)ATP isomers were also modelled into the active site of PGK. The most interesting result was obtained with the Δ axial isomer, which binds to the enzyme best and probably most closely mimics the natural MgATP structure. Figures 28B and C illustrate a possible binding mode of the best inhibiting isomer, when it was superimposed (according to the adenosine rings) on MgAMP-PNP bound in the crystal structure of yeast PGK [70]. The different phosphate chain conformations of the nucleotides may be attributed to their different metal coordination: bidentate in CrATP and tridentate in MgAMP-PNP. In spite of these differences, surprisingly large number of molecular contacts can be formed with the monoammine chromium ATP. The possibility of interactions with Lys Y217/219, Asn Y334/336, Asp Y372/374) and Thr Y373/375, as well as with the peptide N-atoms of Gly-s Y371/373, Y372/374 and Y373/375 (partly through the water molecules coordinating chromium), characteristic of nucleotide binding of PGK are well seen with this isomer. Thus, the tightly interacting axial bidentate chromium ATP complexes more closely mimic the compact tridentate structure of the analogue AMP-PNP due to the strong H-bond from the non-bridging oxygen of the α-phosphate to the coordinated syn water of the coordination sphere. Overall, the agreement between kinetics and the crystal structure is fairly good and supports the validity of probing kinase active sites with monoammine chromium ATP.

2.4. CD-, Fluorescence- and Infrared Spectroscopy

As for other proteins, the characteristic spectral properties of PGK have reported about various structural features of the molecule.

The far-UV CD spectrum of pig muscle PGK, shown in Figure 29 exhibits two minima centred at 207 and 222 nm and a maximum at 193 nm [189]. A closely similar spectrum was recorded later for the human PGK [126]. The spectrum is characteristic of an α-helix rich protein, entirely consistent with the X-ray structural data (cf. Chapter I. 1.2.) Very often, far-UV CD-spectra of proteins conform to the empirical relationship given by Provencher [190] and analysis according to this method gives information about the secondary structure content compatible with the measured spectra. In fact, such analysis has provided data about the amount of α-helices and β-structure of PGK in good agreement with the known X-ray crystallographic PGK structures.

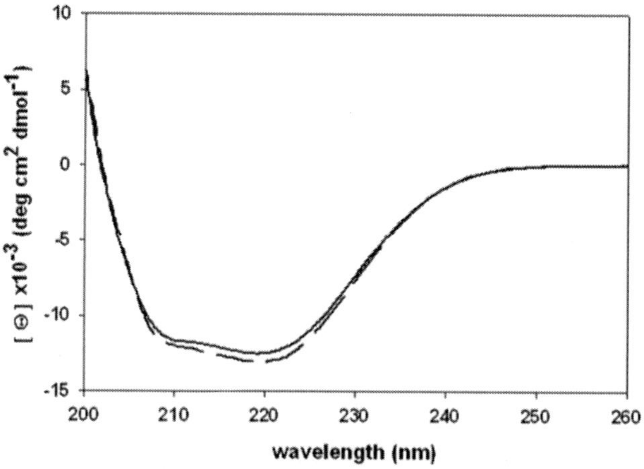

Figure 29. The far UV CD spectrum of wild type PGK The measurements were carried out at 25 °C at a protein concentration of 50 μM, a cell path length of 0.1 mm both for the pig muscle (continuous line) [189] human (dashed line) [126] PGKs.

The protein fluorescence originates from the side-chains of tryptophan and tyrosine residues and therefore reports about the nature of micro-environment of these residues within the protein molecule. Usually the fluorescence spectrum of protein is dominated by that of Trp because of the efficient energy transfer from Tyr to Trp. For this reason in many proteins the fluorescence spectrum appeared around 335 nm (characteristic of Trp) with a small shoulder, if any, at 304 nm (characteristic of Tyr). In the case of yeast PGK which contains 8 Tyr and 2 Trp residues, as published much before the high resolution structural data [12], the emission maximum was found at 308 nm when excited at 280 nm (curve A in Figure 30). Thus, the emission peak was unusually displaced to the blue side and, in fact, was close to that of free Tyr that is 303 nm. However, when the protein was excited at 295 nm, the emission maximum was shifted to 329 nm (shown by curve B in Figure 30), i.e., more close to 335 nm (characteristic of Trp). The fluorescence spectrum of PGK in 4M GuHCl, where the enzyme was presumably in a random coil state, is shown by curve C in Figure 30. Here the Tyr and Trp peaks can be identified at 303 and 343 nm, respectively, as in many other denatured proteins. In the native state of PGK, however, the efficiency of energy transfer from Tyr to Trp residues was profoundly smaller than in many other proteins, where Tyr fluorescence is usually quenched by energy transfer to Trp, and therefore, the protein fluorescence is dominated by the Trp fluorescence. From this observation, it was concluded by Nojima and coworkers [191] that Tyr and Trp residues are located relatively far from each other in the three-dimensional structure of PGK, which is against the occurrence of the efficient transfer. This conclusion was, indeed, confirmed much later, by publication of the high resolution crystal structure of yeast PGK that shows location of Tyr and Trp residues in separate domains (i.e., in the N- and C-terminal domains, respectively) [5].

The emission characteristics of the two Trp residues located in the C-terminal domain of yeast PGK have been determined by Privat and coworkers [192] using photo-counting pulse fluorimetry. They could decompose the fluorescence spectra of yeast PGK into two distinct spectra, which were attributed to the two tryptophyl residues of the protein molecule. These spectra are shifted by 6 nm with respect to each other (λ_{max1}= 332 nm and (λ_{max2}= 338 nm). On the other hand, the fluorescence decay was found to be the sum of three exponential functions. The smallest lifetime (τ_1=0.4-0.6 ns) belongs to Trp_1 with the blue shifted λ_{max}. The fluorescence of this Trp was strongly quenched, and therefore it was predicted to be buried in the protein core. The other two lifetimes (τ_{21}=3.1 ns and τ_{22}=7.0 ns) have been attributed to Trp_2. Since multiexponential decay of a Trp residue generally can be related to several configurations of the local environment of that residue, the same could be applied to Trp_2 of yeast PGK. From its λ_{max} value a more hydrophilic environment of Trp_2 could be concluded compared to that of Trp_1.

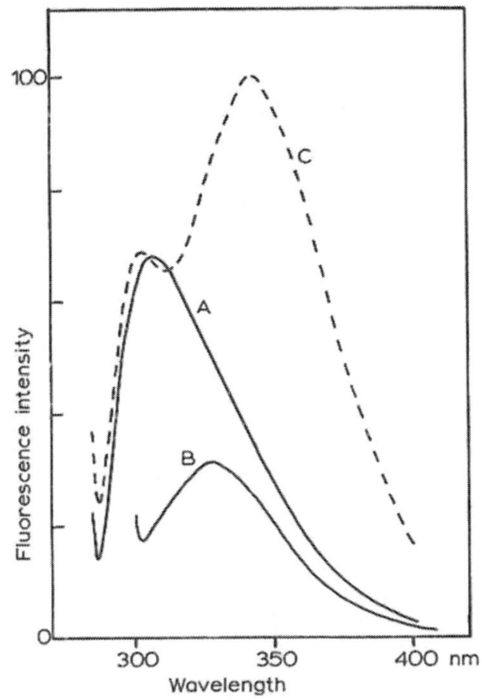

Figure 30. Corrected fluorescence spectra of yeast PGK (adapted from [191]): Native enzyme was exited at either 280 nm (A) or at 295 nm (B) in 50 mM Tris-HCl buffer, pH 8.0. The denatured PGK was exited at 280 nm in the presence of 4 M GuHCl. The protein concentration was 0.1 mg/ml.

In another, frequency domain fluorescence studies with yeast PGK, carried out by Wasylewski and Eftink only a biexponential decay was observed for Trp fluorescence emission, but otherwise, the results were consistent with those of the above study [193]. The short lifetime (τ_1=0.4 ns) component is associated with a spectrum having λ_{max1}= 329 nm and an 18.4 KJ/mole activation energy (E_a) for thermal quenching. The long-lifetime (τ_2=3.5 ns) component has a λ_{max2} = 338 nm and E_a of only 7.9 KJ/mole. The assignment of the two decay times to the individual residues was difficult, but a tentative assignment could be offered. From the X-ray structure [5] Trp Y308/310 appeared to be more solvated of the two. Therefore the redder-emitting long-lifetime component was assigned to Trp 308, while the bluer-emitting short-lifetime component could be assigned to Trp Y333/335. In this work interesting observations have also been made about the effects of substrates. It was found that binding of ATP and PG led to significant decrease in the fluorescence lifetime, a red shift of the emission spectrum and in a decrease in the E_a for both components. These results indicated that binding of the substrates to PGK might influence the conformational dynamics around the two Trp residues located in the C-terminal domain.

Later on in a more detailed steady state and time-resolved fluorescence quenching studies using acrylamide, iodide and caesium ions as quenchers, Dryden and Pain observed an even higher complexity of the heterogeneous fluorescence of yeast PGK [194]. The buried Trp Y333/335 had a blue-shifted heterogeneous emission spectrum characterised by three fluorescence lifetimes, and was inaccessible to quenchers. It implies that this well conserved Trp is well shielded from the solvent and is located in very well packed and stable surroundings. The surface Trp Y308/310 also had a heterogeneous emission with multiple lifetimes. The emission of this Trp could be separated into a blue-shifted emission accessible to acrylamide and cesium only, and a red-shifted emission emission accessible to all three quenchers. Using another quencher, succinimide, the same team demonstrated fully selective quenching of the exposed Trp Y308/310 [195], thus allowing yeast PGK to be studied as a single Trp system.

A study of tryptophan phosphorescence of yeast PGK could sensitively report about the conformational dynamics of the C-terminal domain, in which Trp Y333/335 is located [196]. At room temperature in buffer the phosphorescence is due to only this Trp residue and the biphasic nature of its decay (τ_1=196 and τ_2=25 ms, respectively) revealed the presence of stable, slowly interconverting (in 10-100 ms timescale), conformers with profound differences in the internal fluidity of the C-domain. It could also be stated that complex formation with the substrates affects protein structure in a very selective way. While PG, which binds to the N-domain, has practically no influence on the average lifetime, MgATP and MgADP, which binds to the C-domain, increases it by a factor of 1.9 and 5.3, respectively. The change in lifetime implies a remarkable stiffening of the C-domain which is partly relaxed in the ternary complexes with PG. It is remarkable that the internal conformation of the C-domain is altered in the ternary complex, which is the functioning form. - On the other hand, fluorescence anisotropy data on the horse muscle PGK, possessing 4 Trp residues in the C-domain, reported a decrease in the degree of polarization in the binary and ternary complexes with Mg-nucleotides [197]. This decrease of polarization was attributed to a greater mobility of Trp 335/Y333 upon nucleotide binding, i.e., in contrast to the assumed stiffening of the C-terminal domain. The rotational path of this tryptophan, in the presence and absence of the nucleotide, were established by potential energy minimisation calculations. The results indicated that MgADP induces a displacement of the α-helix 13 (in agreement with the crystallographic data, cf. Chapter I. 1.4.) that decreases the rotational energy barrier of Trp 335 from 16 kcal/mole in the free protein to 8 kcal/mole in the complex. The increased mobility of Trp 335 due to nucleotide binding, otherwise, is approved later by the crystal structure of pig muscle PGK. Namely, rotation of the indole-ring of Trp 335 around its Cβ-Cγ bond was observed upon diffusing the ATP-analogue MnAMP-PNP into the crystal of the PG binary complex [69]. Figure 31 illustrates this flipping of the aromatic ring.

The CD-spectrum of a protein in the near-ultraviolet region also reflects interactions of the aromatic residues (mainly Trp-s) in the tertiary structure. When a His residue in the waist region between the two domains (His Y388/390/Tb393/Tm371/Bs368) has been replaced into the uncharged Gln by mutation of yeast PGK (H388Q) a well-defined change have been observed in the near UV CD-spectrum [198]. The difference spectrum between the wild type and mutant enzymes is characteristic primarily of tryptophan and indicates a loss of asymmetry of the environment of one or both Trp residues (Figure 32A). Thus, perturbation of the waist region interestingly caused a conformational change in the C-domain. Moreover, the authors raised the possibility of close relationship of the conformational change and the enzyme function, since they observed similarly unusual temperature dependence of both the near UV CD ellipticity and the enzyme activity (Figures 32B and 32C).

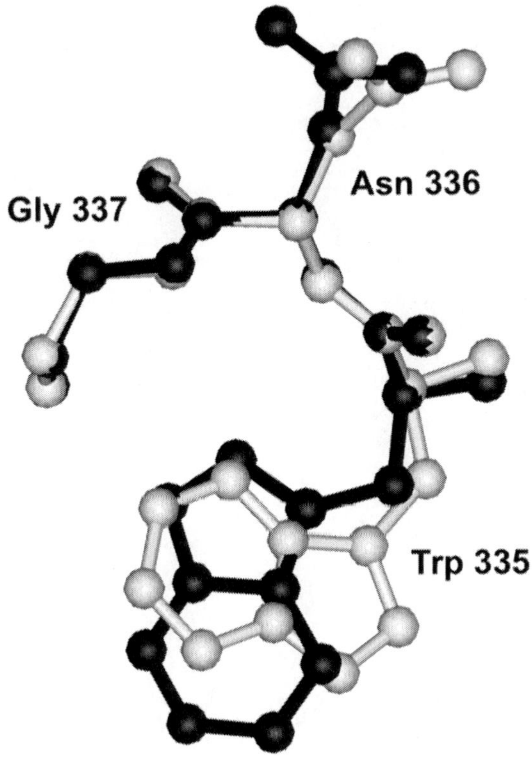

Figure 31. Rotation of the indole-ring of Trp 335 in the structure of pig muscle PGK (adapted from [69]): Ball and stick model of the Trp 335-Asn 336-Gly 337 peptide segment in the structure of PG binary (light) and PG·MnAMP-PNP (dark) ternary complexes of pig muscle PGK.

The substrate-caused conformational changes have been monitored more systematically in yeast PGK by studying the fluorescence of various single Trp mutants [199]. In this study, several PGK mutants with single tryptophans placed in various locations were used as intrinsic fluorescent probes to examine the extent and delocalisation of conformational changes induced by binding of the various substrates (PG, bPG, MgADP, MgATP) and AMP-PNP the non-hydrolysable analogue of ATP. The results showed that only the probes situated in the hinge-region and in parts of each domain close to the hinge reflect substrate-induced conformational changes. Binding of substrates to one domain was found to induce spectral perturbation of the probes in the opposite domain, indicating a transmission of conformational changes between the domains. A combination of both substrates generated much larger fluorescence changes than the individual substrates. This picture fits well to all the structural and functional features of PGK and to the suggested molecular mechanism of domain closure (cf. Chapters I. 1.10., and X.).

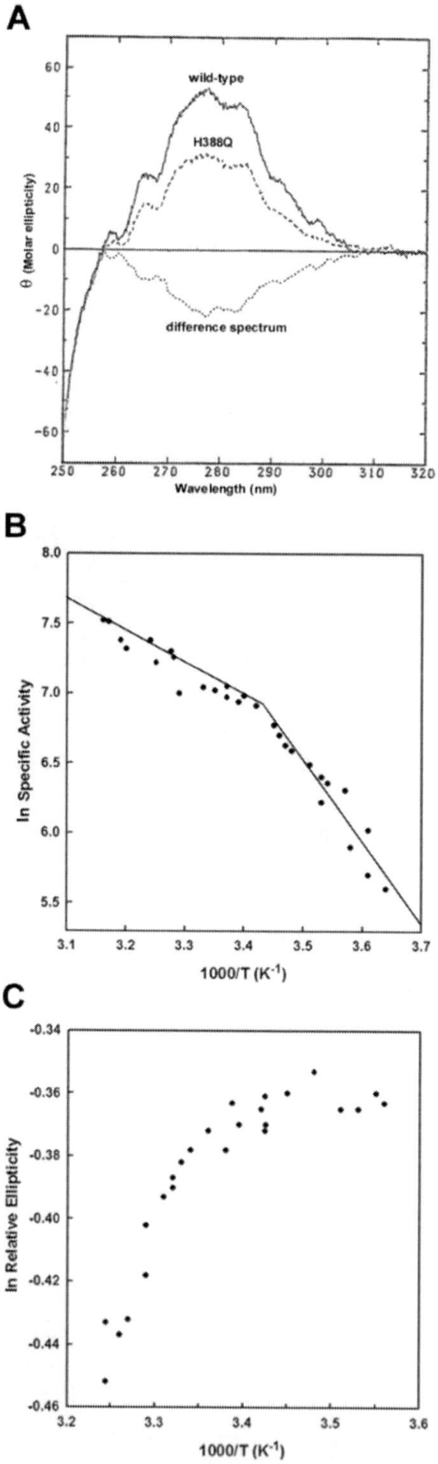

Figure 32. Relationship between the conformation of the waist region of yeast PGK molecule and the enzyme activity (based on the work by Dryden et al. 198]): (A) Sensitivity of the near UV CD spectrum of yeast PGK on the mutation in the waist region (His Y388/390 into Gln); (B) and (C): Unusual temperature dependences of the enzyme activity and the near UV ellipticity of yeast PGK, respectively.

In another work with a similar purpose cysteinyl residues have been introduced in different positions in the yeast PGK molecule by site-directed mutagenesis [200]. By this way several mutants were constructed, each containing a unique cysteinyl residue. Neither the conformation nor the enzyme activity was effected by the substitutions. The reactivity of the thiol-groups and the fluorescence of the probe of N-acetyl-N'-(5-sulfo-1-naphtyl)ethylene-diamine covalently linked to these thiols were used to monitor the conformational changes induced upon ligand binding. It was found that the observed changes mainly involve the part of the protein located in the interdomain cleft, particularly the environment of residues Y35/35 and Y183/185. Interestingly, only PG induced these conformational changes. However, when the fluorescence probe was attached to the residue 377, the binding of both substrates was required to induce modification in the fluorescence of the probe, whereas the reactivity of the thiol-group decreased on binding of each ligand separately. - On the other hand, no alteration was observed on the external side of the protein: for the mutant proteins C97A-S140C and C97A-E249C whose cysteinyl residues are located far away from the cleft, the ligand binding did not affect significantly the thiol-group reactivity. Although, some uncertainty still remain concerning the nature of the ligand triggering the conformational changes, it is likely that the substrates separately or together induce discrete molecular motions in PGK, which may or may not be observed depending on the experimental approach. Furthermore, using the above listed spectroscopic investigations that evidenced the occurrence of ligand-induced conformational changes, it could not be clearly shown to what extent these changes are required for the catalytic process.

A more sophisticated fluorescent approach, determination of interdomain distance distribution by site-specific labelling and time-resolved non-radiative fluorescence energy transfer technique, was applied to characterize the large scale hinge-bending domain motions of yeast PGK [201]. In order to achieve this purpose two genetically engineered cysteines (Cys 135 and Cys 290), one in each of the two domains (illustrated by Figure 33), were covalently labelled with a donor, 5-[2-(2-iodoacetamido)ethylamino]-1-naphthalenesulfonic acid (IAEDANS) and acceptor, 5-iodoacetamidofluorescein (IAF) pair of fluorescent probes. Analysis of sub-nanosecond fluorescence decay curves yielded the equilibrium distribution of interdomain distances. In the absence of substrates, the distribution of distances between the two labelled sites was very broad, with a full width at half maximum estimated as 20 Å or brooder, indicative of a large number of conformational sub-states in solution. The mean distance, 31.5±1.0 Å, was interestingly 8 Å smaller than in the crystal structure. Upon addition of ATP alone or of ATP and PG together, the average distance increased to 38±1 and the width of distribution is decreased. Addition of PG alone induced a similar, but smaller change (the mean distance of 35.2±0.5 Å was derived). The rate of conformational state fluctuations (interconversion between states) was found to be slow on the nanosecond time scale, as expected for a protein with a relatively large interdomain contact area. - It is notable, however, that the contribution of conformational flexibility of the labelled chain segments (loops) cannot be ruled out in the present experiment. The average conformation of these loops in solution can conceivably be different than the conformation observed in the crystal structure, which may be resulted in shorter interdomain distance in solution than that found in the crystal. Further, mutation and the labelling procedure may also cause some perturbation on the enzyme structure and function. Thus, it is possible that the interdomain distance determined in these experiments is somewhat larger or different than what would be found in a non-treated enzyme. At any case, much in advance to publication of the domain closed PGK structures, the experiment demonstrated that the action of PGK is likely to involve a series of interdomain conformational changes.

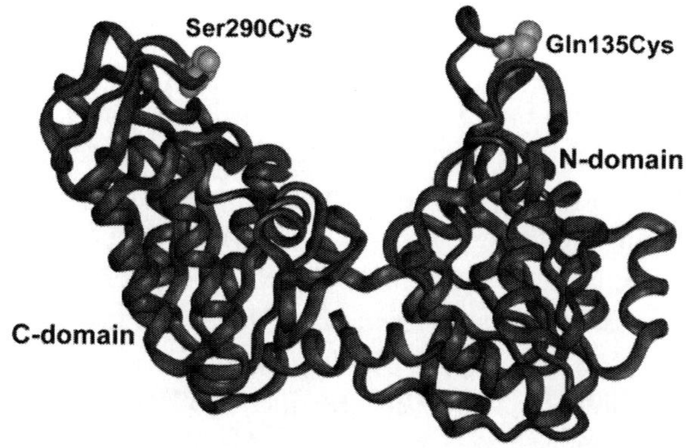

Figure 33. Molecular graphical illustration of introduction of Cys residues into the N- and C-terminal domains of yeast PGK by site-directed mutagenesis (based on the work by [201]):
The mutated side-chains of Gln135Cys (N-domain) and Ser290Cys (C-domain) are shown as grey space-filling models in the molecule of yeast PGK represented by black ribbon diagram.

Infrared spectroscopic studies provided further information complementary to those obtained with other spectroscopic methods. Namely, difference infrared spectroscopy was used to better understand the conformational changes that are unique to the functional PGK·ATP·PG complex in case of yeast PGK [202]. Difference FTIR spectra obtained on PGK·nucleotide·PG complexes show distinct bands that may result from amino acid side-chains as well as structural changes in the hinge-region and/or increased interactions such as salt bridges forming between the two domains. The infrared data obtained on the active ternary complexes show evidence of changes in α-helix and β-structures as well as provide signals consistent with unique perturbation of Arg, Asn, His, Lys, Asp, Glu and other side-chains in the active ternary complex as compared to the other enzyme–substrate complexes. This information was found to be consistent with the findings obtained from the comparison of the closed ternary and the open binary complex structures (cf. Chapters IX. and X.).

2.5. Chemical Modifications

Studies of chemical modification of various reactive side-chains of enzymes with specific reagents often report about their steric position in the protein molecule or about their essential role in the catalysis and in substrate binding. Much before the presently widespread technique of site-directed mutagenesis, chemical modification was used successfully to identify important protein residues.

Modification of Arg and Lys Side-chains

An early example of such studies is the chemical modification of the arginyl residues of yeast PGK with 2,3-butanedione [203]. Modification of one arginyl residues per enzyme molecule has led complete loss of enzyme activity, indicating the essential character of this Arg. Moreover, each of all substrates, MgADP, MgATP or PG largely protected against Arg modification and the accompanied inactivation. In a more extended work with yeast PGK, among other residues, the role of lysines were tested by modification with *O*-methylisourea, 2-methoxy-5-nitropone and pyridoxal phosphate [204]. In all cases several number of lysyl side-chains have been modified, but extrapolation to zero activity indicated the existence of three essential lysines per mole enzyme. From the special protection against modification by MgITP, the role of these lysines in interacting with the nucleotide phosphates has been postulated. It is notable that at the time of these studies no structural information of the protein was available. Much later, participation of one Arg (38/Y38/Tb39/Tm36/Bs36) and two Lys-s (215/Y213/Tb219/Tm197/Bs197 and 219/Y217/Tb223/Tm201/Bs201) in the catalysis have been indeed established (cf. Chapters II. 2.7. and IX.).

Modification of Tyr Side-chains

The latter extended study also demonstrated that nitration of a single Tyr residue of yeast PGK using tetranitromethane led to enzyme inhibition and the substrates have provided partial protection against this process [204]. In agreement, iodination of yeast PGK by other authors [205] has led identification of a critical Tyr residue, the modification of which has accompanied by a complete loss of activity. Further, no significant conformational change could be established upon iodination, as indicated by H-D exchange and CD spectral measurements. The spectral effect characteristic of PG binding was shown to disappear when the enzyme is inactivated. On this basis one might expect the Tyr to be situated at the PG site. In addition, although the nucleotide substrates still bind normally to the modified enzyme, the dissociation constant of MgATP is affected in contrast to that of MgADP that remains unchanged. This seems to indicate proximity between the γ-phosho-group of ATP and the essential tyrosine.

All tyrosyl residues of yeast PGK have been characterized in detail by a more systematic pH-titration study [206]. In the native state one tyrosyl residue has pK_{app}=9.3, another has pK_{app}=12.9, and five have pK_{app} values close to 11.0. In denatured state (in the presence of 6 M GuHCl) all seven tyrosyl residues with pK_{app} values equal to10.0 appear. Under native conditions nitration of three Tyr residues occurs easily when tetranitromethane is used in excess. Four Tyr residues appear to be masked or buried in the protein structure. The Tyr residue having pK_{app}=9.3 can be selectively nitrated that is accompanied by 40% loss of the original enzyme activity. No change in the K_m value of either PG or MgATP could be detected. On the other hand MgATP protects the Tyr residue from nitration, whereas PG appears to be ineffective. The latter finding is in contrast to some previous observation [204]. Taken together, the results suggested the low ionising Tyr residue to be situated close to the binding site of MgATP, possibly in a pocket just behind. Later on efforts have been put forward to identify in the primary sequence of this reactive Tyr. Chymotryptic fragmentation of the nitrated protein followed by peptide mapping revealed approximately fifty peptides, one of which was shown to contain a nitrotyrosine residue. Isolation of this unique peptide was accomplished by gel-filtration and by paper electrophoresis. The sequence as established by Edman degradation and carboxypeptidase hydrolysis is: Lys-NO_2Tyr-Phe-Phe-Lys [207]. This sequence was successfully fitted into the low resolution electron density map of yeast PGK available at that time [60]. The pentapeptide sequence is believed to start at residue 117 and is near the active site of the enzyme [207].

Modification of Glu Side-chains

In another analytical study one glutamyl residue essential for yeast PGK activity has_been also identified [208]. The enzyme is inactivated by 1-cyclohexyl-3-(2-morpholinoethyl)carbodiimide metho-*p*-toluensulfonate and nitrotyrosine ethyl ester. The rationale of the coupling of the two reagents is the following. The primary step for modification of carboxyl-groups in a protein is the activation of these groups by a soluble carbodiimide, the second step being the attack of the acyl-urea derivative by a nucleophile, the nature of which can vary. In order to purify the peptide containing the modified amino acid it is more advantageous to select a nucleophile bearing a chromophore and possessing antigenic properties. Such properties can help in selective purification of the labelled peptide(s). To this end, the nucleophile nitrotyrosine ethyl ester was selected in place of the usually used glycine ethyl ester. This nucleophile appeared to be as reactive and specific as glycine ethyl ester, since the loss of enzyme activity was concomitant with the incorporation of the chromophore; total inactivation was obtained when 1 mol nitrotyrosine was incorporated per mol PGK. A weak protection against inactivation is observed with each substrate added separately. In contrast, the functioning ternary complex with MgATP and PG affords a considerable protection. The critical residue is identified as a glutamyl residue after isolation by immuno-affinity chromatography of nitrotyrosyl peptide resulting from exhaustive proteolytic digestion of the modified protein. In addition, the determination of the primary sequence of the C-terminal part of the protein leads to the location of the modified Glu residue at position eight from the C-terminus: Tyr(NO_2)Met-Glu-Gly-Glu-Leu-Phe-Val-Ala-Lys. On this basis and taking into account the X-ray structural information available at that time [60] the authors concluded that this Glu residue is situated in the domain, which does not bind the nucleotide substrate.

Modification of Cys Side-chains

It was also established that sulfhydryl reagents, as well as mild hydrogen peroxide oxidation, do not inhibit the enzyme activity, indicating that the single thiol-group and three metionine residues present in the molecule are not essential for activity [204]. In agreement, using various alkylating agents (NEM, IAA, IAOH) and the thiol-specific Nbs_2 others have concluded that the single thiol-group of yeast PGK is fairly unreactive, does not seem to be directly located in the active site region, and the modification does not appear to influence the substrate binding and the activity [209]. In order to characterize better the local surroundings of this single thiol in the steric structure of yeast PGK, it was reacted with a chromophoric organomercurial reagent, 2-chloromercuri 4-nitrophenol [210]. Only one mole of mercurial has reacted per mole of enzyme. The modified enzyme appeared identical with the native enzyme in its catalytic properties. The p*K* of the phenolic hydroxyl of free mercurial is 6.60, but became 8.30 when the compound is associated with the single thiol of yeast PGK. This association also resulted in shifts in the wavelength of maximum extinction and decreased the extinction at these maxima (hypochromicity). There are further changes of the p*K* of the phenolic hydroxyl group upon substrate binding: in the presence of PG or MgATP the p*K* is increased to 8.72 or 8.54, respectively. Furthermore, the rate of the relatively slow reaction between the mercurial and the enzyme is increased by MgADP and MgATP and decreased by PG and phosphate ion (the latter is a known activator of the PGK reaction, cf. Chapter VII.). Considering the p*K* increase of the mercurial upon binding to PGK and the relatively slow reaction of this thiol of yeast PGK with this and other thiol reagents suggests a strongly hydrophobic pocket in the surrounding of the thiol-group. It appears that the thiol group of yeast PGK is not freely accessible towards the reagent and is located relatively far from the active site. The effects of substrates on the rate of thiol modification are possibly only due to their indirect effects on the global enzyme conformation.

In contract to yeast PGK the mammalian PGK molecules generally possess more than one thiol. PGK from rat muscle has been published to possess four thiols per mole enzyme: one fast-reacting and three slow-reacting ones towards Nbs_2 [211]. The fast-reacting thiol can be also alkylated with IAOH that is accompanied by complete activity lost. Inactivation by Nbs_2, however, can be reversed by incubation with a thiol-compound, such as MET or DTT.

From the pH-dependent kinetics of iodoacetate inactivation the p*K* value of the reactive thiol group could be evaluated to be 9.1, corresponding to the normal value of the unprotonated from of a free cysteinyl residue. Among the substrates MgADP or ADP completely protect the enzyme against alkylation, whereas MgATP, ATP and PG have no any effect on the reaction. It was therefore concluded that the fast-reacting thiol is located in or close to the MgADP binding site. It was also observed that on storage of rat muscle PGK in the absence of thiol compounds there is a loss of enzyme activity and a concomitant loss of the fast-reacting thiol. Gel-filtration and ultracentrifugation studies have revealed an increase of the molecular weight during storage: a dimeric form is detected due to formation of a disulfide linkage between the fast-reacting thiols of the two monomeric molecules. The dimeric protein is catalytically inactive but can be fully reactivated by reducing agents.

Other mammalian PGKs, such as rabbit [12], pig [92] and horse muscle [212] as well as human [107] enzymes have seven cysteine residues: two of them are similarly reactive towards Nbs_2 and the other five thiols react very slowly. With pig muscle PGK the second order rate constant of the two readily accessible thiols is around $10^3\ M^{-1}\ s^{-1}$, i.e., comparable with the reactivity of a free thiol compound, while the reaction of the slowly reacting five thiols are limited by a first order protein structural change (0.0002 s^{-1}), independent of the reagent concentration [92].

In case of rabbit muscle PGK modification with organothallium reagents (by analogy with the abovementioned organo mercurials) have been found to rapidly label the enzyme and caused its inactivation [213]. The rate of the reaction was found to be variously affected by the substrates. On the other hand, modification of the single thiol of yeast PGK with organothalliums was not accompanied by activity loss. Considering the different behaviours of yeast and rabbit PGKs towards thiol modification suggest that thiols themselves may not be essential residues even for functioning of the mammalian enzymes.

Numerous experimental results have, indeed supported this suggestion. First of all, blocking of the fast-reacting thiols did not abolish the substrate binding ability. In fact, all substrates inhibit the reaction of the slowly reacting thiols of the pig muscle PGK and the order of their protecting effect is very similar to their known structure stabilizing effects [92]. Furthermore, it has been proposed from experiments of the pig PGK with the bifunctional thiol-reagents (either mercuric chloride ($HgCl_2$) or 1,4-bis(bromomercuri)butane), that the two fast-reacting cysteines in the pig enzyme could correspond to Cys 378 and Cys 379 in the horse enzyme, the structure of which was known much earlier than the pig one [91]. The crystal structures of pig muscle PGK have, indeed, shown later the presence of cysteinyl side-chains in the positions of 378 and 379, located in helix 13. Introducing Cys side-chains into the equivalent positions of yeast PGK by site-directed mutations resulted in fast reacting thiols of the yeast enzyme with similar reactivity to those of the mammalian PGKs [212]. These mutational experiments have proved directly that the fast-reacting thiol groups are identical to the non-conserved side-chains of Cys 378 and Cys 379 of the mammalian PGKs.

Knowing that these two cysteinyl residues belong to helix 13, that is close to, but separate from the substrate sites, the question arises: why blocking of these thiols with either Nbs_2 or JAA is accompanied by the loosing of enzyme activity? A possible answer is the following: Since helix 13 is situated in the interdomain region of the enzyme, introducing a bulky group here by chemical modification might prevent important conformational changes needed for the function. In order to test the validity of this hypothesis the fast-reacting thiols were alkylated by reagents bearing groups of different sizes. Namely, the effects of iodoacetamide and iodomethane on pig muscle PGK has been compared [214]. It was found surprisingly that while both reagents can alkylate the fast reacting thiols, introduction of the more bulky carboxamidomethyl-group leads to complete activity loss, whereas methylation of the same groups does not affect at all the catalytic efficiency (Figure 34). Full activity of the methylated PGK on one side approved the non-essential character of the fast-reacting thiols, on the other side clearly confirmed the hypothesis about that the activity loss is due to steric prevention of the activity-related conformational changes (possibly the hinge-bending motion) by a bulky group attached to the thiols. Further SAXS measurements (cf. Chapter II. 2.9.) with both the methylated and carboxamidomethylated PGKs, indeed, supported the occurrence of the hinge-bending motion in case of the methylated active enzyme, while indicated its absence in case of the carboxamidomethylated inactive enzyme, upon addition of the substrates.

Affinity Labelling

Besides the above-discussed selective modification of various protein side-chains, a different type of modification, namely the „affinity-labelling" has to be also mentioned. An affinity labelling reagent is generally a structural analogue of a substrate bearing also a reactive group. Thus, it binds in the substrate site and can react with the reactive side-chains in its neighbour. Thus, this type of reagent can report about the location of reactive groups in the vicinity of the substrate binding sites. An ATP-analogue affinity label, 5'-[*p*-(Fluorosylfonyl)benzoyl]-1,N^6-ethenoadenosine was used to label yeast PGK. The

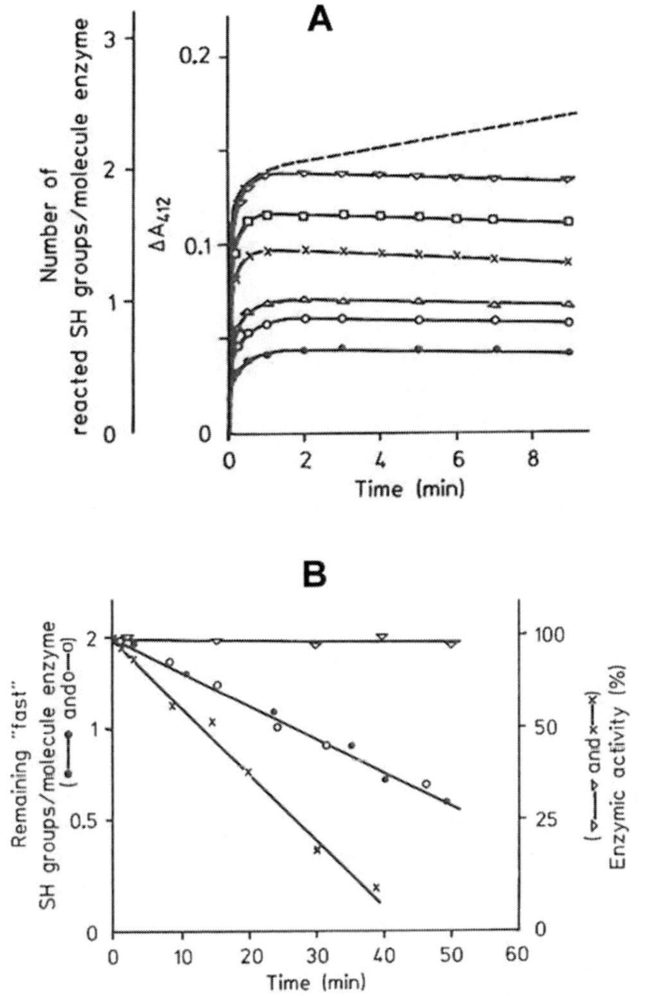

Figure 34. Kinetics of the reaction of the fast-reacting thiol-groups with either iodomethane or iodoacetamide (adapted from [214]): (A) Decrease of the fast-reacting thiol content in the course of the reaction of 15μM pig muscle PGK with either 8 mM iodomethane or 1.9 mM iodoacetamide as determined upon reaction with Nbs_2 in the aliquots withdrawn from the reaction mixtures at different time intervals. (B) Semilogarithmic plots of thiol-modification with 1.9 mM iodoacetamide (•-•) or with 8 mM iodomethane (o-o). Enzyme activity was also determined during the reaction with 3.75 mM iodoacetamide (x-x) and 8 mM iodomethane (∇-∇).

enzyme was irreversibly inactivated upon incubation with the reagent [215]. All substrates markedly protected the enzyme against inactivation. Stochiometric studies showed that the relationship between moles of incorporated reagent per mole of enzyme and the residual enzyme activity exhibited non-linear behaviour. An extrapolated value of 1.5 moles of bound label/moles of enzyme corresponded to complete inactivation. Thus, the authors have suggested a two-site model of the nucleotide binding to PGK. Furthermore, the apparent overall pseudo first-order rate constant for the reaction exhibited saturation behaviour with respect to the concentration of the reagent, indicative of a rapid reversible binding of the reagent to the enzyme prior to modification. Sulfonyl fluoride has a reactive group that can act as an electrophile in covalent reactions with several classes of amino acid residues, including Ser, Tyr, Lys, His and Cys. From the crystal structure of PGK two Lys-s, namely Y213/215 and Y217/219 have suggested as possible candidates for modification. However, analytical identification of the residues labeled with this affinity-reagent has not been carried out in case of yeast PGK.

There are other two examples of affinity labelling of PGK with ATP-analogue reagents. In both cases yeast PGK has been modified and the results similarly led to the conclusion about significant ATP-induced conformational changes. In one case [^{3}H] pyridoxal 5'-diphospho-5'-adenosine (PLP-AMP) has been applied and found to be a potent and highly specific inactivator [216]. 1.2 moles PLP-AMP/ mole enzyme was found to be sufficient to give complete inactivation and MgATP exerted near complete protection. The modified residue was identified as Lys 131 following reduction with $NaBH_4$, digestion with endopeptidase, separation and sequencing the labeled peptide. In the open crystal structure of the yeast enzyme, available at that time [5], Lys 131 is located in an outer loop of the N-domain, approximately as far as 20 Å from the position of the γ-phosphate of ATP bound on the C-domain. From these data the authors have concluded that the structural element containing Lys 131 has undergone substantial movement during the ligand-induced conformational change. – In the case of the other work, another affinity labelling reagent, pyridoxal 5'-diphospho-5'-adenosine ($AdoP_2Pxy$) was used to modify yeast PGK [217]. Incorporation of 1 mole label/mole enzyme was sufficient for complete inactivation of the enzyme. The substrate ATP afforded substantial, while PG exerted partial protection against modification. Analytical identification of the modified residue revealed specific reaction of Lys 385. This residue is located in the β-strand L, immediately following the α-helix 13 in the interdomain region of the enzyme molecule. Model-building studies showed that when $AdoP_2Pxy$ is built into the open form of the enzyme, the pyridoxal group of the reagent can not reach Lys 385 for Schiff-base formation. Thus, labelling of Lys 385 of yeast PGK also suggested occurrence of significant conformational changes upon substrate binding.

2.6. Proteolytic Susceptibility

The classical method of proteolytic digestion can still be considered as a sensitive probe of the conformation of any given protein. Namely, susceptibility of proteins towards proteolytic attack greatly depends on the conformational state of a given protein. It can report e.g., about location of the side-chains on the surface that are sensitive to proteolytic attack. The method can also reveal the effects of ligands/substrates on the protein conformation.

In a study with yeast PGK great resistance of the native enzyme was observed towards either trypsin or yeast proteinase A, but modification of the single non-essential thiol either Nbs_2 or 2-chloromercuri-4-nitrophenol greatly increased the susceptibility of the kinase towards proteolyis [218]. In both cases limited proteolysis occurred. The 28 kDa product that appeared after extensive proteinase A digestion is the same size fragment that accumulated with the trypsin degraded kinase. However, the intermediates between the full size 48 kDa protein and the 28 kDa species were different with the two proteases, and this probably reflects their different specificities. Trypsin hydrolyzes peptide bonds adjacent to Arg or Lys residues, while proteinase A shows properties similar to those of pepsin and hydrolyzes peptide bonds adjacent to aromatic residues. Considering the amino acid composition of yeast PGK, and the specificities of the two endopeptidases used for this study, there should be susceptible residues within the 28 kDa fragment. From the relative resistance of the 28 kDa fragment towards proteolysis it follows that the fragment must have enough structure to render these potential substrate residues sterically inaccessible. It is possible that this type of limited proteolysis is due to the domain structure of PGK (cf. Chapter I. 1.2.), the hinge region that connects the two lobes is being more susceptible to proteolytic cleavage. However, this conclusion remained speculation since it has not been supported by analytical identification of the cleavage sites. Apart from this drawback, other interesting observations of the work that substrates, PG and MgATP, as well as various anions (citrate, succinate, chloride) significantly protected the enzyme against proteolysis. Thus, the effects of these ligands on PGK conformation have been clearly demonstrated by these experiments.

A different aspect of the relationship between proteolytic sensitivity and the domain structure has been established in another work by comparing proteolytic susceptibilities of PGKs from yeast and *E. coli* [219]. It was found surprisingly, that *E. coli* PGK is resistant to proteolytic cleavage while the yeast homolog from *Saccharomices cerevisiae* is not. The two proteins have extremely similar tertiary structures and their global stabilities (as determined by equilibrium denaturation) are also very similar. However, the unfolding rate of *E. coli* PGK is much slower than that of the yeast homolog that can be related to the much stronger domain-domain interactions of the former one. Moreover, these differences are consistent with and explain the differences in the proteolytic sensitivities of the two enzymes.

There is another interesting example of demonstration of the conformational effect of the substrate PG by occurrence of limited trypsinolysis of pig muscle PGK in its enzyme-substrate complex [220]. By following the time course of proteolysis using various methods (Figure 35) it was found that in the presence of PG both the decrease of enzyme activity and the release of trichloroacetic acid-soluble peptides occur after a pronounced lag-period. During this lag phase the native enzyme molecule is split into two fragments with molecular masses of about 30 and 18 kDa, as detected by SDS-PAGE. Under non-denaturing conditions, however, these fragments are held together by non-covalent forces (they cannot be separated by gel-chromatography under native conditions) and constitute a fully active, nicked enzyme molecule. Determination of the N-terminal sequence of the large proteolytic fragment has led to identification of the tryptic cleavage site at Lys155, i.e., at the border of a sub-domain within the N-terminal domain. These proteolytic fragments have also been used in complementation refolding experiments (cf. Chapter III.). - In the absence of PG or in the presence of MgATP the kinetics of tryptic digestion of pig muscle PGK is apparently a single first order reaction leading to the formation of peptides with molecular masses of less than 10 kDa.

As mentioned above (cf. Chapter I. 1.2.) the C-terminal tail of PGK molecule constitutes an integral part of the N-domain. There is a conserved H-bonding pattern between the two termini of the polypeptide chain. These interactions are assumed to contribute in regulating the functionally important domain motions. An interesting limited trypsinolysis study with yeast PGK seems to support this assumption. In the course of this work a single Lys residue (Lys 415) from the C-terminus was removed and this modification was accompanied by a fourfold increase of the enzyme activity. It was concluded that removal of the C-terminal Lys strengthen the interaction between Lys 5 and Glu 413. Thus, this salt bridge possibly stabilizes the yeast enzyme in the higher activity form, while the presence of Lys 415 reduces the strength of that interaction. Although these residues are not conserved, this finding has supported the importance of the C-terminal region in mediating the conformational changes required during catalysis.

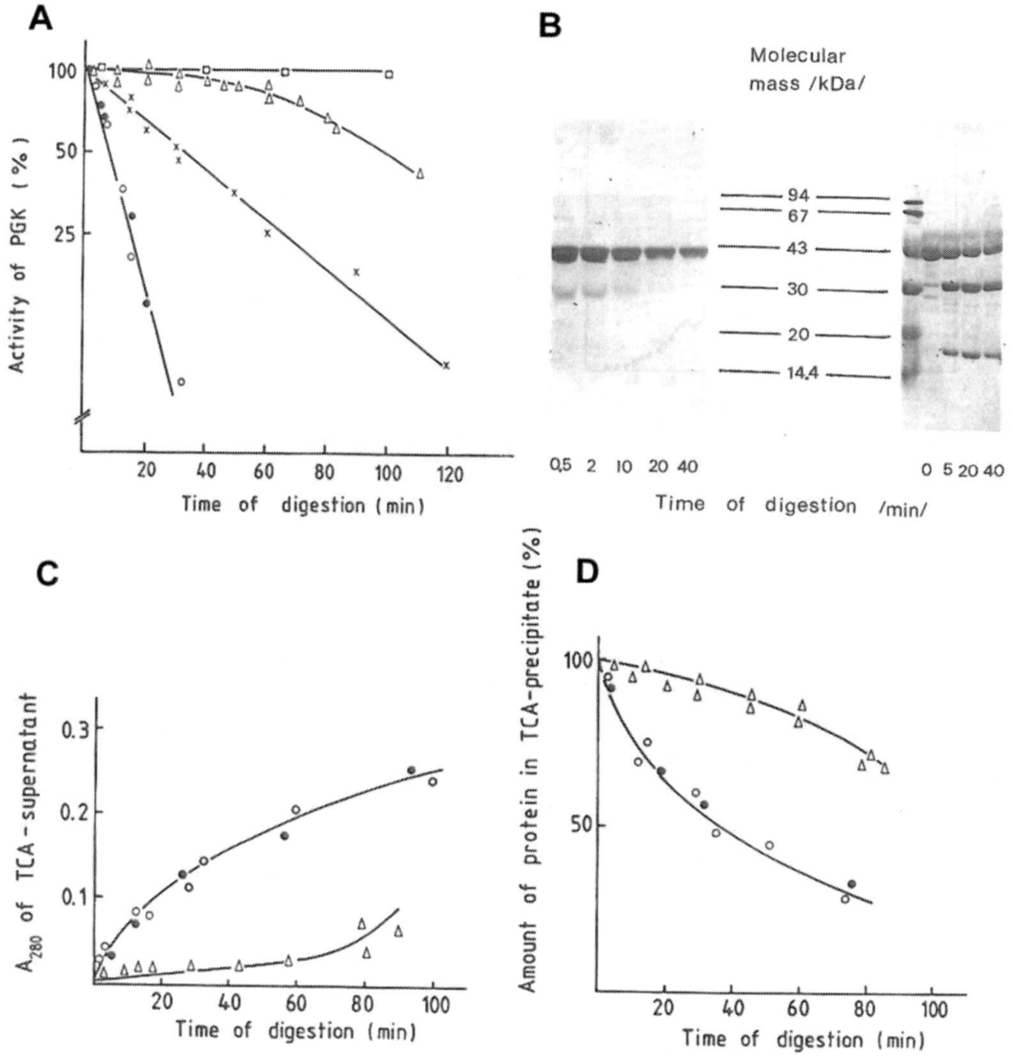

Figure 35. Limiting tryptic digestion of pig muscle PGK (adapted from [220]): (A) Enzyme activity tests of the digestion mixture of 50 μM substrate-free enzyme in the presence of 2.2 (x-x) and 10.9 (o-o) TU_{BAEE}/ml trypsin and in the absence (□-□) of trypsin. The experiment with 10.9 TU_{BAEE}/ml trypsin was repeated with either 6.6 mM PG (△-△) or 6.6 mM MgATP (•-•). (B) SDS-PAGE analysis of the digestion mixtures of 0.3 mM PGK and 10.9 TU_{BAEE}/ml trypsin in the absence of substrates (1) and in the presence of 6.6 mM PG. C and D: Increase of the TCA-soluble peptides during digestion of 0.5 mM PGK in the absence of substrates (o-o) or in the presence of 6.6 mM MgATP (•-•) with 2.2 TU_{BAEE}/ml trypsin as well as in the presence of 6.6 mM PG with 10.9 TU_{BAEE}/ml trypsin (△-△).

2.7. Mutagenesis Studies

Site-directed mutagenesis, in some features, is very similar to chemical modification of selected side-chains. Namely, mutation allows replacement of a selected side-chain by another one having different chemical nature or different size. From the modification caused changes in substrate binding, catalytic activity or in the physico-chemical properties, conclusions can be drawn about the role of the mutated residue in substrate binding, catalysis, enzyme conformation or in the catalytically important conformational changes.

Mutations within the N-Terminal Domain

The first mutational studies were carried out with yeast PGK much earlier than its X-ray structure has been determined, and aimed to identify side-chains of the basic patch of the inner surface of the N-domain that was assumed to bind the substrate PG. As mentioned above (cf. Chapter II. 2.1.) from the three histidines of the basic patch two conserved ones, namely His Y62/62 and His Y167/169 have been assigned by NMR [131]. Later mutational studies have indeed been identified them together with the non-conserved His Y170/172 as constituents of the basic patch [132, 221]. Their role in PG binding has been also revealed by increasing the K_m or K_d values of PG upon their mutation into a neutral residue (cf. Table 10).

Besides the histidines, several arginines (Y21/21, Y38/38, Y65/65, Y121/122, Y168/170) have been exchanged by various type of mutations as summarized in Table 10. These works have been suggested their role in PG binding [107, 125, 156, 222-225]. These suggestion has been confirmed by later X-ray studies, except for Arg Y21/21, that is not involved in substrate binding [15]. The increase of the Km values of its mutants is probably due to partial abolishment of the H-bond network involved in substrate binding [125].

In general, mutations in the N-domain mainly destroy the interactions with PG and only slightly influence those of with MgATP. The results have been suggested binding of PG to the N-domain much in advance of the structural identification of its binding site. Moreover, kinetic data of the mutant Arg Y38/38 [107] clearly demonstrated the essential role of this side-chain in the catalysis, which was clearly approved only later by the closed crystal structure of human PGK with a TSA (Chapter IX.).

Mutations within the C-Terminal Domain

Mutations have been carried out at the nucleotide binding site of yeast PGK [226] following determination of the X-ray structures with the nucleotides. The first mutated residue was the conserved Asp 374/Y372. The X-ray structure suggested that one of the carboxyl oxygens of this Asp is H-bonded to the 3' hydroxyl of the ribose and that the other interacts with the Mg-ion complexed with ATP [5]. Later on, higher resolution crystal structure of MgADP binding to BsPGK has been determined and showed clearly the interaction of Mg-ion with both the equivalent Asp and the phosphates of ADP [6] (cf. Figure 4A). Thus, this acidic residue appears to help to desolvate the magnesium ion bound to the nucleotide. The substitution of the charged carboxyl group with the neutral amide in the AspY372Asn mutant affects the binding of the nucleotide substrate, but not the binding of PG.

Table 10. Summary of kinetic effects of the mutations within the N-terminal domain of PGK

Amino acid residue	Type of mutation	K_m^{PG} (factor of increase)	K_m^{MgATP} (factor of increase)	k_{cat} (%)[#]	Anion activation	K_d^{PG} (factor of increase)
21/Y21	Arg→Ala [125]	**21**	4	60%	abolished	—
	Arg→Lys [222]	8	3.6	55%	abolished	—
	Arg→Met [222]	4.3	4	15%	abolished	10
38/Y38	Arg→Ala[107]	**45**	8.5	**0.05-0.1%**	abolished	**27**
	Arg→Ala [156]	7.2	5.8	**0.2%**	abolished	**50**
	Arg→Lys[107]	2.8	1	66%	unaffected	3
62/Y62	His→Gln [125], [221]	2	2	200%	unaffected	3
	His→Ala [156]	5	4.2	85%	abolished	**43.6**
65/Y65	Arg→Gln [223]	6	2.2	165%	abolished	6
	Arg→Met [223]	5.4	2.2	165%	abolished	—
	Arg→Met [224]	**21**	4.9	10-15%	abolished	**>50**
	Arg→Ala [223]	5.7	2.2	165%	abolished	—
	Arg→Lys [224]	2.2	1	slight increase	decreased	4
122/Y121	Arg→Met [224]	2.6	1	50-60%	decreased	—
	Arg→Lys [224]	2.6	1	50-60%	abolished	—
169/Y167	His→Ser [125]	**10.5**	1,5	40%	abolished	—
170/Y168	Arg→Met [225], [132]	**7.8**	3,5	20%	decreased	**15**
	Arg→Lys [225], [132]	2.7	2.7	40%	abolished	2.3
	Arg→Gln [125]	**19**	2.3	55%	abolished	—
172/Y170	His→Asp [132]	—	—	100%	—	4

The largest effects of the mutations on the kinetic constants are shown in bold and underlined.
[#]Activity of the wild-type enzyme is considered as 100%.

Moreover, the overall velocity of the enzymic reaction is reduced 10 fold (Table 11). This reduction of velocity is considerably less than one would expect from its known structural position. Thus, this conserved residue has no direct role in the catalysis in spite of its possible importance of the charge balance in the active site cleft (cf. Chapter IX.).

There is another interesting conserved residue located close to the phosphate chains of the bound nucleotides, namely Lys 215/Y213/Bs197/Tm197/Tb219. However, as discussed in Chapter I. 1.4., its side-chain does not interact with the nucleotide phosphates in any of the open PGK structures [6, 69, 80]. Moreover, a characteristic feature is a considerable flexibility of the side-chain of Lys 215 in both the AMP-PCP- and the MgATP-binding open structures of pig muscle PGK [7, 72]. In contrast, in the closed structure of TmPGK Lys 215/Tm197 simultaneously interacts with both the γ-phosphate of AMP-PNP and the carboxylate of PG [73] (Figure 9C). This observation has suggested the essential role of this residue in the chemical phospho-transfer. In fact, a large (more than 10Å) movement of this residue has taken place between the open and closed enzyme conformations (Figure 8). Thus, the right geometry of the active site with the proper position of the catalytic residues can apparently be formed during domain closure. This hypothesis and the catalytic role of Lys 215 have been confirmed by site-directed mutation of Lys 215 into either the neutral Ala or another positively charged residue (Arg) (cf. Table 11 [107]). It is notable that both the binding and the catalytic interactions with MgATP, but not with MgADP is weakened upon its mutation into Ala. Based on these findings an initial weak interaction of Lys 215 with the γ-phosphate of MgATP already in the non-functioning open conformation and simultaneous movement of Lys 215 together with this transferring phosphate have been concluded. It is indeed, plausible that during this movement the transferring phospho-group is positioned properly for catalysis. The essential catalytic role of the highly conserved Lys 215 in human PGK has been fully explored by a transient and equilibrium kinetic as well as crystallographic study with the active site mutant K215A and it has been shown that this side-chain is indeed directly involved in the chemical phospho-transfer step [126]. For more detailed discussion cf. Chapters V. 5.2. and V. 5.6.

Table 11. Summary of kinetic effects of mutations within the C-terminal domain of PGK

Amino acid residue	Type of mutation	K_m^{PG} (factor of increase)	K_m^{MgATP} (factor of increase)	K_m^{MgADP} (factor of increase)	k_{cat} (%)[#]	Anion activation	K_d^{PG} (factor of increase)	K_d^{MgATP} (factor of increase)	K_d^{MgADP} (factor of increase)
215/Y213	Lys→Ala [107]	4.1	**22**	0.8	**0.06%**	abolished	1.1	4.4	1.5
	Lys→Arg [107]	3.5	5.2	0.9	**0.2%**	abolished	1.05	2.5	0.85
218/Y216	Asp→Asn [107]	4.2	1.4	—	32%	largely decreased	1.2	0.9	—
219/Y217	Lys→Ala [228]	**17**	**40**	**12**	**0.1%**	abolished	1.2	**8**	**68**
	Lys→Ala [29]	—	1.5	5	**0.9%**	—	—	—	—
	Lys→Arg [29]	—	1.3	2.3	**1.7%**	—	—	—	—
336/Y334	Asn→Ala [228]	5.4	**12.2**	3.5	**0.7%**	abolished	1.5	3	**13**
	Asn→Asp [29]	—	1	0.3	**0.9%**	—	—	—	—
	Asn→Gln [29]	—	1.1	2.4	40%	—	—	—	—
343/Y341	Glu→Ala [228]	3.1	**15**	**7.6**	**1.7%**	abolished	0.9	4.9	**19**
	Glu→Ala [29]	—	—	3	12%	—	—	—	—
374/Y372	Asp→Asn [226]	1.0	2.5	—	10%	unaffected	—	—	—
375/Y373	Thr→Ala [228]	6.3	**9.5**	**9.3**	15%	abolished	—	0.8	10.3

The largest effects of the mutations on the kinetic constants are shown in bold and underlined.
[#]Activity of the wild-type enzyme is considered to be 100%.

Beside Lys 215 the largely conserved Asp 218 at the N-terminus of helix 8 has been suggested to be involved in the interaction with the ATP phosphates. Namely, in the MgAMP-PCP bound structure the γ-phosphate of AMP-PCP interacted with Asp 218 through Mg-ion [72] (Figure 4H). The importance of Asp 218 has been assessed by its mutation into the neutral Asn, but no evidence has been obtained in favor of the interaction of Asp 218 with Mg-ion (cf. Table 11 [107]). However, involvement of this residue in activation by anions has been supported. It has to be added that during the time of preparing this book further mutagenesis studies with Asp 218 as well as Asp 374 of HsPGK (not indicated in Table 11) have revealed their important roles in the interaction with the catalytic metal ion (cf. Chapter VIII.).

Among the nucleotide binding residues, Lys 219, which binds the nucleotide α-phosphate (Figure 4), is approved to be at least as important as Lys 215, which raised the possibility of direct participation of this side-chain in the catalysis [228] (cf. Table 11). NMR and X-ray crystallographic studies of the complex with a transition state analogue of human PGK have been clearly approved this assumption [76] (cf. Chapter IX.).

The other three residues, Asn 336, Glu 343 and Thr 375 are also involved in binding of the nucleotide substrates as discussed in Chapter I. 1.4. (Figure 4). They are also important in transmission of the nucleotide effect towards the main hinge (cf. Chapter X.). However, they are not essential in the catalysis, as reflected by the kinetic and binding data of their single site mutants (Table 11).

It may be notable that in Table 11 the data of Krishnan and coworkers are not supported in all aspects by the data of Szabó and coworkers, which is possibly due to the different methodology used by these authors for kinetic characterisation of the mutants.

Site-directed mutagenesis studies (summarised in Table 12) were also applied to test the operation of the assumed main hinges of PGK, located in the interdomain region, namely, at both termini of the α-helix 7 and within the β-strand L.

According to an early hypothesis the conserved salt-bridge between Glu 192/Y190 and His 390/Y388 (that thereby connects α-helix 7 and β-strand L, cf. Figure 36A) might contribute to the operation of the hinge-bending mechanism [230, 233]. However, various mutations of these two side-chains did not led to dramatic changes in the kinetic constants (Table 12), providing no evidence in favor of the responsibility of these residues. On the other hand, the interdomain interactions seems to be altered in the mutant of this His into Gln, resulting in the appearance of independent thermal transition for the two domains, in contrast to the single cooperative transition observed for the wild-type enzyme [235] (cf. also Chapter II. 2.8.). This experiment clearly demonstrated the role of the interactions of this His (including the salt-bridge with Glu 192/Y190) in the interdomain communication of PGK. In accordance, calorimetric and thiol-reactivity measurements with the mutants (Glu 192/Y190 and Phe 165/Y163 into Ala) indicated considerable destabilization effects on the whole molecule [228]. Thus, the hydrophobic and electrostatic, as well as H-bonding interactions operating between the abovementioned three side-chains (cf. Figures 36B and 36D) are essential for the structural integrity of the hinge-region as well as for the whole PGK molecule. Among the contributors to the hydrophobic interactions, Phe 196/Y194 or both the open and the closed conformational states of the molecule, thus they are not directly involved in the molecular mechanism of domain closure. It can be proposed, however, that multiple interactions of these side-chains located in the interdomain region may be important Leu Tm178/Bs178 (shown in the same figure) has to be also mentioned that is not a conserved residue, but always hydrophobic in its nature. These interactions are present in for transmission of the substrate-induced conformational changes in PGK. Mutations of Phe 165, Glu 192 and Phe 196 of HsPGK into Ala, indeed have led to substantial decrease of all k_{cat}/K_m values [228].

Mutations at the Hinge-Regions of PGK

Table 12. Summary of kinetic effects of mutations at the hinge-regions of PGK

Amino acid residue	Type of mutation	K_m^{PG} (factor of increase)	K_m^{MgATP} (factor of increase)	K_m^{MgADP} (factor of increase)	k_{cat} (%)#	Anion activation	K_d^{PG} (factor of increase)	K_d^{MgATP} (factor of increase)	K_d^{MgADP} (factor of increase)
165/Y163	Phe→Ala [228]	**19**	7.5	1.0	32%	abolished	1.3	2.3	0.2
184/Y183	Ala→Pro [229]	1.1	0.9	—	121%	diminished	—	—	—
192/Y190	Glu→Gln [230]	2.2	1.7	—	26%	abolished	—	—	—
	Glu→Asp [230]	2.2	0.8	—	36%	abolished	—	—	—
	Glu→Ala [228]	**10**	2.2	0.8	18%	abolished	1.3	1.0	0.4
196/Y194	Phe→Leu [231]	0.3	0.3	—	60%	decreased	2.3	0.3	—
	Phe→Trp [231]	1.0	1.0	—	50%	increased	0.9	1.0	—
	Phe→Ala [228]	**14**	3.0	—	73%	abolished	2.1	2.5	0.9
206/Y204	Pro→His [232]	3.2	3.8	—	**1%**	decreased	1.5	3	**13**
	Pro→Phe [232]	—	—	—	**0.02%**	unaffected	—	—	—
390/Y388	His→Gln [233],[234], [235]	1.0	0.2-0.3	—	20-30%	unaffected	**33**	—	—
	His→Lys [236]	0.6	0.4	—	**6%**	unaffected	—	—	—
	His→Ala [236]	0.4	0.5	—	13%	unaffected	—	—	—
392/Y390	Ser→Ala [228]	6.4	0.7	—	69%	unaffected	3.4	2.0	1.4
393/Y391	Thr→Ala [228]	**20**	8	0.8	**7%**	abolished	4.6	2.0	1.1
	Thr→del [228]	2	1.2	—	**0.08%**	abolished	1.2	2.0	0.3
392-393/ Y390-391	Ser-Thr→ Ala-Ala [228]	**33**	6	—	**1.9%**	abolished	3.4	1.3	2.1
398/Y396	Ser→Ala [228]	1.2	1.2	—	26%	unaffected	1.0	1.3	—

The largest effects of the mutations on the kinetic constants are shown in bold and underlined.
#Activity of the wild-type enzyme is considered as 100%.

The K_m values for PG in these cases are better increased than those for MgATP, thus these residues may better contribute to mediating the PG caused conformational effect towards the main hinge at β-strand L. Depicting the structural details of this transmission from the PG binding site to the bending region is shown by Figures 36A and 36B. PG is linked to α-helix 5 through its interaction with Arg 170 and the conserved Phe 165 of α-helix 5 interacts with Glu 192 (HsPGK numbering) of α-helix 7 of the interdomain region in all known PGK structures. Furthermore, Glu 192 interacts permanently with the side-chains of Ser 392 and His 390, the constituting residues of βL. Operation of the hinges in terms of substrate-caused changes of the atomic interactions will be discussed in more detail in Chapters IX. and X.

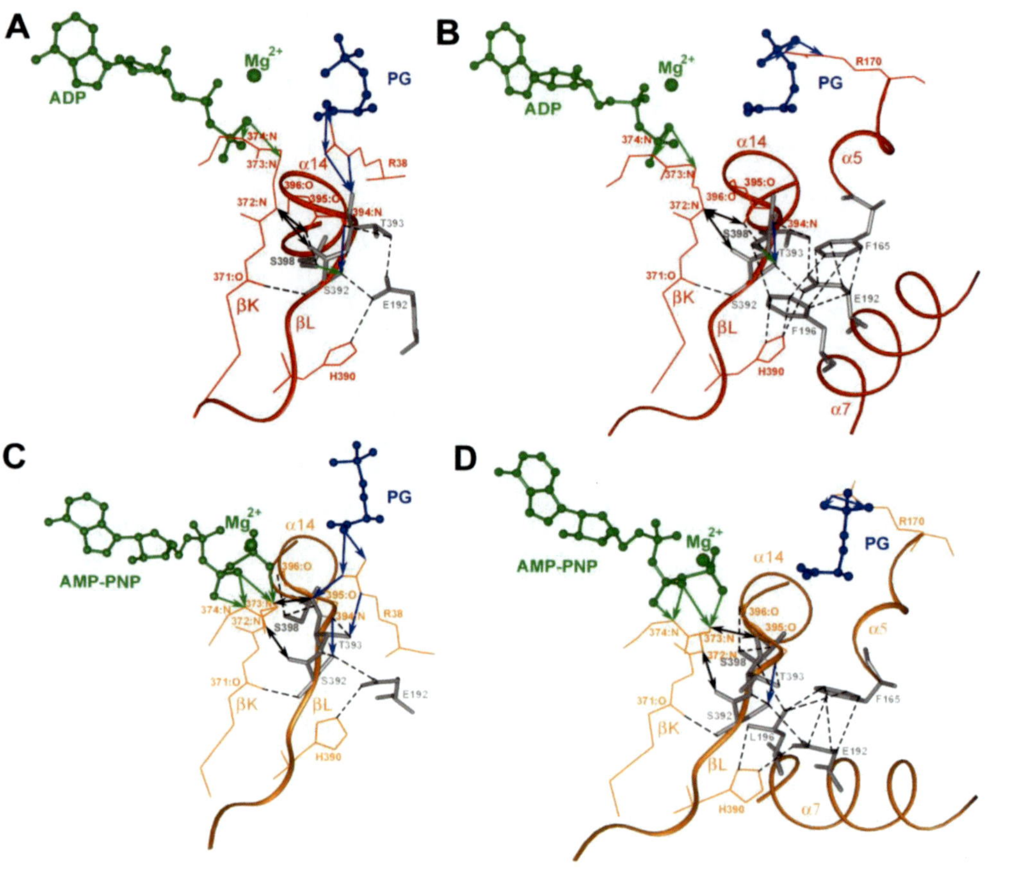

Figure 36. Details of the main hinge (βL) of PGK (adapted from [228]): The atomic interactions in the closed conformations of the ternary complexes of MgADP·PG of TbPGK [84] (A and B) and of MgAMP-PNP·PG of TmPGK [73] are illustrated (C and D). Substrates and the interacting side-chains are shown by ball and stick and stick models, respectively. The atomic contacts formed upon PG and nucleotide binding are shown by arrows of the respective colours. Other structural parts of the protein are shown in red (TbPGK) and orange (TmPGK). Side-chains are numbered according to the HsPGK sequence. The mutated side-chains are highlighted in grey colour. The permanent interactions are shown by dashed black lines. The H-bonds formed only in the ternary complexes are indicated by black double-sided arrows.

The hinges at the N- and C-termini of α-helix 7 have been also probed by mutations. Introduction of a Pro in place of Ala 184/Y183, i.e., artificially increase the rigidity of this N-terminal region, did not cause any appreciable change in the catalytic properties. On the other hand, mutation of the conserved *cis*-Pro 206/Y204 at the C-terminal region into either His or Phe caused severe impairing the enzymic activity, indicating that possibly the hinge-bending motion is hindered [232].

Operation of the assumed main hinge at the β-strand L have been similarly tested by systematic mutational studies [228]. The side-chains of Ser 392 and Thr 393 (mammalian sequence numbering) are highly conserved ones at the C-terminus of β-strand L, where the shape of this structural element is substantially changed upon domain closure, i.e., where the assumed main hinge is located (cf. Chapter I. 1.11., Figure 16A). Upon preparing single Ala mutants of these residues, T393A exhibited very large (about two orders of magnitude) loss in the catalytic efficiency (k_{cat}/K_m) compared to the wild type enzyme, while S392A possesses much less diminished efficiency (Table 12). These results indicated the importance of Thr 393 in operation of the assumed hinge at βL. The double mutant S392A-T393A, however, accumulated the properties of both single mutants, indicating the cumulative effects of these residues in the mechanism of domain closure. On the other hand, mutation of the less conserved Ser 398 (that interacts with Ser 392 in one of the closed crystal structures) into Ala did not cause large decrease of the catalytic efficiency compared to the wild-type enzyme. Thus, the assumed role of Ser 398 in operation of the main hinge should be excluded.

The results with T393A mutant support the suggestion about the role of Thr 393 in communication between the two domains. As observed in PGK X-ray structures, containing bound PG, the side-chain of Arg 38, besides participating in PG binding, creates a contact with Thr 393, either with its backbone [15, 84] (Figure 36A) or with the side-chain itself [73] (Figure 36C). This would be a plausible way of communication between the two domains and of the transmission of the conformational effects of PG from its binding site on the N-domain to the hinge at βL, as illustrated by the blue arrows in Figures 36A and 36C. The greatly decreased k_{cat} of T393A and its increased K_m value for PG (Table 12) clearly support the participation of Thr 393 in this process.

In addition to the above mutations a deletion mutant with lacking of the residue Thr 393 was also prepared and this modification indeed caused a dramatic activity loss [228]. This finding underlines the importance not only the side-chain of Thr 393 itself, but also the proper shape and length of the βL in the catalytic function. The deletion caused at least four orders of magnitude loss in the catalytic efficiency, but the interaction with the substrates (as expressed in their respective K_d values (Table 12), have not changed. Thus, βL has no direct role in substrate binding, but its intact main-chain is essential for the function. These findings clearly support the previous hypothesis [80] about the role of βL in functioning of PGK.

The more detailed description of operation of the main molecular hinge will be given in Chapter X. The question of a possible relationship or cooperation between the main hinge at βL and the hinges at the termini of α-helix 7 will be also addressed below.

2.8. Differential Scanning Calorimetric (DSC) Measurements

Differential scanning calorimetry (DSC) has frequently been used in the study of multidomain proteins since it allows individual domains to be seen as they undergo cooperative thermal unfolding. If the domain transitions are well separated on the temperature axis, then thermodynamic parameters and the transition temperature values (T_m) may be obtained directly for each domain. If the interactions between the domains are strong, their thermal unfolding transitions overlap severely and are merged into an apparently single transition.

There are numerous DSC calorimetric studies with the two-domain containing PGK. Heat-or cool-induced unfolding of yeast [235, 237-243], thermophilic bacterial [244], cold-active bacterial [245] and pig muscle [108] PGKs in the absence [235, 239-244] and presence [237, 238] [108, 245] of substrates have been monitored by DSC technique. Disruption of native PGK structure upon cooling occurs in two distinct stages, corresponding to independent and reversible unfolding of the individual domains [239, 242, 243]. It has been suggested later that the uncoupled unfolding of the two domains is the result of the presence of low concentration of GuHCl in cold denaturation experiments [246]. On the other hand, thermal unfolding of PGK invariably proceeds in an apparently single DSC transition both at low GuHCl concentration [241] and in diluted buffer [108, 237, 238, 240]. Strong interdomain stabilization has been claimed, but the slightly asymmetric thermal unfolding profile may be accounted for by assuming partially separated unfolding of the C- and N-terminal domains. However, with pig muscle and yeast PGKs the existence of strong cooperativity between the two domains has been demonstrated by following the heat transitions with both calorimetry and fluorimetry. Since in both proteins the Trp residues that provide the fluorimetric signal, are located exclusively in the C-domain, and furthermore, the calorimetric and the fluorimetric heat transitions are entirely coincided, a highly cooperative unfolding of the two domains have been evidenced (Figures 37A-C) [108]. The structural explanation of this cooperative behaviour is the conservative nature of the H-bonding, hydrophobic and electrostatic interactions operating in the interdomain region of all PGKs (Figures 38A-E).

On the other hand, separated DSC transitions attributable to the individual PGK domains have been observed in the case of engineered mutants of yeast PGK that contain modifications in the hinge-region between the two domains [235, 238] and in the case of cold-active PGK [245], possibly due to weaker interdomain interactions in these cases. It is notable that the substrate PG in one case [235], while PG and MgADP together (the non-functioning ternary complex) in the other case [245] caused merging of the two transitions into a single one. This indicates increased interdomain cooperativity upon substrate binding in the mutants.

Domain coupling and its dependence on substrate binding are important not only in the stabilizing mechanism of PGK structure, but also in interdomain communication during the catalytic cycle. For the wild-type PGK, after having very scarce experimental data [237, 238, 245], for the protective effects of substrates against thermal unfolding of PGK, a systematic study have been devised with pig muscle PGK, as well as with the yeast enzyme, to characterize the extent of domain coupling and its regulation by each substrate [108]. It was found that a similarly cooperative mechanism between the domains operates in the binary complexes with either substrate, i.e., their stabilizing effect is not restricted to the N- or C-terminal domain to which they bind.

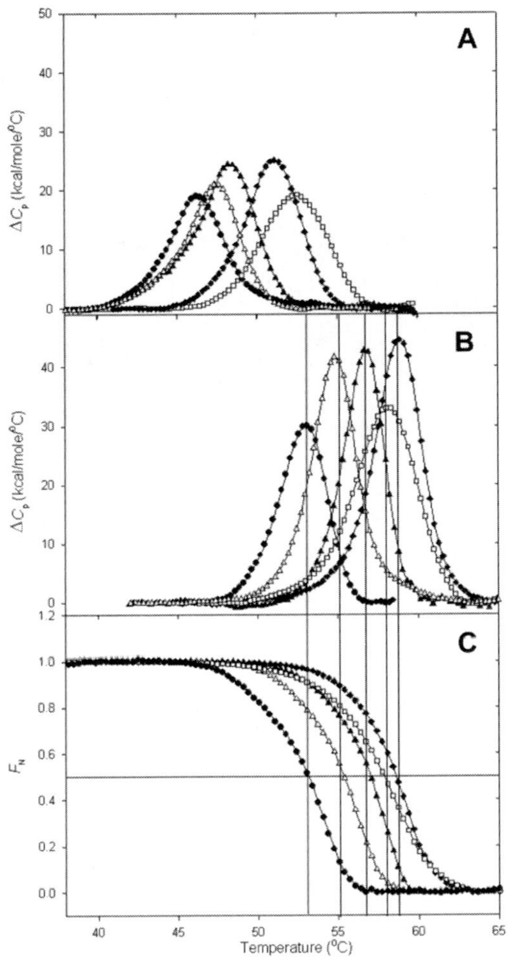

Figure 37. Effect of substrates on the temperature-dependent unfolding of pig muscle PGK: DSC (A and B) and fluorimetric (C) heat denaturation curves were determined with CM-PGK (A) and unmodified (B and C) pig muscle PGK in the absence of substrates (•), in the presence of 10 mM MgATP (△), 10 mM MgADP (▲), 10 mM PG (□) and 10 mM MgADP plus 10 mM PG (◆) (adapted from [108]).

The largest protection was shown by PG (cf. Figure 37 and Table 13), which could be explained by the presence of the atomic contacts that are known from the X-ray structure of the binary complex with this substrate (Figure 39A). Namely, all side-chains interacting with PG belong to separate structural elements of the N-domain, i.e., βA, βB, βD and βE as well as α helices 1 and 5. Thus, these structural elements are strongly fixed together by the bound PG and this may result in an increased stability of the whole N-domain. The calorimetric experiments, however, indicated that not only PG, but any of the other substrates stabilizes the whole molecule, not only the domain to which they bind. This effect is most probably

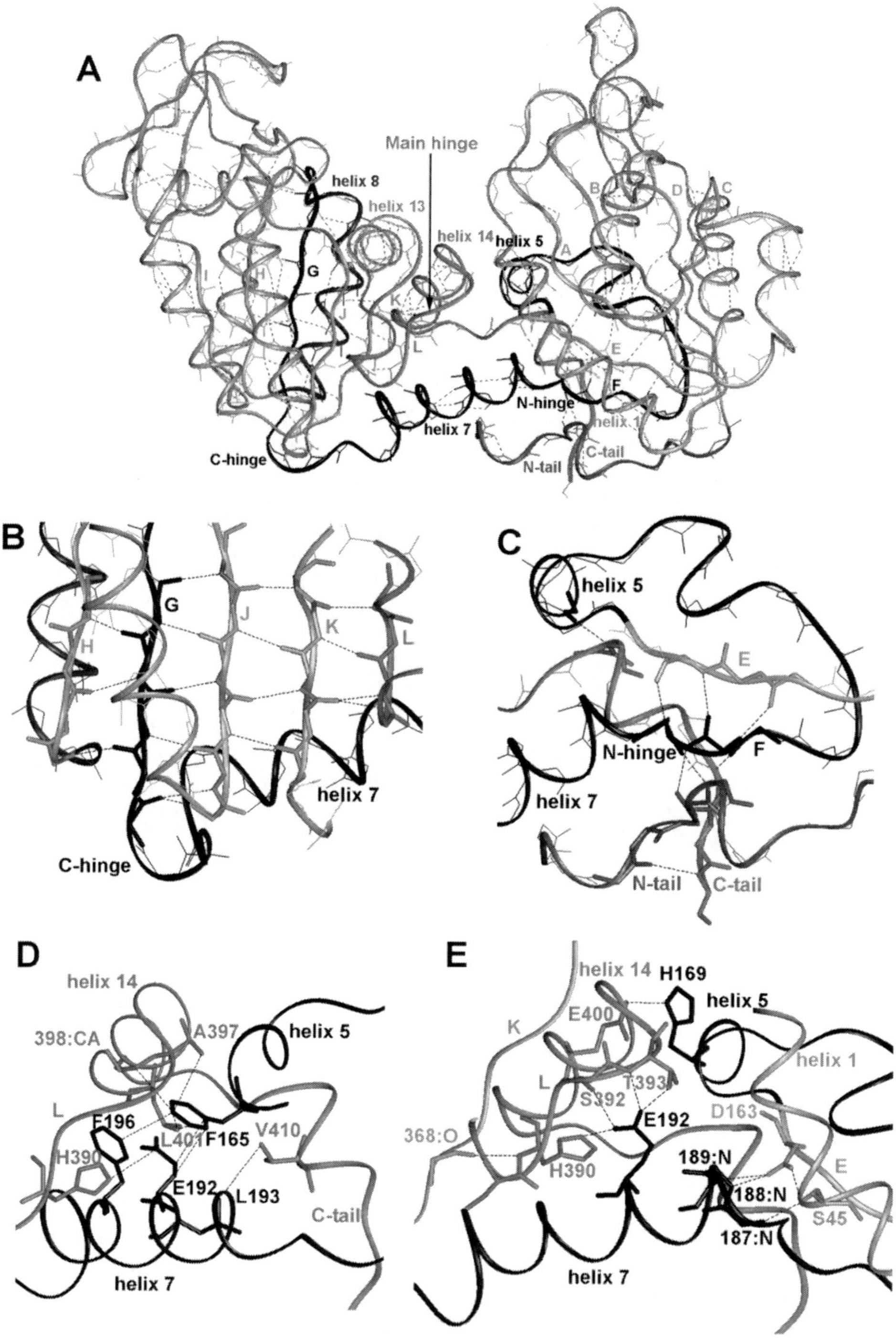

Figure 38. Non-covalent bonds in the PGK molecule responsible for interdomain interactions: The ribbon diagram of the open conformation of the substrate-free pig muscle PGK [72] (A) and its details at the C-hinge (B) at the N-hinge (C) as well as in the interdomain region including the main hinge at βL (D and E) are shown. The backbone and the conserved side-chains are illustrated by stick model. The H-bonds (A, B and C) and the hydrophobic (D) and electrostatic (E) interactions are illustrated by dashed lines (adapted from [108]).

Table 13. Mid-point temperatures (T_m) and calorimetric heats (Q_t) of thermal transitions of pig muscle and yeast PGKs (adapted from 108])

Ligand	Pig muscle PGK					Yeast PGK		
	Unmodified			CM-		Unmodified		
	$T_{m,cal}$	$T_{m,fluor}$	Q_t	$T_{m,cal}$	Q_t	$T_{m,cal}$	$T_{m,fluor}$	Q_t
No	53.0	53.1	114	46.5	85	56.4	56.2	110
MgATP	54.8 55.6[a]	55.3 —	125 —	47.6 —	91 —	59.2 —	59.8 —	127 —
MgADP	56.9 57.6[a]	57.0 —	139 —	48.7 —	98 —	60.7 —	61.4 —	140 —
3-PG	58.2	57.5	147	52.4	110	60.1	60.0	150
3-PG+MgADP	59.1 59.8[b]	58.6 —	175 —	51.1 —	105 —	62.8 —	63.4 —	157 —

T_m and Q_t values are given in °C and kcal/mole, respectively. $T_{m,cal}$ and $T_{m,fluor}$ have been determined by DSC and fluorimetric experiments, respectively [108]. The experimental errors of T_m and of Q_t are varied between ±0.2-0.3 °C and ±5-10 kcal/mole, respectively.

[a]published values, taking into account the decreasing effect of Mg^{2+} on PGK stability [7]

[b]extrapolated to zero concentration of free Mg^{2+} using the method described earlier [7]

promoted by the initially existing interactions (Figures 38A-E) between the two domains. The transmission of PG induced effects from the N- to the C-domain can be visualized by observing the newly formed interactions in the complex with this substrate (coloured as violet in Figure 39A). The new H-bond of Arg 38 with the backbone O-atom of Thr 393 in one side, and the permanent electrostatic interaction between the side-chains of Arg 38 and Asp 23 on the other side forms a new connection between the two domains. This connection is characteristic of all PG bound structures, whereas binding of MgADP or MgATP to the substrate-free enzyme does not induce formation of this bond.

As for the stabilization effects of MgADP and MgATP, the common characteristics of their binding are H-bonding and hydrophobic interactions of the adenine ring to βq (Trp 310/Y308/Bs290/Tm291/His Tb312, Gly 312/Y310/Tm293/Ala Bs292/Tb314), and of the ribose ring to the loop between βJ and α helix 12 (Pro 338/Y336/Bs318/Tm319/Tb340) as well as the ionic interaction of the α-phosphate with Lys 219/Y217 /Bs201/Tm201/Tb223(NZ) as illustrated in cases of Bs/pig muscle PGKs (Figures 39B and 39C) at the N-terminus of the α helix 8. Thus, the secondary structural elements may be similarly fixed together in the C-domain either by MgADP or MgATP. There are, however, further interactions that are different for the two nucleotides and account for their different protecting effects against heat-denaturation (Figure 37 and Table 13). These differences, as shown in detail in Chapter I. 1.4. (Figure 4), are the followings. For MgADP, in contrast to MgATP, the α- and β-phosphates are linked through the bound Mg^{2+} to the carboxylate of Asp Bs352/374 located in α helix 13 (Figure 39B). Furthermore, the β-phosphate of MgADP is linked to Ser Bs353/Thr375 (OG) from the same α helix. These multiple interactions of MgADP are further strengthened by H-bonds and electrostatic forces between the positively charged N-terminus of this α helix and the negatively charged β-phosphate. The above interactions also result in a complete ordering of the α helix 13, as discussed earlier (Table 1). In addition, a β-phosphate O-atom of MgADP is H-bonded to the ND2 atom of Asn Bs316/336 from the β-strand J (Figure 39B). Neither of these interactions is formed with the bound MgATP (Figure 39C). These structural differences are in complete agreement with the greater stabilization effect of MgADP compared to that of MgATP (Figure 37 and Table 13). The new interactions, characteristic of the complexes with MgADP and MgATP are coloured as violet in Figures 39B and 39C, respectively and explains the extension of the nucleotide effects from the C- to N-domain. These structural data are in line with the observed cooperative unfolding of the two domains also in the binary complexes with the nucleotides.

The thermal stability of PGK is further increased in the ternary complex, i.e., in the simultaneous presence of the two substrates (Figures 37B and C and Table 13). In order to avoid the heat effect due to the enzyme reaction, an unproductive ternary complex was investigated in which the substrate, MgATP, was replaced by the product, MgADP. The increased stability of the ternary complex can equally be resulted by an assumed strictly additive conformational effect of each substrate or by assuming further, much larger conformational changes, such as formation of the catalytically competent closed conformation in the ternary complex. The calorimetric behaviour of the native PGK ternary complex itself, thus, did not allow to distinguish between these two possibilities. However, further calorimetric experiment with a modified PGK that lacks the ability of domain closure provided an appropriate experimental test for the possible further stabilization effect of domain closure on PGK conformation. As reported above (Chapter II. 2.5.), pig muscle PGK contains two reactive Cys residues that belong to helix 13 in the hinge region between the two domains and these side-chains have no any direct role in the enzyme activity. Yet, their chemical modification with a relatively bulky reagent (such as carboxymethylation) leads to complete loss of enzyme activity, due to steric prevention of domain closure, as evidenced by SAXS data with this modified CM-PGK [103].

DSC runs of CM-PGK show significantly wider peaks compared to the unmodified enzyme (Figure 37A) indicating a less cooperative heat transition. In addition, there is a remarkable loss of stability of the whole molecule compared to the unmodified enzyme (Figure 37A and Table 13), possibly due to some perturbation of the conformation of helix 13. However, the most interesting part of the results is the protective effects of the substrates against thermal unfolding of CM-PGK. Protection of each substrate correlates well with the full substrate binding ability of this modified enzyme. The order of effects by each substrate in the binary complexes is the same as found for the unmodified active PGK (Figures 37A-B). There is, however, a remarkable difference between the modified and unmodified enzymes: the ternary complex of CM-PGK does not exhibit the highest stability among the enzyme-substrate complexes. This can only be due to lack of the domain closure, since substrate binding to CM-PGK has failed to cause domain closure. Thus, DSC experiments with the unmodified pig muscle PGK are compatible with the highest conformational stability of the native, functionally competent ternary complex. It follows further that the domain-closed conformation is an exclusive characteristic of the ternary complex, i.e., both substrates are required for inducing domain closure.

The different effects of substrates on the calorimetric behaviour of CM-PGK and on the wild-type enzyme also provided a suitable tool for testing the domain closure abilities of the various site-directed mutants compared to the wild-type enzyme. These experimental results are summarized in Table 14. When both substrates together in the ternary complexes protect better the investigated mutant compared to any of the single substrate in the binary complexes, similar to the wild type enzyme, the mutation does not influence domain closure,

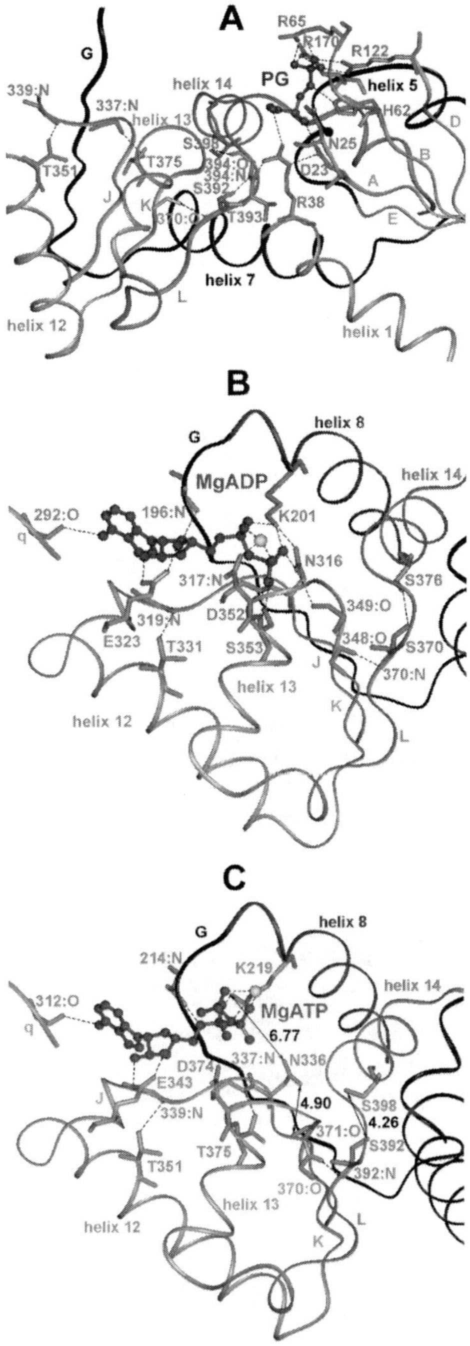

Figure 39. Details of the interdomain region in the binary complexes with substrates: (A) PG binding to pig muscle PGK [15], (B) MgADP binding to Bs PGK [6] and (C) MgATP binding to pig muscle PGK [7]. In each case sequence numbering refers to the corresponding species. The important secondary structural elements are highlighted by ribbons. The side-chains and backbone atoms of the residues interacting with the substrates are indicated. Arrows connect the eqivalent non-interacting atoms in (C) and the distances between these atoms are indicated in angstroms (adapted from [108]).

Table 14. Summary of the DSC experiments with the site-directed mutants

Enzyme	Substrate-free	+3PG	+MgADP	+3PG+MgADP	ΔT_m (terner-biner)
Wild-type HsPGK	54.7	59.1	58.9	60.5	**+1.4**
R38A	49.0	50.1	53.8	53.6	**–0.2**
K215A	53.5	58.2	57.4	59.2	**+1.0**
K219A	53.7	58.4	54.9	57.3	**–1.1**
N336A	58.2	61.4	59.4	60.9	**–0.5**
E343A	53.3	58.5	55.9	57.6	**–0.9**
T375A	53.6	58.1	56.5	58.4	**+0.3**
F165A	49.0	51.9	53.8	54.8	**+1.0**
E192A	49.9	53.2	55.6	56.5	**+0.9**
F196A	51.3	54.4	56.6	57.7	**+1.1**
S392A	51.6	55.1	56.7	57.3	**+0.6**
S392A-T393A	51.9	54.8	56.5	57.2	**+0.7**
T393del	53.4	56.5	58.4	58.4	**0**
S398A	53.1	58.2	58.9	60.9	**+2.0**

The calorimetric T_m values were determined both in the absence and presence of substrates (part of the results are published in [227]) and are given in °C. The experimental error is ± 0.5 °C.

i.e., the investigated residue (such as Ser 392, Ser 398) possibly has no any role in the domain closure. On the other hand, if the protecting effects of the two substrates are additive or smaller than the sum of the two substrates, then the investigated residue (such as Arg 38, Lys 219, Asn 336, Glu 343 and Thr 393) possibly has an important role in the domain closure.

2.9. Analytical Ultracentrifugation and SAXS-Measurements

The cooperation between the two substrates in inducing domain closure, as strongly suggested by the DSC experiments, was finally approved by more direct physical methods.

First of all, as early as 1980, Roustan and coworkers have demonstrated characteristic changes in the shape of PGK molecule upon formation of the various enzyme-substrate complexes by their analytical ultracentrifugation studies [167]. Accordingly, yeast PGK was found in two conformational states: the unligated enzyme as well as the enzyme complexed with either PG or with the nucleotide substrate (binary complexes), characterized by s_{20} of 3.10-3.20 S; and the ternary complexes either with PG and MgATP (functioning) or with PG and MgADP (non-functioning), characterized by s_{20} of 3.80-3.90 S. It is also notable, that sulphate ion (the known activator of PGK, cf. Chapter VI.) alone also caused similar increase in the s_{20} value. Such variation of this value could result either from a conformational change or from an association such as dimerization. The second possibility is unlikely, since no protein concentration dependence was found and moreover, the molecular weight, estimated by ultracentrifugation equilibrium, is identical for the unliganded and liganded species. Thus, the changes observed for the fully liganded species should be related to a change of the shape of the molecule and could be interpreted as a transition of the molecule from an asymmetric to a more compact globular shape. These data, therefore gave evidence for the presence of two conformational states in equilibrium upon ligand control. The change in the shape of the molecule would correspond to a motion of the two domains closing together.

SAXS is another effective method to study the overall structure of macromolecules and, in particular, to validate the crystal structure in solution (see e.g., Koch et al., [247] for a review). Occurrence of large scale conformational changes of PGK in the presence of both substrates (PG and MgATP) have been directly supported by SAXS measurements with solubilized yeast enzyme [102] even at the early time when only initial X-ray structure determination of the substrate-free enzyme [4, 5] was available. Pickover and coworkers have detected an about 1 Å decrease in the radius of gyration (R_g) of PGK molecule upon binding of both substrates. This decrease of R_g value corresponds well to the estimated decrease, upon modelling of domain closure, starting from the X-ray structures existing at that time. If only one of the two substrates was added to the enzyme they did not observe any detectable decrease in the R_g value, while others reported a similar decrease already in the binary complex with CrATP, a non-reactive MgATP-analogue, using neutron-scattering [115]. Although at that time this problem was not further clarified, the decrease of R_g essentially reflected the closure of the deep cleft between the two domains upon binding of substrate(s). SAXS measurements with HK have led to similar results [248]. Thus, the nature of the ligand-induced conformational changes and the magnitude of the hinge-bending motion may be similar for PGK and HK [249].

A strong correlation between domain closure and enzyme activity was also shown by SAXS experiments with an active and an inactive, chemically modified derivatives of pig muscle PGK [103]. In this work also a decrease of 1 Å in the radius of gyration of the native enzyme was detected upon addition of both substrates. Similar decrease has been found in the R_g of the completely active PGK form that has been methylated on its two freely accessible thiols. However, when the same experiments were carried out with an inactive PGK form, carboxamidomethylated on the same two reactive thiol groups, no decrease could be observed in R_g upon addition of the substrates. Since carboxamidomethylation of PGK does not affect substrate binding, the lack of its activity can only be due to the absence of domain closure (cf. also Chapter II. 2.5.). Therefore, these SAXS data clearly demonstrated that domain closure is an exclusive property of the active, functioning enzyme. Prevention of domain closure and the consequent activity loss upon carboxamidomethylation may be due to the presence of a bulky carboxamido-group introduced into the interdomain region. At the same time, introduction of a much smaller methyl-group into this site has no effect either on enzyme activity or the occurrence of domain closure.

As noted above, earlier SAXS experiments with solubilized PGKs, however, either investigated only the ternary substrate complexes of the pig muscle enzyme [103] or provided ambiguous results concerning the binary complexes of the yeast enzyme [102, 114, 115]. This question has been reassessed by a comprehensive SAXS study of various enzyme-substrate complexes of the wild type human PGK [109]. The sequence of the latter enzyme is identical up to 97 % with that of pig muscle PGK, several crystal structures of which were known at that time.

Table 15. Comparison of SAXS experimental parameters of the wild type human PGK with those of the crystallographic models

(adapted from [109])

SAXS experiments			Discrepancy values ($\chi 2$)between the scattering from crystallographic models and experimental data					
PGK complexes with various ligands	*Rg* (experimental), Å		Open crystal structures				Closed crystal structures	
			Pig PGK	BsPGK	Pig PGK		TmPGK	TbPGK
	GNOM	Guinier	substrate free	MgADP binary	MgATP binary	PG binary	[a]ternary1	[a]ternary2
No ligand	23.9±0.5	24.2±0.1	**2.75**	4.33	3.52	3.16	9.14	9.56
PG	23.3±0.7	23.8±0.2	2.68	5.33	3.30	**1.96**	6.05	6.13
bPG	23.7±0.6	24.0±0.2	1.71	2.59	2.00	**1.58**	3.32	3.64
MgATP	23.2±0.6	23.4±0.2	3.86	2.85	**2.41**	3.39	3.18	3.91
MgADP	23.5±0.7	23.7±0.2	1.49	3.24	1.63	**1.14**	5.15	6.19
MgATP·PG	22.5±0.5	23.0±0.2	6.14	6.04	4.66	5.31	2.25	**1.61**
MgADP·bPG	23.6±0.6	23.9±0.2	1.76	1.83	1.54	**1.52**	1.58	1.69
MgADP·PG	23.4±0.5	23.9±0.2	2.30	3.52	2.80	2.05	**2.02**	2.92
Rg(theoretical), Å [b]			24.25	24.34	24.02	23.97	23.26	22.64
Molecular mass, kDa [b]			43.7	43.2	43.6	43.8	43.7	45.3

The minimum values of discrepancy $\chi 2$ (bold) indicate best correlation between SAXS and crystallographic data.

[a]ternary1 and ternary2 denote MgAMP-PNP·PG and MgADP·PG ternary complexes, respectively.

[b]The radius of gyration and the molecular mass of the high resolution models as retrieved from the PDB entries: 1PHP for the Bs PGK·MgADP complex, 1VJC for the pig PGK·MgATP complex, and 1VPE for the Tm PGK and 13PK for Tb PGK ternary complexes; the PDB coordinates of the pig substrate-free structure and the PGK·PG binary complex [15].

The scattering curves of HsPGK in the absence of substrates and of the binary complexes with PG, MgATP, MgADP or bPG, as well as of the ternary complexes with PG plus MgATP or with bPG plus MgADP (the functional complexes) and with PG plus MgADP (a non-functional complex) are determined [109, 250] and presented in Figures 40A-B. The experimentally determined radii of gyration (R_g) are summarized in Table 15. Independent of the direction of the catalyzed reaction, a significant decrease of R_g is detected in the presence of both substrates, similar to previous observations with the pig muscle PGK [103]. This decrease is compatible with the change expected upon domain closure, as seen from the R_g values computed from the crystal structures (Table 15). On the contrary, none of the binary complexes exhibited significant changes of the R_g compared to the substrate-free protein. Thus, neither of the substrates alone is able to induce domain closure.

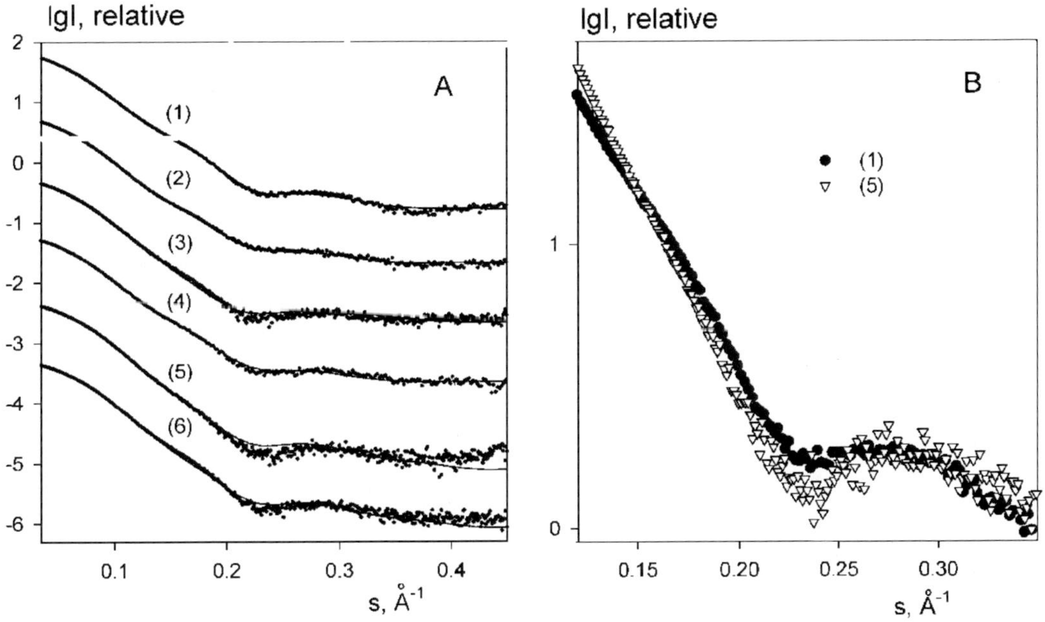

Figure 40. Comparison of the experimental and the calculated SAXS scattering curves (adapted from [109]): The experiments were carried out in the absence of substrates (1), as well as in the presence of 20 mM PG (2), 10 mM MgATP (3), 2.5 mM MgADP (4), 20 mM PG plus 10 mM MgATP (5) and 20 mM PG plus 2.5 mM MgADP (6). The logarithm of the scattering intensity (*I*) is plotted as a function of the momentum transfer *s*. In (A) the dots (with error bars) are the experimental data, while the solid lines are the best fits from the crystallographic models computed by CRYSOL [251]. In (B) the scattering data of the open substrate free form (1) and the closed form of the PG·MgATP ternary complex (5) are compared. The curves (2), (3), and (4) are similar to (1), while the curve (6) is similar to (5).

To validate the experimental scattering curves further, scattering profiles were computed from the available crystallographic data and were systematically screened against the experimental scattering data. For the substrate-free-enzyme and all binary complexes, the best fits (presented in Figures 40A-B) were obtained for the crystal structures, representing open domain conformation. This is expressed numerically by the discrepancy χ^2 values between the experimental scattering data and the corresponding crystallographic models (Table 15).

Table 16. Effects of mutations on the experimental SAXS scattering curves

(The table summarizes the results published by Szabó et al. in [227] and [228].)

SAXS experiments			Discrepancy values ($\chi 2$) between the scattering from crystallographic models and experimental data					
PGK forms (complexed with 3-PG and MgATP)	R_g (experimental), Å		Open crystal structures				Closed crystal structures	
			Pig PGK	BsPGK	Pig PGK		TmPGK	TbPGK
	GNOM	Guinier	substrate free	MgADP binary	MgATP binary	3-PG binary	[a] ternary1	[a] ternary2
Wild type	22.5±0.5	23.0±0.2	6.14	6.04	4.66	5.31	2.25	**1.61**
E192A	23.7±0.3	23.9±0.2	2.89	3.01	**1.99**	2.27	2.79	3.43
S392A	23.8±0.2	24.1±0.3	1.45	1.78	1.68	**1.43**	2.44	2.40
T393A	23.7±0.3	24.0±0.2	2.77	3.72	2.35	**2.11**	2.93	2.94
S392A-T393A	23.9±0.2	24.1±0.2	2.71	3.99	2.51	**1.97**	3.55	3.76
T393del	23.8±0.2	24.1±0.3	**1.27**	1.45	1.29	1.28	1.83	1.95
S398A	22.7±0.3	23.0±0.2	3.99	4.02	3.25	3.45	1.96	**1.40**
R38A	24.0±0.3	24.3±0.3	1.41	1.35	**1.26**	1.33	1.62	1.91
K215A	22.6±0.2	22.9±0.3	3.48	3.74	2.79	2.93	2.26	**2.21**
K219A	23.6±0.4	23.9±0.2	1.49	1.83	1.58	**1.44**	2.32	2.37
N336A	23.7±0.3	24.0±0.2	2.71	3.70	2.89	**2.35**	4.26	3.71
E343A	23.7±0.3	23.9±0.2	1.53	1.75	1.41	**1.37**	1.55	1.66
T375A	22.7±0.3	22.9±0.3	1.92	1.95	1.64	1.69	1.40	**1.37**
Rg (theoretical), Å [b]			24.25	24.34	24.02	23.97	23.26	22.64
Molecular mass, kDa [b]			43.7	43.2	43.6	43.8	43.7	45.3

The minimum values of discrepancy χ^2 (bold) indicate best correlation between SAXS and crystallographic data.

[a]ternary1 and ternary2 denote MgAMP-PNP·3-PG and MgADP·3-PG ternary complexes, respectively.

[b]The radius of gyration and the molecular mass of the high resolution models as retrieved from the PDB entries: 1PHP for the BsPGK·MgADP complex, 1VJC for the pig PGK·MgATP complex, and 1VPE for the TmPGK and 13PK for TbPGK ternary complexes; the PDB coordinates of the pig substrate-free structure and the PGK·3-PG binary complex [15].

There is only one apparent exception for the MgADP binary complex, where the BsPGK model [6] does not yield the best fit to the data [109]. The reason is most probably the extra loop between Glu 128/Bs124 and Ile 146/Asp Bs127, which is present in pig or in human, but missing in BsPGK. The scattering data of the MgADP binary complex, however, correlate rather well with all of the three other open crystal structure models. These results clearly demonstrate that all the binary complexes exhibit an open domain conformation, both in crystal and in solution. Since in solution no conformational restriction by the crystal lattice forces exists, the open structures observed in the crystals of PG binary [15] or of MgATP binary [7] complexes of pig muscle PGK and of the MgADP binary complex of BsPGK [6] are not artifacts, supporting their usage in the above structural comparisons.

Concerning the functional MgATP·PG ternary complex of human PGK, the SAXS-based analysis provided unequivocal result in favor of domain closure showing the best fit to the completely closed TbPGK [84] crystal structure (Table 15) [109]. This is in good agreement with the earlier SAXS results with yeast [102] and pig muscle [103] PGKs. In the case of the other functional MgADP·bPG and non-functional MgADP·PG ternary complexes, however, the data are equivocal due to the marginal differences between the respective discrepancy (χ^2) values in Table 15, which are almost within the accuracy limit of the experimental data [250]. Taken the data formally, the best fit was obtained to the largely (but not completely) closed TmPGK [73] structure and to the partially closed PG binary complex of pig muscle PGK in the cases of MgADP·PG and MgADP·bPG complexes, respectively (Table 15). This means that in these ternary complexes domain closure does occur, but its extent is difficult to determine. The characteristic features of the scattering profiles of the functioning (curve 5) and the non-functioning (curve 6) ternary complexes in Figure 40A are the less pronounced shoulder at s = 0.18 $Å^{-1}$ and the more pronounced maximum at s = 0.25-0.30 $Å^{-1}$. These alterations in the scattering profile upon domain closure are better visible in Figure 40B, where the scattering patterns of the MgATP·PG ternary complex (curve 5) and the open form of unligated PGK (curve 1) are directly compared.

In summary, SAXS and crystallographic data correlate well and support the view that simultaneous binding of both substrates to PGK are required for complete domain closure. This analysis, using a systematic screening of the entire computed scattering patterns against the experimental data, is much more conclusive than previous SAXS studies, based solely on the values of R_g (which might be disturbed *e.g.*, by aggregation or concentration effects).

SAXS measurements were also applied to specify better the contribution of certain side-chains to the catalysis, studied previously by mutagenesis, and the results are summarized in Table 16. As reported above, Arg 38 is a catalytic residue, which participates in the binding of the substrate PG or bPG and interacts with the transferring phospho-group. At the same time, Arg 38 interacts (through the contact with Thr 393:O) with βL (cf. Figure 39A), thus its participation in domain movements is expected and clearly supported by the SAXS data, suggesting the absence of domain closure for the R38A mutant. The way of transmission of the PG-caused effect towards the main hinge βL was illustrated above in Figure 36A and will be summarized in Chapters IX and X.

Similar SAXS experiments revealed that the other catalytic residue, Lys 215 at the nucleotide-site does not fulfill a similar role (Table 16). The large movement of this residue during domain closure is possibly dictated by the other nucleotide binding residues, Lys 219, Asn 336 and Glu 343, which may be directly involved in domain closure, as suggested by the

SAXS data (Table 16). Namely, the mutants K219A, N336A and E343A do not exhibit domain closure, even in their ternary complexes with both substrates. On the other hand, T375A exhibits domain closure, indicating that the side-chain of Thr 375, although important for the proper interactions with the nucleotide phosphates, has no direct role in the molecular events leading to domain movements.

All these data clearly demonstrate the importance of Lys 219, Asn 336, Glu 343 residues in domain closure. These residues interact with the α- and β-phosphate of the nucleotide, respectively and Glu 343 interacts with the ribose OH-groups. Chapter X. describes in detail the possible route of the conformational transmission from the nucleotide site to the main hinge-region at βL.

In summary, the importance of Arg 38, but not of Lys 215 (i.e., only one of the two known catalytic residues) in the domain movement has been demonstrated. Thus, the large movement of Lys 215 during catalysis is not the reason, but a consequence of domain closure. Naturally, assistance of this movement by the flexible phosphates of MgATP or bPG (cf. the description of this hypothesis in Chapters I. 1.4. and IX.) cannot be excluded, but this is not the primary driving force of the motion. The SAXS data have shown that among the nucleotide binding residues Lys 219, Asn 336 and Glu 343 contribute to the transmission of the nucleotide effect towards the main hinge, while Thr 375 seems to be irrelevant in this respect.

As discussed (Chapter II. 2.7.) a deletion mutant of human PGK that lacks the residue Thr 393 has led to a dramatic activity loss. This underlines the importance not only the side-chain of Thr 393 itself, but the proper shape and length of the βL in the catalytic function, too. The deletion caused at least four orders of magnitude loss in the catalytic efficiency, but the interaction with the substrates seemingly has not changed. Thus, βL has no direct role in substrate binding, but its intact main-chain is essential for the function.

The results of SAXS measurements carried out with the above mutants correlated well with the kinetic and binding data. Only the S398A mutant exhibited complete domain closure upon binding of both substrates, similar to the wild-type human PGK. For all other mutants (except T393del) an intermediate situation, partial domain closure can be deduced from the analysis of the SAXS scattering curves (summarized in Table 16) and this coincides with the partial catalytic activities of these mutants (cf. Tables 10-12), indicating that neither of the corresponding side-chains is responsible alone for functioning of the molecular hinge at βL. The results with the mutant T393del clearly suggest that domain closure did not occur, in agreement with its highly reduced catalytic activity. It is notable that these results are nicely compatible with the suggestions derived from the above discussed DSC experiments (Chapter II. 2.8.).

In: Phosphoglycerate Kinase: A Hinge-Bending Enzyme
Editor: Mária Vas
ISBN: 978-1-62808-836-6

Chapter III

Folding Mechanism of a Two-Domain Protein as Examplified by PGK

Jeannine Yon-Kahn
Institute de Biochimie et Biophysique Moleculaire et Cellulaire,
Université de Paris-Sud, Orsay, France

3.1. The Folding Problem

Milestone Studies

Protein folding is a fundamental problem since it completes the information transfer from DNA to the active protein product. It represents the final part of the genetic message according to the following:

DNA → RNA → polypeptide chain → active folded protein

The mechanism by which a nascent polypeptide chain acquires its three-dimensional and functional structure has intrigued scientists for several decades. As early as 1930, the reversibility of the unfolding process was demonstrated for several proteins, provided the disulfide bridges were not disrupted, implying the concept that protein folding is a spontaneous process. The denaturation process and the recovery of the native state were analysed at that time by Wu [252] in a review. In the 1950's significant contributions underlined the importance of non covalent interactions in the stability of proteins. In 1959, Kauzmann [253] proposed, on thermodynamic considerations, that the hydrophobic effect was the driving force of protein folding.

Starting in the 1960's, the determination of three-dimensional structure of proteins by X-ray diffraction, joined in the 1980's by NMR spectroscopy, provided a new basis for structure analysis and for studying the folding process. The remarkable success of Anfinsen [254] in refolding denatured and reduced ribonuclease into a 100% active enzyme marked the beginning of intensive research into the protein folding. The author concluded that "*All the information necessary to achieve the native conformation of a protein in a given environment is contained in its amino acid sequence.*" [255]. As an inference to this statement, referred to as the Anfinsen's postulate, the thermodynamic control of protein folding was accepted, meaning that the native protein is at the minimum of the Gibbs free energy. This inference was questioned by Levinthal [256]. Indeed, an unfolded polypeptide chain can adopt an astronomically large number conformations; thus a random search of the native structure corresponding to the minimum of free energy through all the possible conformations, would require a considerable time, incompatible with the folding time which is on the order of seconds or minutes *in vitro* as well as *in vivo*. For Levinthal, protein folding would be under kinetic rather than thermodynamic control. During the folding process, the protein would be trapped in an energy minimum but not the global one, a high energy barrier preventing it from reaching the latter. Such a situation, referred to as the Levinthal paradox, has dominated discussions during the last three decades, leading to different models for the folding pathway. The attempts to solve the Levinthal paradox have stimulated kinetic studies. It was clear that a random search of the native structure was not realistic and that evolution has chosen an efficient solution to solve this combinatorial problem.

Folding Models

In order to understand how the polypeptide chain could overcome the Levinthal paradox, different folding models have been proposed, arising from experimental observations, theoretical considerations, or folding simulations.

The classical nucleation-propagation model, which applies to helix-coil transitions, involves a nucleation step followed by rapid propagation, the limiting step being the nucleation process. It has been proposed to explain the folding of ribonuclease A [257]. But this initial model was forsaken after new kinetic data of the refolding of ribonuclease were obtained [258]. More recently a nucleation-condensation model, different from the classical one, was proposed by Fersht [259]. This model is based on a mechanism involving a weak local nucleus which is stabilized by long range interactions.

A stepwise sequential and hierarchical folding process, in which several stretches of structure are formed and assemble at different levels following a unique route, was supported by several authors and was generally accepted for many years [260, 261]. According to this model the first event, nucleation, is followed by the formation of secondary structures which associate to generate supersecondary structures, then domains, and eventually the active monomer; the association of domains induces the final conformational refinements which generate the functional properties. Such a hierarchy of protein folding corresponds to the hierarchy of protein structure. Similarly, the framework model assumes that secondary structure is formed in an early step of folding, before tertiary structure, emphasizing the role of short range interactions in directing the folding process [262]. According to this model, the folding pathway could be represented as shown in Figure 41.

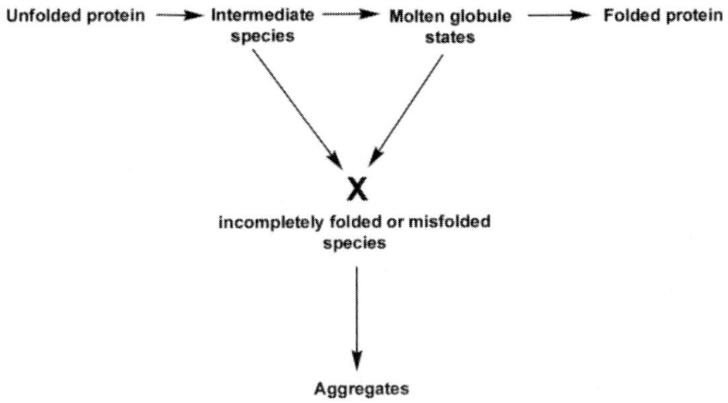

Figure 41. The classical sequential model of protein folding including a side-reaction giving rise to aggregates.

A modular model of folding was suggested on consideration of the three-dimensional structures of proteins. This model assumes that not only domains, but also sub-domains can be considered as folding units which fold independently into a native structure, forming structural modules that assemble to yield the native protein [65, 263].

The diffusion-collision model of folding was developed in 1976 by Karplus and Weaver [264] on theoretical considerations and reconsidered in 1994 in the light of more recent experimental data [265]. In such a model, nucleation occurs simultaneously in different parts of the polypeptide chain, generating microstructures which diffuse, associate and coalesce to form sub-structures with a native conformation. These microstructures have a lifetime controlled by segment diffusion, so the folding of a polypeptide chain containing 100 to 200 amino acids can occur in a very short time, less than a second. According to this model, folding occurs through several diffusion-collision steps.

The hydrophobic collapse model implies that the first event of protein folding consists of a collapse via long range hydrophobic interactions which occurs before the formation of secondary structure [266]. The idea originates from the work of Kauzmann considering the hydrophobic effect as the driving force in protein folding and stabilisation. Later, Chan and Dill [267] proposed that the formation of stretches of secondary structures could occur simultaneously with the hydrophobic collapse.

The jigsaw puzzle model was introduced in 1985 by Harrison and Durbin [268]. This model admits the existence of multiple folding routes that can be taken in reaching a unique solution. According to this hypothesis, the identification of folding intermediates represents a kinetic description rather than a structural one, each intermediate consisting of heterogeneous species in rapid equilibrium. It was the subject of controversy until recently, as described further on in this chapter, but presents some similarities with the diffusion-collision model proposed by Karplus and Weaver [264].

Detection and Characterisation of Intermediates on the Pathway of Protein Folding

The unfolding-refolding transition under equilibrium conditions has often been described as a two-state process indicating a very cooperative transition. The two-state approximation is often verified for small proteins; the intermediates are generally unstable and poorly populated at equilibrium.

As mentioned above, the attempts to solve the Levinthal paradox stimulated many kinetic studies. For many proteins, unfolding kinetics are monophasic while refolding kinetics are multiphasic. These investigations have demonstrated the existence of intermediates in the folding pathway, even when the two-state approximation describes the overall process. The structural characterisation of the intermediates is a prerequisite to solving the folding problem. However, two major impediments to characterising these species are the high cooperativity and the rapidity of the process. Such intermediate species generally have a very short lifetime, making the analysis of their structural properties difficult. In spite of these inherent difficulties, much effort has been devoted to characterising these transient species. Technological advances allowed some of the intermediates to be structurally characterised. Furthermore, for some proteins, transient folding intermediates were accumulated by the use of low pH, protein engineering techniques, or by working on protein fragments unable to complete the folding process, making possible their characterisation. More recently, submillisecond kinetic methods have improved the resolution time of kinetic studies allowing detection of the very early events of folding.

The occurrence of secondary structures in the early steps of the folding process has been detected for several proteins (for a reviews see Yon [269] and references therein). The formation of folded but flexible intermediates was proposed by Ptitsyn [262] from theoretical considerations. Ohgushi and Wada [270] termed such a species the molten globule. Ptitsyn et al. [271] suggested the molten globule to be a general intermediate in the folding pathway of proteins. Goldberg and coworkers introduced the term "specific molten globule" with a precise definition [272]. The commonly accepted definition of the molten globule is a rather compact intermediate with a high content of native secondary structure but a fluctuating tertiary structure. It contains accessible hydrophobic surfaces which bind a hydrophobic dye, usually anilinonaphtalene sulfonate (ANS, cf. abbreviations). In this intermediate, the aromatic residues can rotate in a symmetrical environment, as shown by the absence of the near UV circular dichroism spectrum. The occurrence of a molten globule state has been reported for several proteins. The rapid formation of transient intermediates with a significant secondary structure content is supported by a great number of observations. Thus, it has been inferred that the folding process originates from short range interactions. However, the molten globule is a transient species which appears in the dead-time of a stopped-flow device and it is possible that its formation might be preceded by an earlier event. Furthermore, the secondary structures observed in the early folding steps are not always native-like structures. For several proteins non-native secondary structures are formed in the early folding steps, requiring further reorganisation of their conformation [272]. The use of rapid kinetic techniques allows detection of very fast events which occur on a time scale of less than a microsecond. Thus, the upper limit for the rate of protein folding has been evaluated to be $1\mu s^{-1}$ [273] in agreement with theoretical estimates [274].

The Role of Domains and Subdomains in the Folding Process

In large proteins, the polypeptide chain is folded in several compact structures, referred to as domains, within the molecule. A domain has more contacts with itself than with the rest of the molecule. These domains have been considered by Wetlaufer [263] as folding units, forming structural, folded modules that then assemble to generate the native protein. Several reviews concerning the role of domains in the folding process have been published, among them those of Ghélis & Yon [275], and Jaenicke [276]. Many experimental data have shown that domains obtained either by limited proteolysis or by genetic engineering behave as folding units, giving rise cooperatively to a native-like structure. However, the ability of independently folded domains to associate to yield an active protein has been observed in only a few cases, such as thioredoxin [277], elastase [278], and methionyl-t-RNA synthetase [279]. The correct folding of isolated domains is not generally sufficient to yield a functional complementation of a protein. This will be shown below with phosphoglycerate kinase.

Several investigations of the folding of fragments smaller than domains have been carried out (for reviews see ref. [275, 276, 280]). C-terminal fragments from thermolysin have been reported to fold autonomously and cooperatively when their size is equal three helices or larger [281]. Oas and Kim [282] have deduced from their experiments with BPTI (cf. abbreviations) fragments that subdomains fold autonomously and associate to generate the protein. The results obtained on the folding and complementation of barnase are also consistent with the modular model of folding [283]. However, from a great number of results involving several different proteins, it appears that the modular model does not account for a general mechanism of protein folding.

The "New View": the Energy Landscape and the Folding Funnel

More recently, a unified model of protein folding based on the effective energy surface of a polypeptide chain has arisen from theoretical studies principally introduced by Wolynes and colleagues [284]. This is referred to as the "new view" and describes the folding process in terms of an energy landscape and a folding funnel. It has evolved during the last decades from both experimental and theoretical results through the use of simplified models. This model describes the thermodynamic and kinetic behaviour of the transformation of an ensemble of unfolded molecules to a predominantly native state as illustrated schematically in Figure 42. Thus, as written by Wolynes et al.., "*to fold, a protein navigates with remarkable ease through a complicated landscape*". According to this view, protein folding consists of a progressive organisation of ensembles of partially folded structures that arise through multiple routes. A wide variety of folding behaviour emerges from the energy landscape depending on the energetic parameters and conditions. Towards the bottom of the folding funnel, the number of protein conformations decreases as does the chain entropy. The steeper the slope, the faster is the folding.

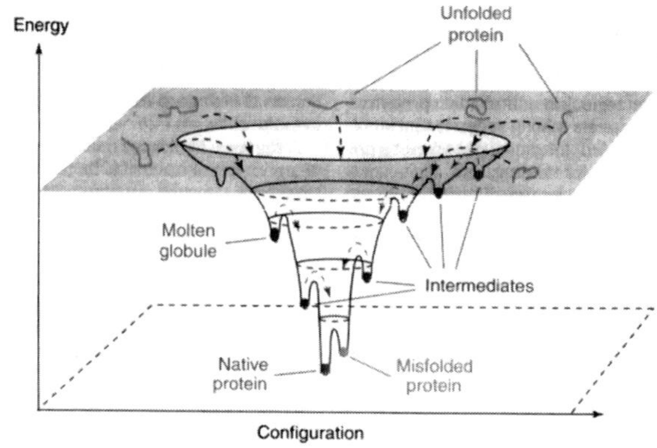

Figure 42. The folding funnel: The width of the funnel represents the entropy and the depth, the energy. Q is the fraction of native contacts (based on [284]).

The energy landscape of protein folding resembles a partially rough funnel whose local roughness gives rise to transient trapping of the protein in local free energy minimum. When local barriers are high enough, protein molecules can be trapped and, depending on the nature of the states, eventually aggregate.

As written by Dobson and Karplus [285], "*An important aspect of the new view is that it provides a simple way of understanding why the Levinthal paradox is not a real problem.*". The energy landscape and the folding funnel metaphor provides a conceptual framework for understanding the different scenarios of protein folding and also protein misfolding and aggregation. This view is in fact a rationalisation of the jigsaw puzzle model [268]. Thus, the new view has progressively replaced the classical model of a hierarchical and sequential pathway of folding and is now quite generally accepted. Experimental results are generally consistent with this view. There is an increasing amount of evidence showing that the initially extended polypeptide chain folds through a heterogeneous population of partially folded intermediate species in fluctuating equilibrium. Recent interpretations of protein engineering experiments also support the idea that the transition state for a folding reaction consists of a distribution of structures rather than of a single structure [286, 287].

3.2. Folding Mechanism of Phosphoglycerate Kinase

Phosphoglycerate kinase (PGK) is the first enzyme of the glycolytic pathway, which produces MgATP through high energy phosphate transfer from 1,3 diphosphoglycerate (bPG) and MgADP, with the formation of 3-phosphoglycerate (3PG) according to Eq. 2 in Chapter IV. 4.1.

The structure of the enzyme is known for various mesophilic [4, 15, 58, 60], thermophilic, archae [71, 84, 288], and psychrophilic [289] species (see Chapter I). The structures of these monomeric proteins are very similar exhibiting the same global fold. The most characteristic feature of the molecule is the presence of two well resolved domains, corresponding to the C- and N-terminal halves of the molecule, separated by a large cleft (Chapter I. 1.2.). Nucleotide substrates bind to the C-domain whereas phosphoglycerate substrates bind to the N-domain through a network of hydrogen bonds to a cluster of basic aminoacids. The binding of specific substrates induces a hinge bending motion, closing the cleft and allowing the chemical reaction to take place. The structure of this enzyme made PGK a good model to study the role of domains in the folding mechanism of large proteins.

Equilibrium Studies

The unfolding-refolding transition of small proteins is a highly cooperative process and has been described as a two-state process. The validity of the two-state approximation means that the intermediates are poorly populated under equilibrium conditions. Deviations from two-state behaviour is often observed for larger proteins.

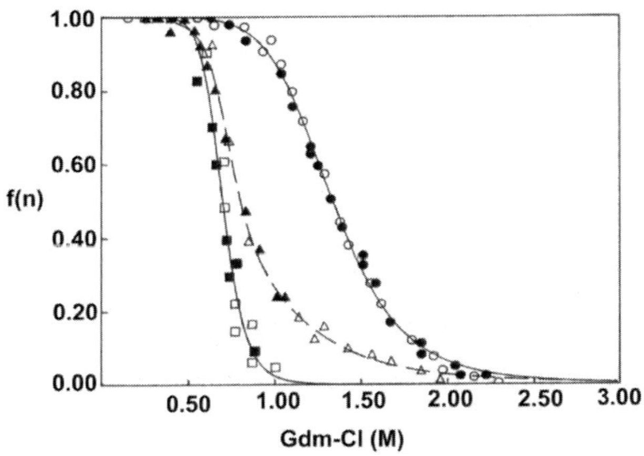

Figure 43. Unfolding (open symbols)-refolding (closed symbols) equilibrium transition of horse muscle phosphoglycerate kinase (adapted with permission from [290]; Copyright 1984 American Chemical Society): Unfolding and refolding transitions have been observed by different methods: o,•: fluorescence intensity at 336nm; ▲,Δ: ellipticity at 220nm; ■,□: enzyme activity.

Equilibrium studies of the unfolding-refolding process induced by GuHCl have been monitored for PGKs from horse muscle and yeast [290, 291]. Different signals, including enzyme activity, circular dichroism, difference spectrophotometry and fluorescence, have been used. As shown in Figure 43 for the horse muscle enzyme, the transitions assessed by these different signals do not coincide, indicating the presence of significantly populated intermediates under equilibrium conditions. Since the tryptophan residues are located in the C-domain, these data indicate than this domain is more stable than the N-domain. Furthermore, the circular dichroism signal is asymmetrical and results from the contribution of two transitions. The yeast enzyme has been found to be more sensitive to denaturant than horse muscle PGK. Nevertheless the transition curves also indicate the presence of intermediates, but they are less populated than in the horse muscle enzyme. Although the yeast enzyme is less stable than the horse muscle PGK, for both enzymes the data are consistent with an independent folding of the domains, the C-domain being more stable than the N-domain.

The existence of intermediates was confirmed by experiments which have revealed the occurrence of a side reaction arising from these intermediates. When horse muscle PGK is exposed to a critical concentration of denaturant under equilibrium conditions, the enzymatic reactivation becomes partially irreversible (not illustrated). The extent of the irreversibility, as indicated by the amplitude of a "trough", was found to be dependent on protein concentration and temperature. This phenomenon was only observed following the return of the enzyme activity, whereas the transitions as assessed by the different conformational probes (cf. Figure 43) were found to be perfectly reversible under the same conditions. However, it was completely abolished at low temperature, suggesting that irreversibility originates from hydrophobic intermolecular interactions between partially folded molecules [292]. The same phenomenon was observed for the yeast enzyme although to a smaller extent, in agreement with the smaller shift in the transition curves [293, 294]. The structural properties of these intermediates have been analysed for the horse muscle enzyme. All the ß structures were recovered but about 30% of the helix content was missing. Furthermore, the C-terminal fragment (residues 193 to 416), isolated by limited proteolysis under critical concentration of denaturant was found to have the same characteristics as this intermediate. It was concluded that in this intermediate, the C-terminal domain is partially folded, the N-terminal domain being completely unfolded [295]. These data supported the idea that the C-domain behaves as a folding unit.

Kinetic Studies: Detection and Characterisation of Intermediates

The kinetics of the unfolding-refolding reactions were studied using three different signals, fluorescence emission intensity, circular dichroism at 220nm and enzyme activity. They confirmed the occurrence of intermediates in the folding pathway of horse muscle PGK [296]. At least two refolding phases were detected by fluorescence as well as by circular dichroism, whereas only the slow phase was detected by the recovery of the enzyme activity. During the fast phase an intermediate was formed with fluorescence intensity higher than that of the native protein, but devoid of enzyme activity. Only the slow phase was detected for the unfolding process, which was not attributable to proline isomerisation. A proline-isomerisation independent slow refolding was also confirmed by studies with pig muscle and yeast PGK-s [297]. In this work the kinetics of refolding were followed by four different signals (Trp fluorescence, ultraviolet absorbance, bound ANS fluorescence and enzyme activity) and found to be very similar (with half time between 80-120 s) for these two different PGK species, as well as similar to the findings with horse muscle PGK [295]. Thus, the large extent of structural similarity of PGK-s of different origin [24] may be reflected by their similar refolding time courses. Since the Trp residues are located in the C-domain (two and four residues per molecule of yeast and mammalian PGK, respectively), it follows that renaturation of the whole protein molecule (restoration of activity and decrease of affinity towards ANS) and refolding of the C-terminal part, including Trp-s (restoration of protein fluorescence and UV absorbance) are closely related processes. In accordance, removal of a large part of the N-terminal domain from the whole molecule by trypsinolysis (about 13 kDa or 18 kDa of the yeast and pig muscle enzyme, respectively) has no effect on the slow refolding rate of the remaining C-terminal part [297]. Although the influence of the small remaining part of the N-domain could not be completely excluded, slow conformational process(es), localised in the C-terminal part/domain was assumed to limit the folding rate of the whole molecule [297]. As for the conformational origin of the slow isomerisation process, it was shown for horse muscle [296] and yeast (Semisotnov, G.V. and Kuwajima, K., unpublished) PGK-s that about half of the far UV CD signal of the intact PGK molecule restored at a similar slow rate as the other investigated structural parameters. This finding raises the possibility that the rate-limiting step during refolding of PGK is some slow stabilisation step at the secondary structural level. As for horse muscle PGK, α helices located at the exterior of the molecule unfold prior to the β-sheets, which form the core of the molecule [296]. From further, more detailed kinetic analysis of reactivation of pig muscle PGK from its isolated C- and N-terminal proteolytic fragments, a sequential domain refolding mechanism was deduced [298] (cf. below).

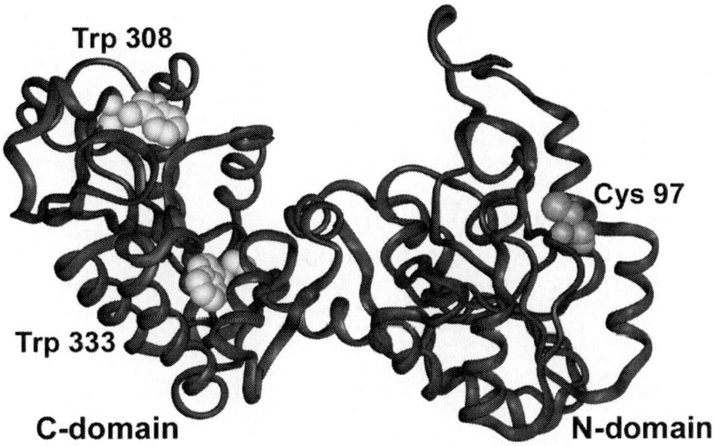

Figure 44. Representation of the three-dimensional structure of yeast PGK: Space-filling models illustrate in grey the unique cysteine residue (Cys 97) in the N-domain and in white the two tryptophan residues (Trp 308 and Trp 333).

Generally, kinetic studies provide only phenomenological information. The use of intrinsic conformational probes located at known positions in the polypeptide chain is useful in obtaining local information. Yeast PGK contains a unique cysteine residue (Cys 97) located in the N-domain, and two tryptophans, Trp 308 and Trp 333 in the C-domain as shown in Figure 44. In order to detect local events in the folding pathway of PGK, several mutants, each containing a unique cysteine in selected buried positions, were constructed from a single mutant devoid of cysteine [299, 300]. All the mutants retained enzyme activity and their overall conformation. In all but one case, three labelling phases were observed during the refolding process, indicating the presence of three species: the unfolded, intermediate and folded PGK. The intermediate species were characterised (Figure 45). In this intermediate, each domain appeared to have a high content of secondary structure, but a fluctuating tertiary structure; this intermediate, which had the characteristics of a molten globule, remained in fluctuating equilibrium with a widely unfolded form [299]. The folding data deduced from the accessibility of the thiol group indicated that the stability of the molecule in the different regions increased in a particular order, but no difference attributable to the location of the residue either in the N- or in the C-domain was detected [299, 300].

These results are apparently in contrast with those obtained for the horse and pig muscle enzymes in which the C-domain was found more stable. The region around Ala 183, buried through its interaction with the C-terminal segment connecting the two domains, seemed more sensitive to denaturant, suggesting that interdomain interactions are destabilised in the first steps of the unfolding process.

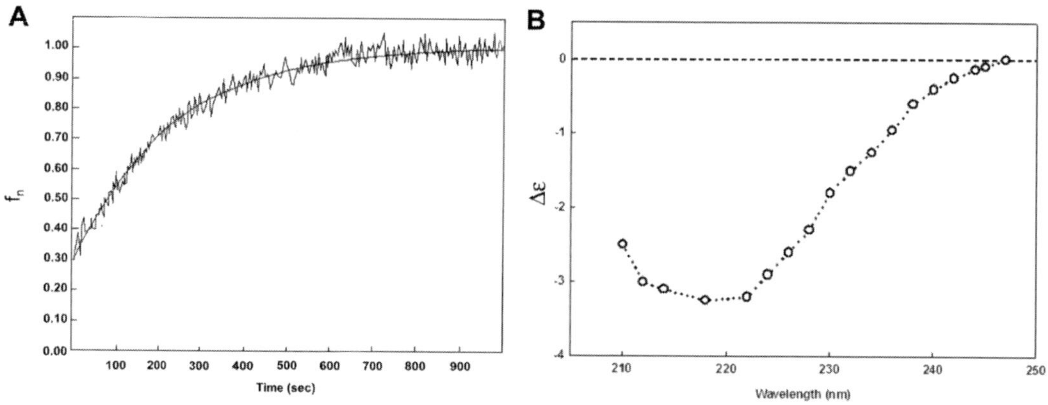

Figure 45. Kinetics of the refolding of yeast PGK as assessed by circular dichroism: (A) Restoration of the native protein fraction during refolding; (B) The CD spectrum of the intermediate occurring during the fast refolding phase. The amplitude was calculated for a ½ intermediate to unfolded species ratio, as indicated by the stochiometric labelling of the SH groups (based on [299]).

Detection of Residual Structures in the Unfolded Form of Yeast Phosphoglycerate Kinase

The unfolding-refolding transition of yeast PGK, as followed by steady-state fluorescence, clearly showed the occurrence of a hyperfluorescent form [301]. In order to determine the contribution of each of the two tryptophans to the fluorescent properties of the enzyme and to characterise the hyperfluorescent forms, two single tryptophan mutants were constructed in which either Trp 308 or Trp 333 was replaced by a tyrosine and a phenylalanine [301]. Neither the catalytic nor the physicochemical properties of the enzyme are significantly altered by these mutations. The unfolding-refolding transition was studied by circular dichroism and tryptophan fluorescence emission. Both tryptophans contribute to the hyperfluorescence in the first transition. For guanidine hydrochloride concentrations higher than 0.9 M, it appears that the second transition results from further unfolding. It is accompanied by a decrease in fluorescence intensity and a 5nm red shift of the maximum emission wavelength. However, when unfolding was induced by urea, the end of the transition corresponds to the hyperfluorescent state. A further addition of guanidine hydrochloride sufficed however to induce complete unfolding. The results are illustrated in Figure 46. Such a situation was attributed to the presence of residual microstructures around Trp 308 and Trp 333 in the hyperfluorescent state which might thereby constitute nuclei in the folding process. Using several single tryptophan mutants of yeast PGK, Mas and her group [302, 303] have presented different conclusions, attributing the differences for mutants Trp 308 and Trp 333 to differences in the slope of the post transition base-line.

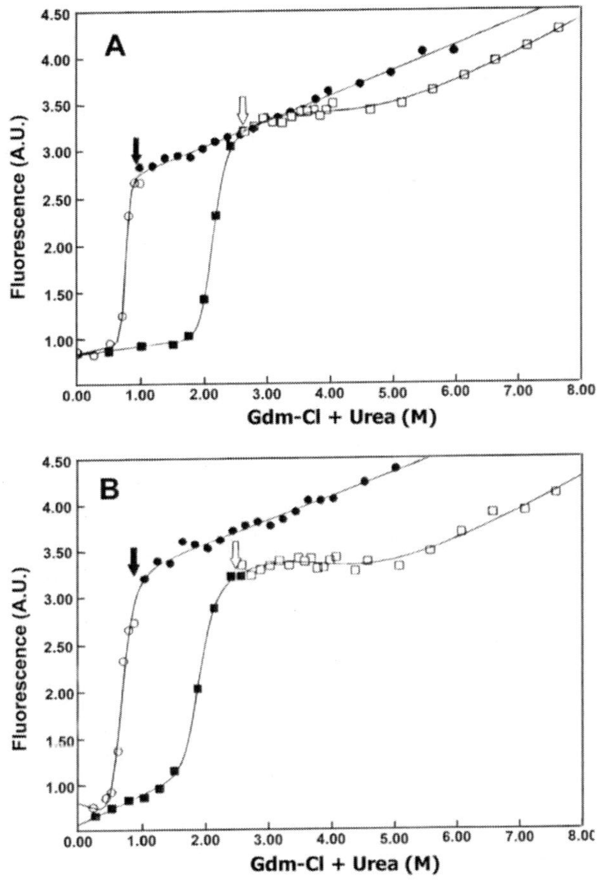

Figure 46. Unfolding transition of (A) wild type PGK and (B) W333F mutant as assessed by fluorescence intensity at 350nm (adapted with permission from [301]; Copyright 1995 American Chemical Society): The first transition was obtained by dilution of the GuHCl induced hyperfluorescent state (o), or of the urea unfolded state (■). The white arrow indicates the urea concentration of the solution to which concentrations of GuHCl were added (□), i.e., 2.7 and 2.6 M for wild type and mutant PGK, respectively. The black arrow represents the concentrations of the solution to which the urea concentrations from 0 to 6 M were added (•), i.e., 1 and 0.9 M GuHCl for wild type and mutant PGK, respectively. The abscissas correspond to the sum of denaturant concentrations.

In a further study, these residual structures and their contribution as nuclei in the folding process were investigated by Garcia et al. [304]. Tryptophan accessibility in the various conformational forms of the protein was assessed by fluorescence quenching using acrylamide. In the hyperfluorescent state, the tryptophans are not fully accessible. The residual structures around Trp 333 are less compact than those surrounding Trp 308. Comparison of static and collisional quenching constants indicated that both tryptophans are involved in residual structures, although Trp 333 seems to be involved in microstructures that are less compact and more dynamic. The microstructures around Trp 333 may be related to the persistence of native secondary structures. The unfolding curves at low temperatures suggest that the residual structure around Trp 308 is mainly stabilised by hydrophobic interactions. Trp 308 would interact with Ile 304, Ala 306 and/or Ile 311 in this structure.

Further characterisation of residual structures in yeast PGK has included a description of the conformational distribution of the denatured state of the protein in 4 M guanidine hydrochloride. This was described by a combination of small angle neutron scattering (SANS) experiments, statistical mechanical modelling and molecular mechanics calculations [305]. The denaturation of the protein produces a clear change in the SANS profile and a large increase of the radius of gyration. In the statistical mechanical model the region of contrast neutron scattering density is pictured as a chain of freely joined spheres (Figure 47). This model was used to fit the SANS data for the unfolded protein; the best fit was obtained for 100 spheres of 8.5 Å radius. This corresponds to a persistence length of about four residues, in agreement with that obtained for a polyalanine coil [306]. The SANS studies were extended to large angles [307]; under these conditions, it was shown that the fit could be improved with a model using 200 spheres, each sphere containing two amino acids.

The dynamic properties of the unfolded and the native states of PGK have been studied using quasi-elastic neutron scattering (QENS) [308]. Data on the native form and on the protein unfolded by 1.5 M guanidine hydrochloride were compared. This work represents some of the first experimental data on dynamics at the picosecond time scale. The momentum and energy transfer dependence of the scattering for both states were fitted by an analytical model, in which the dynamics of a fraction of the non-exchangeable hydrogens were described as a superposition of vibrations with uniform diffusion in a sphere, the rest of the hydrogens undergoing vibrational motions only. The radius of the sphere also changes slightly, from approximately 1.8 Å in the native protein to approximately 2.2 Å in the unfolded state. The lower limit of the free energy variation associated with the dynamic changes was roughly estimated; it was about 1.5 kT per residue. This value is consistent with the theoretical value of 1.7 kT corresponding to the difference between a free and a collapsed chain. These comparisons suggest that the change in such fast diffusive motion may play a significant role in the thermodynamics of the folding process from highly unfolded states.

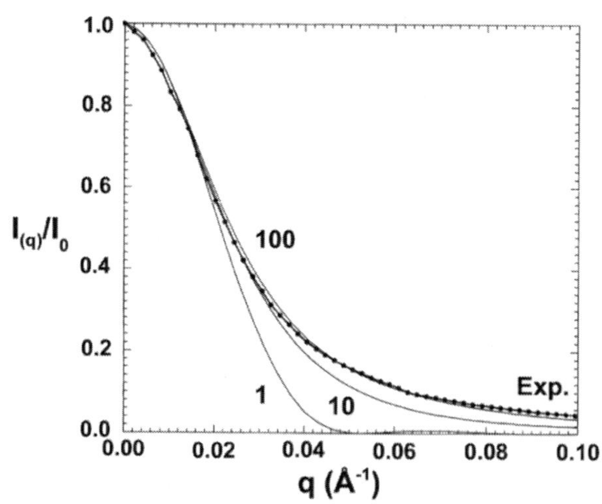

Figure 47. Comparison of the Small Angle Neutron Scattering data for the unfolded yeast PGK: The curves calculated for 1, 10 and 100 spheres models (adapted from [307]).

Folding of the Isolated Domains

The behaviour of domains as independent folding units has been addressed by genetic engineering. Two interspecies hybrids were constructed, each made of one domain from the human enzyme and the other domain from the yeast enzyme [309]. In spite of a 35% difference in the amino acid composition between the two species, the hybrid enzymes have catalytic properties similar to those of the parental proteins. Thus the evolutionary substitutions within these two distantly related proteins do not significantly affect the formation of the active site cleft, the mechanism of domain closure and the enzyme activity. Moreover, it suggests that domain behaves as independent structural units.

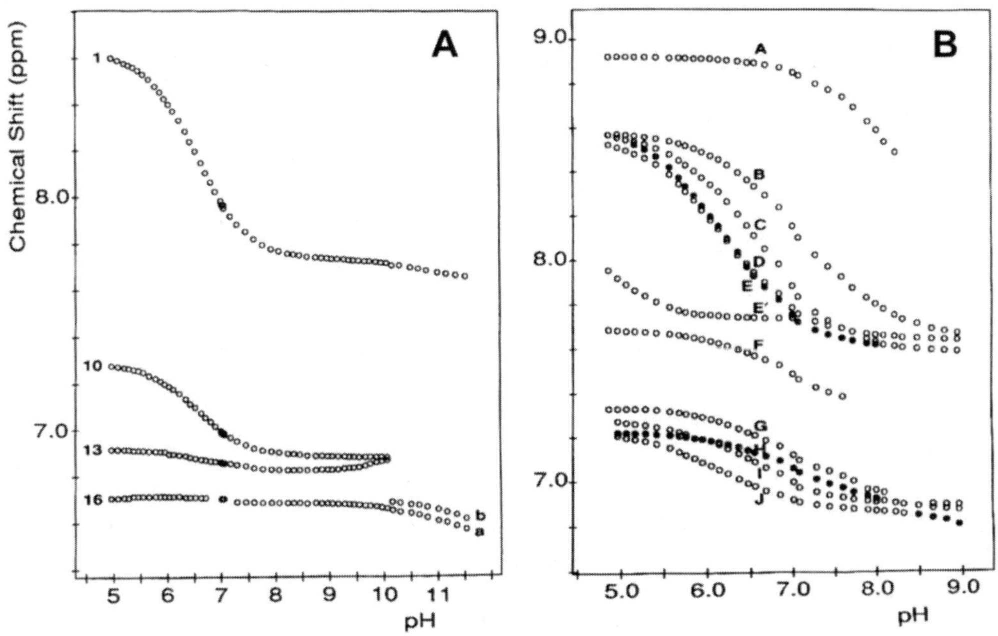

Figure 48. pH dependence of chemical shifts for some aromatic resonances of the isolated C-domain of yeast PGK (adapted with permission from [315]; Copyright 1996 American Chemical Society): (A) Peaks 1 and 10 are assigned to C2-H and C4-H of His 388, respectively. (B) pH dependence of chemical shifts for His C2-H and C4-H resonances of the isolated N-domain.

To study the role of domains in folding, several approaches were used to isolate these domains. The first approaches using limited proteolysis have provided significant results for horse muscle PGK [295], and for the pig [189, 220, 297, 298, 310] enzymes, although the size of the proteolytic fragments did not correspond to the size of domains. Genetic engineering provides an unambiguous method for obtaining isolated domains in large amounts. It was first achieved for the yeast enzyme [311, 312], then for the enzyme from Bs [313, 314] and from Tm [244]. All the domains thus obtained were shown to have a quasi-native structure as assessed by various criteria. Proton NMR spectroscopy has provided direct evidence that the isolated domains from yeast PGK are able to fold independently [312]. The C-domain was shown to be more stable than the N-domain, which shows greater conformational flexibility. The basic patch in the N-domain is clearly perturbed by the removal of the C-domain. While the C-domain binds ATP (although with 30-fold reduced affinity), the isolated N-domain lost completely its ability to bind PG [311]. The helix content of the isolated C-domain is however slightly lower than expected from the X-ray data. This is likely due to the absence of stabilising interactions between the C-terminal peptide including helices 13 and 14 (yeast PGK numbering, which correspond to helices 14 and 15 of mammalian PGK-s) and the N-domain. Figure 48 shows the NMR data for the isolated domains. Similar conclusions showing the integrity of the N-domain from Bs PGK were obtained by heteronuclear NMR spectroscopy [127]. In contrast, Mas et al. [127] concluded from their data that the tertiary fold of the N-domain of PGK requires the complete polypeptide chain including the C-terminal domain. Although the isolated domains are capable to fold into a quasi-native structure, it was shown that neither functional, nor structural complementation was obtained [311]. This indicates that the modular model does not apply to yeast PGK.

In order to determine the role of the C-terminal helix in the folding and stability of yeast PGK, a mutant lacking the last 12 amino acids corresponding to the C-terminal helix 14 (PGK Δ404-415), was constructed [187, 315]. The mutant folds into a conformation very similar to that of the wild type enzyme, but exhibits a very low activity. Apart from the deleted region, the C-domain was found to be correctly folded. The structural properties of the truncated protein are very similar, at least qualitatively, to those of the isolated domains. The addition of the missing peptide which has a dissociation constant of 80 μM provokes a 40-fold increase in enzyme activity at saturation. Moreover, this peptide, which displays a random structure in solution, folds into a helical structure on binding, as assessed by NMR (TRNOESY, cf. abbreviations). This fold is in good agreement with the corresponding helix 14 of the crystallographic structure of wild type yeast PGK. These results show that the C-terminal part of yeast PGK is not necessary for most of the initial folding, but acts to lock the C-domain to the N-domain, thus ensuring the expression of full enzyme activity, possibly by mediating domain motions (cf. Chapter I. 1.2.). The importance of the C-terminal peptide (401-415) as well as the interdomain (185-199) in the folding and stability of the N-domain was elegantly demonstrated by producing a mutant lacking of the whole C-domain. This mutant contained only the C-terminal peptide attached to the amino-terminal domain (plus the interdomain helix 7) through a flexible peptide linker [316]. By this way a native-like N-terminal domain has been produced as demonstrated by various physico-chemical measurements and its PG binding ability.

Another argument for an autonomous folding of the domains was provided by studies of cold denaturation by differential scanning calorimetry [239]. The unfolding proceeds in two distinct stages, that probably correspond to the independent unfolding of the domains. This was confirmed by further studies using dynamic light scattering [242, 317]. The data obtained by differential scanning calorimetry concerning the denaturation of the isolated domains under high- and low-temperature conditions have shown that the C-domain was the more stable on heating but the less stable in the cold conditions [243]. This is shown in Figure 49.

The unfolding and refolding of domains has been studied both in the native yeast PGK and isolation [291, 318]. Tryptophan residues (Trp-s 308 and 333) where used as conformational probes for the C-domain. A fluorescent probe, N[[iodoacety]aminoethyl]-8-naphtylamin-7-sulfonic acid (AEIDANS, cf. abbreviations) was bound to the unique cysteine residue Cys 97 to observe the conformational changes in the N-domain. The refolding kinetics of each domain from guanidine was found to be multiphase. However, it was found that a refolding step occurring during the refolding of the isolated domain is markedly masked during the refolding of the whole protein. The N-domain was found to refold faster when isolated than when integrated in the molecule.

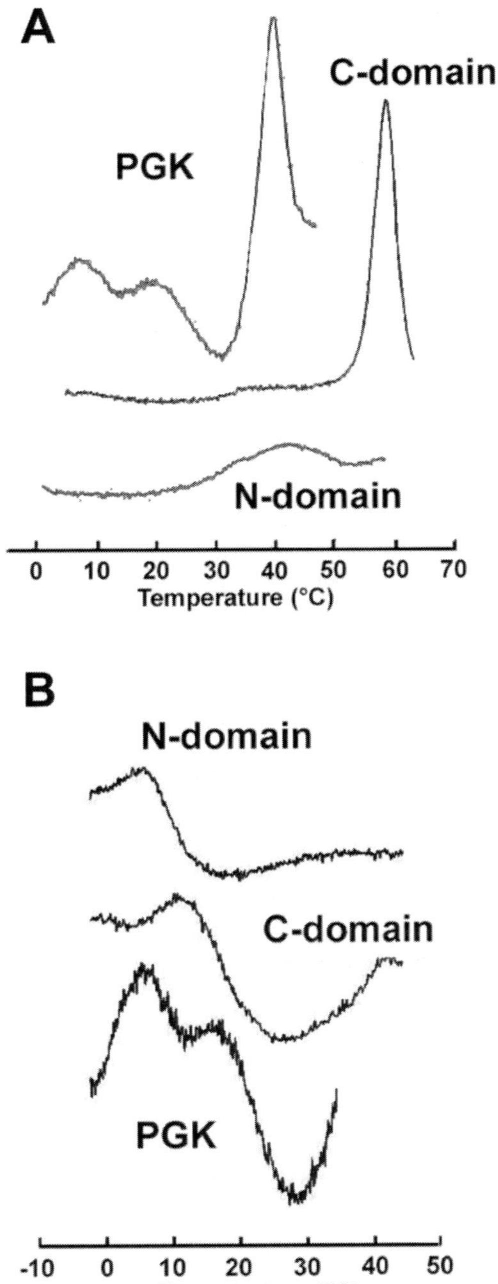

Figure 49. Differential scanning calorimetric curves of yeast PGK (adapted from [243]): (A) Heat denaturation of the isolated N- and C-terminal domains in the absence of GuHCl, and of entire PGK in the presence of 0.7 M GuHCl. (B) Cold denaturation of the isolated N- and C-domains and of entire PGK under destabilising conditions.

A kinetic study of yeast PGK mutants each containing a unique tryptophan residue, either in the N-domain (Trp 122), or in the C-domain (Trp 333) in which Trp 308 was replaced by a phenylalanine, was reported by the group of J. Fidy [319]. This protein system and its individual domains were studied. Refolding kinetics from the guanidine-unfolded state was monitored by stopped flow or manual mixing from 1ms to 1000 s. As shown previously [291, 318], it was confirmed that the folding of the N-domain was significantly faster when isolated than in the complete protein. On the contrary, folding of the individual C-domain was slower than in the complete enzyme. The presence of the C-domain directs the folding of the N-domain along a different pathway than that of the isolated N-domain, while folding of the C-domain follows the same route as within the entire protein.

The recombinant isolated domains of TmPGK exhibit a native structure, but no complementation of the domains to form the nicked protein has been observed, as for the yeast enzyme [244]. In contrast to the yeast enzyme, that is inaccessible to reversible thermal denaturation in the absence of GuHCl, this PGK and its domains show reversible thermal denaturation at pH 3-4. The isolated domains unfold via partially structured intermediates following a three-state model. The complete protein unfolds via three well-defined unfolding transitions which were attributed to the disruption of domain interactions and subsequent sequential unfolding of the domains.

A sequential refolding of the domains of the pig muscle enzyme has been suggested on the basis of the complex reactivation kinetics with a pronounced lag phase. This lag disappears when the C-terminal proteolytic fragment (residues 156-416) is pre-renatured before mixing with the remaining terminal part (residues 1-155) [298]. Thus, it could be concluded that the folding of the C-domain might precede that of the N-domain. A sequential process was also described for the refolding of BsPGK. It involves three stages. In the first stage, each domain rapidly and independently collapses to form an intermediate. Then the N-domain converts to a folded state followed by the C-domain. In the reverse process, the N-domain unfolds faster than the C-domain. A comparison with the isolated domains suggests that the interdomain contacts have an inhibitory effect on the folding kinetics. The presence of the folded C-domain appears to reduce the effective rate of folding of the N-domain [313].

To summarise, the domains in the different PGKs, when isolated, acquire a quasi native structure, but are inaccessible to a functional or structural complementation. The formation of the active enzyme with the correct contacts between domains occurs during some intermediate refolding step.

Folding of Fragments Smaller or Larger than Domains

Since isolated domains are not capable of associating to generate a functional enzyme, several other complementary fragments, different from structural domains, were constructed by genetic engineering [320, 321]. The aim of this work was to study their folding mechanism and then to determine the requirements for their reassociation and functional complementation. The method used for producing rigorously defined fragments involved the introduction of a unique cysteinyl residue into the protein followed by specific chemical cleavage at this residue. In a first study of this type [320] two pairs of complementary fragments were thus obtained, residues 1 to 96 and 97 to 415 on one case, and residues 1 to

248 and 249 to 415 in the other. The short fragments 1-96 and 249-415 were found to contain some secondary structure but had a low stability. Each large fragment showed high structural content, close to that of the isolated structural domain. In contrast to the studies on the isolated domains, a weak but significant complementation, as measured by enzyme activities of ~0.25 % and 8 %, was observed for the two pairs of fragments. An independent refolding of the complementary fragments before reassociation decreased the yield of complementation for one pair of fragments (1-96/97-415) but did not affect the complementation for the other pair. Thus, the correct folding of the isolated fragments, as well as for the domains, is not a prerequisite for their complementation.

In a further study, several fragments corresponding to the limit between globular subdomains in the protein were produced according to the same method [321]. Fragments corresponding to globular subdomains recovered the expected helix content, but not the cooperativity observed in the isolated domains. In this respect the structural domains appear to be minimum units of stability in PGK. The cooperativity in the folding transition was only observed for fragments larger than a domain. Furthermore, the presence in a fragment of a sequence extending beyond that of the C-domain modified the folding process.

According to this and other studies, it seems that an isolated fragment with a low or an undetectable native conformation can recover a native conformation upon mixing with a complementary fragment. This might reflect that this protein reaches its functional conformation through the other parts of the molecule involving long-range interactions. The polypeptide chain can explore interactions outside its own folding unit.

Pig muscle PGK, when submitted to limited tryptic proteolysis, yields a nicked enzyme without loss of catalytic activity. Under non-denaturing conditions, these fragments are held together by non-covalent interactions insuring the active conformation [220, 310]. The unfolded, isolated fragments were mixed under conditions of renaturation and the reactivation process was observed to occur up to 60-85% recovery of enzyme activity. The rate of reactivation occurred with the half-time of 70-110 s and did not differ substantially from the reactivation rates of either the denatured intact and nicked PGK. At that time no detailed kinetic analysis has been made, but it was found that a separate preincubation of one or the other or both fragments under conditions of renaturation before complementation did not cause substantial acceleration of the reactivation process [310]. The high extent of activity recovery may indicate that the position of cleavage in the protein structure is crucial for obtaining an efficient complementation. For the pig muscle enzyme the cleavage site by trypsin has been determined to be at Lys 155, i.e., exactly at the border of a subdomain at the end of helix 4 in the N-domain [297], as shown in Figure 50. Later, a more detailed analysis of complementation from these fragments of pig muscle PGK was carried out (based on reactivation, restoration of Trp fluorescence and ANS release) in a wide protein concentration range (from 10 nM to few μM [298]). The results unambiguously demonstrated that the bimolecular association of the fragments is not rate limiting as the process is a unimolecular one. Moreover, reactivation is a complex process with a pronounced lag phase (either from the intact enzyme or from the two fragments) indicating the presence of an inactive intermediate. The lag phase, however, disappeared when the C-fragment was renatured prior to mixing with the N-fragment, and the time course was simplified into a single first order transition. On the other hand, the same lag-phase was observed when the N-fragment was preincubated under renaturing conditions, before mixing it with the C-fragment [298]. From these results the authors suggested that the folding of the C-domain is independent and an inactive intermediate with folded C-domain and still unfolded N-domain may exist. In accordance, in a further study, Szilágyi and coworkers [189] showed that the refolding of the isolated fragment from the N-domain was incomplete. It produces a protease-resistant, globular, but highly flexible intermediate with non-native secondary structure and highly exposed hydrophobic clusters, which even favour dimerisation. This is in agreement with that found previously for yeast PGK, where complementation occurs when at least one of the fragments is unfolded. The isolated N-fragment of pig muscle PGK, with its non-native character under renaturing conditions, can indeed be considered as a potentially productive intermediate, as its refolding can be completed by an association with the complementary C-terminal part of the molecule.

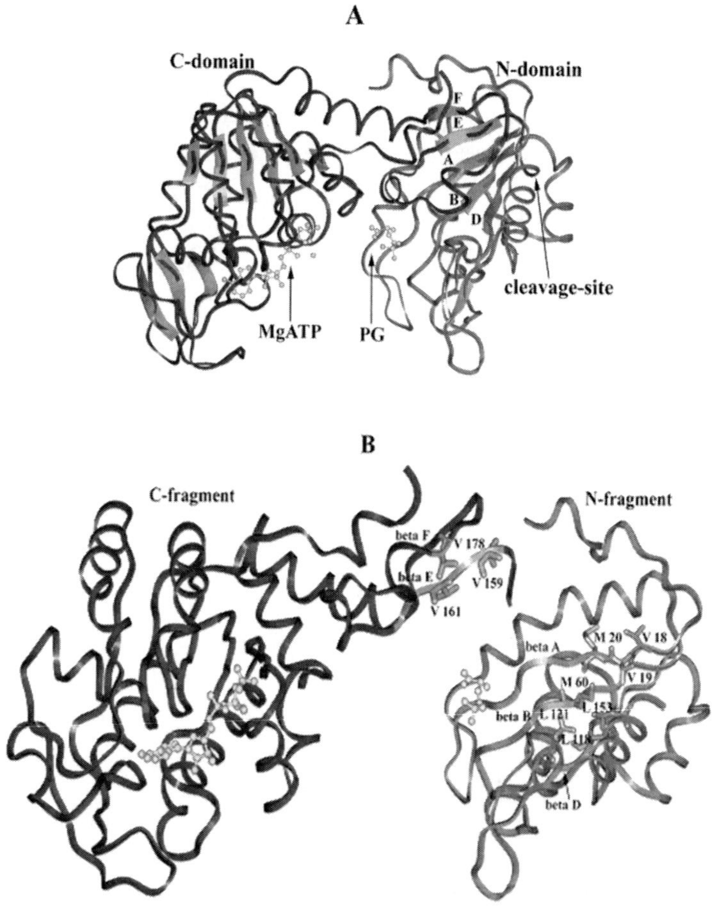

Figure 50. Productive complementation of fragments from pig muscle PGK (adapted from [298]): (A) The N- and C-terminal fragments of the PGK molecule are represented by grey and black ribbons, respectively, and the bound substrates are shown by white ball-and-stick models. The tryptic cleavage site is indicated by an arrow within the N-domain; (B) The hydrophobic side-chains (grey stick models) located at the interfaces of the two fragments may explain the inability of the separated N-fragment to fold into its native structure: the N-domain core ß-sheet system cannot be formed and for the same reason the N-fragment itself tends to aggregate.

According to the results concerning the role of domains in PGK from different organisms, it seems that a strictly modular, sequential and hierarchical view of protein folding does not describe the folding process of this protein adequately.

Role of Circular Permutations in the Folding of PGK

The role of domains as folding units can be understood for large proteins, where they manifest local interactions dominating the first folding steps. But the intuitive concept which associates local interactions and the early folding steps cannot explain the folding of proteins made of discontinuous domains. This leads to the following question: is the continuity of the domains an obligatory requirement for the correct folding of proteins? In order to answer this question, two circularly permuted variants of yeast PGK were constructed by a circular permutation of the polypeptide chain, using a simple polymerase chain reaction methodology [128]. The natural chain termini were directly connected and new extremities were created either in the N-domain (Asn 71 and Glu 72), or in the C-domain (Ala 291and Asp 292) (Figure 51). The two mutant proteins were over expressed and purified, exhibiting 16 and 26 % activity, respectively. The crystallographic structure of the circularly permuted protein in the N-domain indicated that the overall folding pattern is very similar to that of the wild type protein [322]. The effect of this topological mutation on the enzyme activity is more easily accounted for by a restriction of the hinge-bending motion than by structural changes. The unfolding-folding studies have revealed a loss in stability for both mutant proteins, with ΔG^{app} values of 3.3 kcal/mol and 4.2 kcal/mol for the mutant proteins permuted in the N- and in the C-domains, respectively, compared to 7.9 kcal/mol for the wild type PGK.

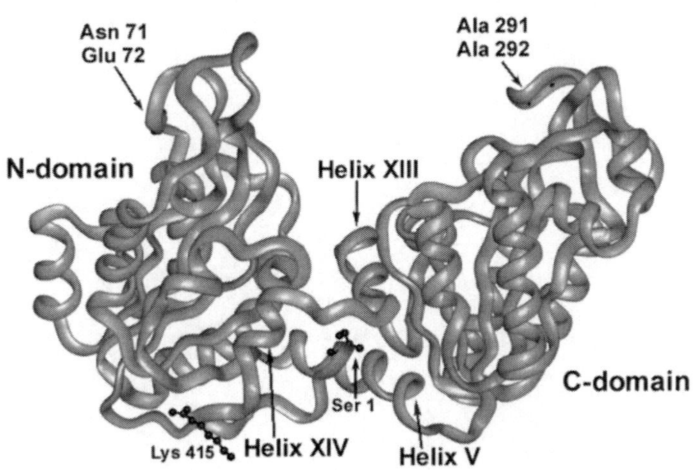

Figure 51. Three-dimensional structure of yeast PGK indicating the cleavage sites for the construction of mutants by circular permutation (adapted with permission from [128]; Copyright 1995 American Chemical Society).

Although they have a lower stability, their fold is not fundamentally affected. Thus, the continuity of the domains might be important for the stability of the folded PGK, but is not required for this protein to find a folding pathway and to achieve a native-like and functional structure. These data suggested that the efficiency of the folding process of a protein made of continuous domains does not strictly depend on the natural connectivity of the native structure.

Role of Insertions in the Folding of Yeast PGK

To further explore the robustness of the structure and the folding of yeast PGK, some short oligopeptide sequences of Gly or Thr were inserted into the loops connecting different regions of secondary structure. For short insertions such as dipeptides, there was no direct relationship between the length of the inserted oligopeptide and the destabilising effect of the insertion. The insertion of a dipeptide is more destabilising per residue than the insertion of longer sequences. Insertions in two long β–α loops (at positions 71 and 129) were found to be more destabilising than an insertion in a short α–β loop (at position 89). However, although these insertions are destabilising, they do not prevent the correct folding of the protein. The effect on stability is mainly due to a large increase in the unfolding rate rather than a decrease of the folding rate [323]. These results showed that large structured insertions can be artificially introduced within a protein.

As a consequence, it should be expected that very large insertions could be accepted by PGK. Indeed, the sequences of entire enzymes, dihydrofolate reductase and β-lactamase, were introduced by genetic engineering in different positions within this enzyme, Asn 71, Lys 89 and Ser 129 in the N-domain and Ser 280 in the C-domain [324]. The resulting chimeric enzymes were all overexpressed and were found to be functional. The structure of the chimeric PGK-β-lactamase is illustrated in Figure 52. Furthermore, functional coupling between the two fused proteins was observed for three chimeric enzymes as shown in Figure 53.

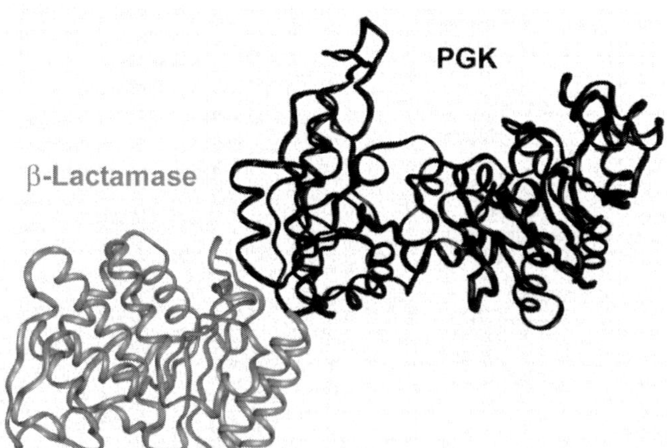

Figure 52. Schematic representation of the chimeric protein PGK-β-lactamase: β-Lactamase was inserted into the sequence of PGK at position 89 (illustration was prepared on the basis of [324]).

These results indicate that the insertion of large proteins may be tolerated, provided that the inserted protein is introduced at positions in which there is no steric hindrance.

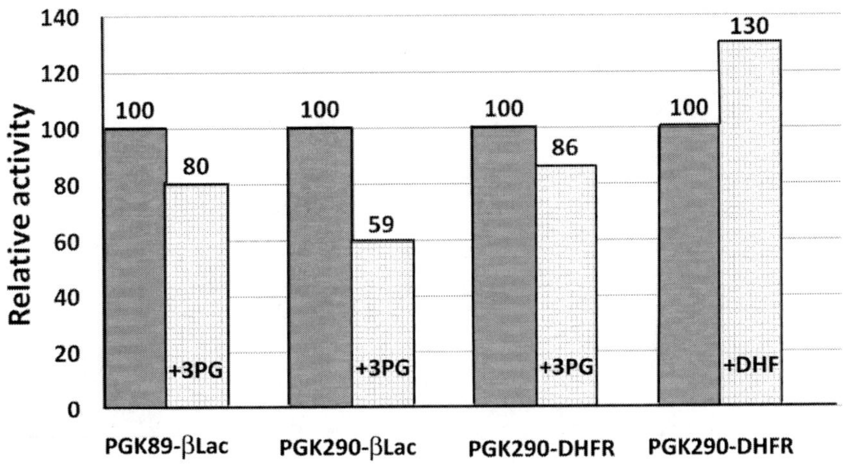

Figure 53. Functional interactions observed between three chimeric proteins (prepared using the data published in [324]): Enzymes are PGK-β-lactamase (inserted at position 89 and 290) and PGK-dihydrofolate reductase (inserted at position 290). Enzyme activities are expressed with respect to that measured in the absence of the tested ligand PG (3-phosphoglycerate), or DHF (dihydrofolate).

Heterogeneity of the Intermediates During the Folding of Yeast PGK

Heterogeneous species have been observed during the refolding of PGK [246, 325, 326]. Under equilibrium conditions, multimeric species, i.e., dimers, trimers and tetramers, have been also detected during the refolding of the N-terminal fragments, where such species were not observed for the whole protein nor for the C-terminal fragments [325]. A kinetic analysis has shown that transient multimeric species are formed during the refolding process of the whole protein, yielding the native monomeric protein in the slow folding step. They are not in equilibrium, but they are rapidly formed and disappear during the slow step. Contrary to classical aggregates, their distribution does not depend on protein concentration, and they are produced for protein concentration as low as 0.05 μM. The distribution of multimeric species is completely stable at the end of the fast refolding phase. To take into account all these observations, a model has been proposed as presented in Figure 54. This model is formally similar to a co-polymerisation between two types of monomers. The multimeric species observed during the folding of the whole protein also occurs during the folding of the N-terminal fragments, but they cannot transform to the native structure in the absence of the interactions with the complementary fragments. These results indicate that during the folding process, the polypeptide chain explores non-native interactions.

In these experiments, the protein was refolded from guanidine hydrochloride. During temperature-jump induced refolding of cold denatured PGK, non exponential kinetics have been reported by Sabelko et al. [326] and interpreted in terms of downhill folding, which points a heterogeneous ensemble of structures en route to the folded state. Further detailed

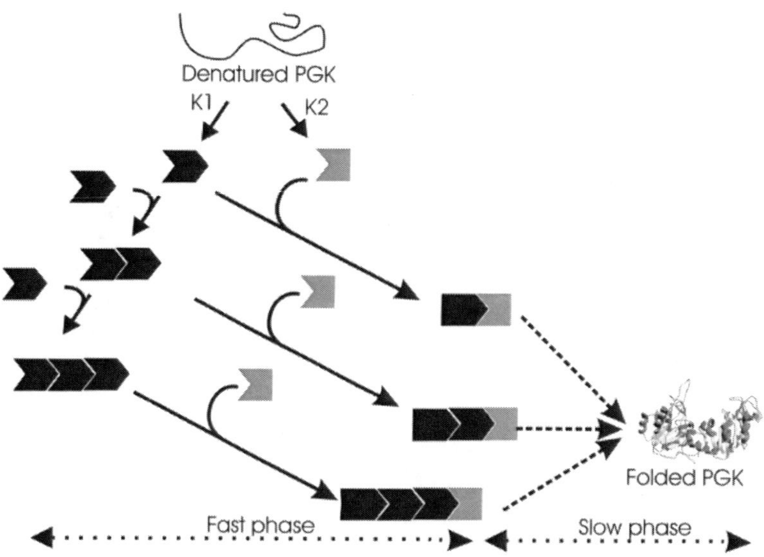

Figure 54. Model proposed for the occurrence of transient multimeric species during the refolding of yeast PGK (based on [325]): The refolding produces two types of conformers, one of these can be extended by association of the same type of monomer or by association of the other type of monomer. In the latter case, the association cannot be extended further.

analysis by the same group [246] confirmed the formation of heterogeneous species in the early folding step. Furthermore the data showed that non-exponential kinetics are intrinsic to the C-domain, whereas the N-domain refolds through an intermediate by single exponential kinetics. Such a discrepancy might be explained by the asymmetric stability of the two domains. Indeed, as previously mentioned, the C-domain is the most stable on heating, but the least stable in cold conditions [243]. In a more recent paper, Osváth et al. [327] analysed the non-exponential yeast PGK refolding kinetics in terms of hierarchic structures in the energy landscape. The result was a hierarchic finite level energy landscape model that quantitatively describes the refolding of PGK from 1 ms to 15 minutes. The early steps of the folding process happen in the upper region of the landscape leading to stretched kinetics in the early phase of the folding. The lower region of the energy landscape is dominated by a trap with the accumulation of molten globule intermediate state. Then, the protein can reach the global energy minimum through a cross-barrier folding step. All these observations, indicating the occurrence of heterogeneous intermediate species and the existence of several pathways during the folding of PGK, are consistent with the new view of protein folding.

Conclusion

An extensive corpus of data has been collected concerning the folding mechanism of a paradigmatic two-domain protein, phosphoglycerate kinase, PGK. Although the domains represent autonomous folding units capable of reaching a quasi-native structure when isolated *in vitro* as well as *in vivo*, they are not capable of complementation to generate the functional enzyme. An ensemble of results indicates that, in the whole molecule, the polypeptide chain explores interactions outside its own folding unit to reach the native structure, contrary to what is expected from a modular model of folding. Furthermore, circular permutations have shown that the continuity of the domains might be important for the stability but is not a prerequisite for the protein to find a folding pathway and to achieve a native and functional structure. Alternative pathways are also demonstrated by the occurrence of heterogeneous species as folding intermediates. PGK exhibits a significant adaptability, with a tolerance to large insertions which is not restricted to very limited regions of the polypeptide chain. The occurrence of a natural fusion protein PGK-TIM from Tm [328] and the possibility of creating various chimeras indicate that the folding process can lead to the same final structure through a variety of pathways. In conclusion, a modular sequential and hierarchical model does not adequately describe the folding mechanism of PGK; on the contrary, all the data are consistent with the new view of protein folding, with the existence of multiple pathways to reach the native and functional structure.

Part II

Physiology: Functional Characteristics

In: Phosphoglycerate Kinase: A Hinge-Bending Enzyme ISBN: 978-1-62808-836-6
Editor: Mária Vas

Chapter IV

Physiological and Non-Physiological Functions of PGK and Its Abundance

Andrea Varga and Mária Vas
Institute of Enzymology, Research Centre for Natural Sciences,
Hungarian Academy of Sciences, Budapest, Hungary

4.1. Physiological Processes

Participation in the Metabolic Pathways of the Cell

PGK catalyses the reversible phospho-group transfer from bPG to MgADP, producing the universal energy storage molecule, ATP. In the reaction equilibrium, the concentration of MgATP and PG is dominating (K_{eq}=3·10^{-4}), because the production of the ATP is thermodynamically more favourable than its hydrolysis.

(bPG) (PG)

Eq. 2

PGK is an essential enzyme for all living organisms. It is involved in both major phases of metabolism. In the aerobic degradation of glucose i.e., in glycolysis the formation of the main energy storage, ATP occurs in the PGK catalysed reaction. Glycolysis is one of several catabolic pathways, sometimes called homolactic fermentation, and occurs in many microorganisms and in the cells of higher animals and plants. In alcoholic (anaerobic) fermentation, characteristic of many types of yeast, the glucose molecule is broken down into ethanol and carbon dioxide. The survival of some anaerobic pathogens like *Trypanosome* or *Plasmodium* species is dependent solely on glycolysis being an essential pathway for their energy metabolism. Therefore PGK as a major flux controller can be an important target to interfere with the life cycle of these pathogens.

PGK has also an important role in the photosynthesis of plant cells. In the dark phase of the process (second step), glucose and other reduced products are formed from carbon dioxide. In this process the PGK catalysed reaction produce ADP from ATP. In the biosynthesis of carbohydrates when the conversion of pyruvate to glucose-6-phosphate occurs, PGK contributes to usage the energy of ATP for this anabolic process. This is the reversal of glycolysis, but most of the high energy required steps are substituted by lower energy-cost steps.

Involvement of PGK in the Invasive Biochemical Processes Leading to Cancer

Although PGK is an essential enzyme in the carbohydrate metabolism, it is also known to be activated transcriptionally by hypoxia inducible factor (HIF-1) in cancer cells [329]. Hypoxia is a central feature of many solid tumours. Transcriptional activation of hypoxia-inducible genes is known to play a critical first step in the adaptation of tumours to hypoxia. Hypoxia-inducible genes also regulate many glycolytic enzymes to facilitate anaerobic metabolism. These molecules are required to provide for the high-energy demands of growing neoplasm and to generate macromolecules critical for growth [330]. Consequently, the shift in energy production from oxidative phosphorylation to glycolysis, the so-called „Warburg effect", is likely to be a fundamental property of cancer cells and is known to correlate with enhanced tumour progression.

Surprisingly, PGK is secreted also extracellularly by tumours despite not having an identifiable signal sequence. While the intracellular PGK assists in tumour growth, extracellular PGK acts against it. In the extracellular milieu it is involved in regulation of angiogenesis by contribution to formation of angiostatin, which is an angiogenesis inhibitor [16]. Angiogenesis, the outgrowth of new capillaries from pre-existing vessels, which can supply oxygen inside the tumour, is necessary for tumour growth and metastasis formation. Angiostatin is formed from the first four disulfide-linked structures of plasminogen, known as kringle domains 1-4 and part of kringle 5 (the latter is shown in Figure 55).

Angiostatin formation is initiated by the reduction of disulfide bonds in the kringle 5 domain of plasmin by PGK. The general mechanism of this reduction, however, is not a thiol-dependent process [17]. This fact was proved by either alkylation/oxidation of the fast reacting Cys or by various site-directed Cys→Ala mutations of the enzyme. It is conceivable, that an acid-based assisted hydrolysis occurs (see in Figure 56) instead of thiol-disulfide redox exchange or alkaline hydrolysis. This may be accomplished by the aid of glutamic or aspartic carboxylic group of PGK, which can polarize the disulfide bond of plasmin and lead to the cleavage of the bond and formation of the angiogenesis inhibitor, angiostatin [27]. Thus, while intracellular PGK assists in tumour growth, extracellular PGK acts against it.

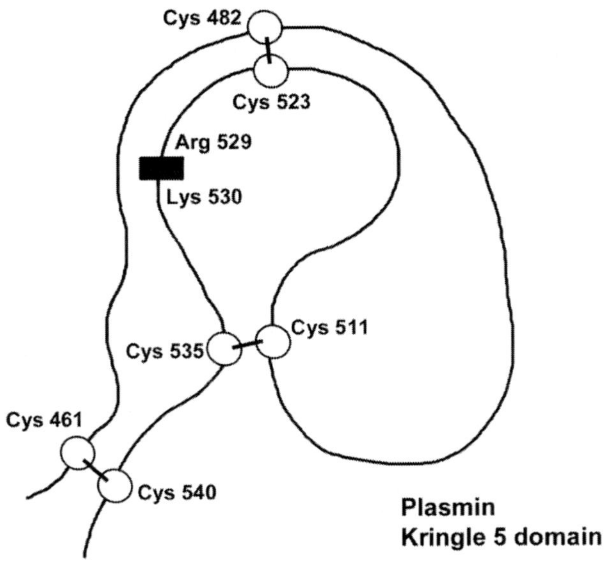

Figure 55. Model for generation of angiostatin by disulfide bond reduction and proteolysis in kringle 5 of plasmin (based on [331]): It is suggested that the Cys 461–Cys 540 and Cys 511–Cys 535 disulfide bonds in plasmin kringle 5 are reduced by plasmin reductase [331]. The reduced kringle 5 is then subjected to proteolysis at the Arg 529-Lys 530 peptide bond (*solid black line*) and also at two other positions (not seen).

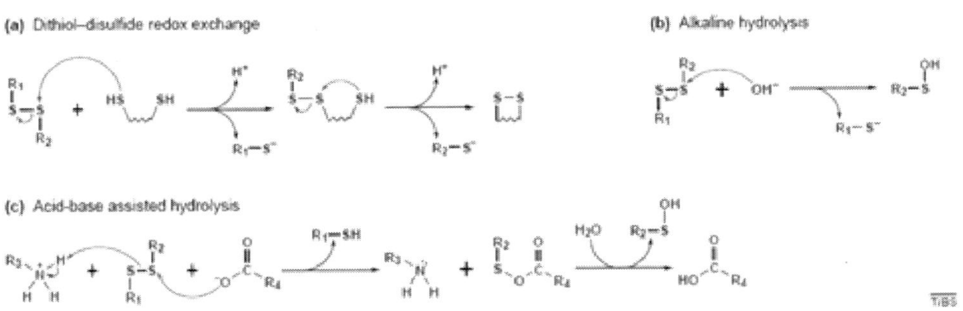

Figure 56. Variuous possible chemical mechanisms for disulphide bond cleavage (adapted from [27]): The last scheme (c) may represent the plasmin reduction by PGK. Namely, a primary amin from the kringle domain of plasmin (Arg 474) may assist in the reduction of the disulphide bond of plasmin by the aid of a glutamic or aspartic residue of PGK [27].

Plasminogen —uPA / uPAR (⊣ PGK)→ Plasmin —PGK→ Reduced plasmin —MMPs→ Angiostatin

Scheme 1. Regulation of plasminogen activation at two different steps by PGK.

In addition, PGK can regulate the angiogenesis of a tumour cell via modulation of the expression of the urokinase plasminogen receptor (uPAR), which is the cell surface receptor of the plasminogen activator (uPA) and thus, an upstream regulator of plasminogen activation [332]. uPAR expression is reduced by PGK via binding to the coding region of the uPAR mRNA (Scheme 1). This reduction, however, is suspended by phosphorylation of PGK on the residue Tyr 76. On the other hand, uPA itself promotes PGK phosphorylation on the residue Tyr 76, which in turn suspends PGK binding to uPAR mRNA, thus resulting in continuous plasminogen activation in lung epithelial cells [333].

In addition, PGK can regulate the angiogenesis of a tumour cell via the modulation of the expression of the urokinase plasminogen receptor (uPAR), which is the cell surface receptor of the plasminogen activator (uPA) and thus, an upstream regulator of plasminogen activation [332]. uPAR expression is reduced by PGK via binding to the coding region of the uPAR mRNA (Scheme 1). This reduction, however, is supposed to be inhibited by phosphorylation of PGK on the residue Tyr 76. On the other hand, uPA itself promotes PGK phosphorylation on the residue Tyr 76, which in turn suspends PGK binding to uPAR mRNA, thus resulting in continuous plasminogen activation in lung epithelial cells [333]. The question arises whether PGK is constitutively phosphorylated on Tyr 76 in other cancer cells as well? Furthermore, tyrosine kinase(s) responsible for phosphorylation of PGK is/are not yet identified. In addition, no mechanism is known that would regulate the ratio of phosphorylated and non-phosphorylated PGK forms, although this ratio would maintain a delicate equilibrium between angiogenesis and angiostatin formation.

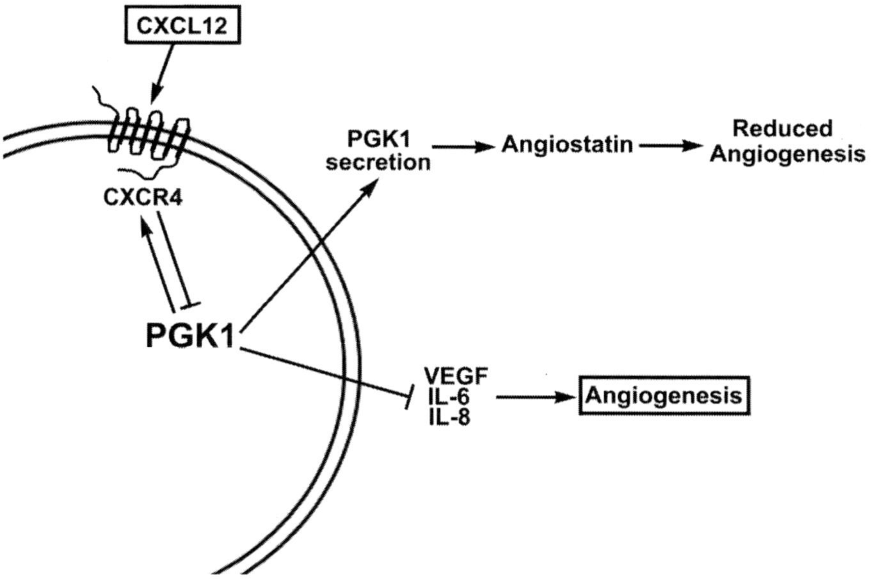

Figure 57. Model of PGK action in prostate cancer (based on [335]): At site of low CXCL12, CXCR4 signalling regulates PGK expression, which reduces angiogenesis by generating angiostatin and by reducing the secretion of the proangiogenic cytokines VEGF and IL-6, Il-8 (interleukines). At sites of high CXCL12 production (e.g., bone, lymph node and liver), however, PGK secretion is inhibited, i.e., favouring angiogenesis.

Further question is what kind of signalling mechanisms operate within the tumour cells for controlling PGK gene expression and secretion of the enzyme? Understanding such mechanisms might allow regulation of cancer cell growing and invasion. It was shown that although PGK production was enhanced by hypoxia, its secretion was inhibited. In contrast, hypoxia did not inhibit secretion of the angiogenesis activator, vascular endothelial cell growth factor (VEGF). PGK secretion was under the control of the oxygen-sensing protein hydroxylases, but direct hydroxylation of PGK was found not to be the mechanism [334].

A linkage between tumour metastasis and angiogenesis through the chemokine CXC-motif receptor, CXCR4 and its ligand CXCL12 signalling pathway was recently explored [335-337]. In the tumour cells, CXCR4 signalling regulates PGK gene expression and secretion and establishes a delicate equilibrium between the hypoxic response and tumour angiogenesis, which leads to tumour progression.

Figure 57 illustrates the operation of CXCL12/CXCR4 directed “angiogenic switch” that is mediated by extracellular PGK levels in prostate cancer cells. Thus, if we were able to regulate PGK secretion by the tumour cells, we might be able to regulate the formation of metastasis.

4.2. A Non-Physiological Role: Involvement of PGK in Drug Action

Activation (Phosphorylation) of Anticancer and Antiviral Drugs

PGK is completely specific for its substrates 3-phosphoglycerate (PG) or 1,3-bisphosphoglycerate (bPG), but not for its nucleotide substrates (ADP or ATP). This fact was published first in 1970 by Bücher and coworker [12], who found that PGK can handle either GTP or ITP as a substrate. The activity of PGK was found to be 56 and 77%, respectively, with these nucleotides compared to ATP (100%). The K_m values measured were in the same order of magnitude as for ATP. These data have been supported and extended towards other nucleotides by recent studies carried out with human PGK [33] using various nucleotide diphosphates as substrates, in place of ADP. Catalytic efficiencies measured are summarised in Table 17. Comparing the efficiency of PGK with purine- and pyrimidine-based nucleotides it is clear that PGK prefers purines (see efficiencies with ADP and GDP) over pyrimidines (see efficiencies with CDP and UDP). In addition, deletion of the 2’-OH group of the ribose (see efficiency with D-2’dADP) is not detrimental for the phosphorylation of the nucleotide by the enzyme. Table 17 also includes data with other nucleotide derivatives including mirror image nucleotide analogues, the significance of which will be discussed below.

The lack of nucleotide specificity of PGK and some other cellular kinases is particularly useful in antiviral therapy. Namely, the nucleoside analogue pro-drugs have to be administrated to the patient as nucleosides, in order to be able to pass the plasma membrane. Thus, the mechanism of action of nucleoside analogues is based upon the intracellular phosphorylation into their 5’-triphosphate form that can interact with the specific polymerases of virus, as a competitive inhibitor or an alternate substrate for these target enzymes, and thereby usually prevents further chain elongation of viral nucleic acid. The intracellular activation of the drug compounds therefore requires sequential steps of phosphorylation. The first two phosphorylation steps of the nucleosides are well-studied. The first step is catalysed by deoxycitidine kinase [338-340], thymidine kinase 2 [341] or deoxyguanosine kinase [342]. Uridine/cytidine monophosphate kinase [343, 344] is responsible for the second phosphorylation step. The last step of the cascade is still largely unexplored. This step may involve kinases with relaxed specificity towards its nucleotide substrates e.g., nucleoside diphosphate kinase (NDPK), phosphoglycerate kinase (PGK), pyruvate kinase (PK) and creatine kinase (CK).

Table 17. Catalytic efficiencies of human PGK measured with various nucleoside diphosphates as substrates

Purine-base analogues				**Pyrimidine-base analogues**			
ADP (analogues)	k_{cat}/K_m ($M^{-1} \cdot s^{-1}$)	**GDP** (analogues)	k_{cat}/K_m ($M^{-1} \cdot s^{-1}$)	**CDP** (analogues)	k_{cat}/K_m ($M^{-1} \cdot s^{-1}$)	**UDP/TDP** (analogues)	k_{cat}/K_m ($M^{-1} \cdot s^{-1}$)
[a]D-ADP	$9.0 \cdot 10^6$	[a]D-GDP	$9.1 \cdot 10^6$	[a]D-CDP	167	[a]D-UDP	4635
[a]D-2'dADP	$2.9 \cdot 10^5$			[a]L-CDP	$2.8 \cdot 10^4$	[c]D-FMAUDP	1173
[a]L-ADP	$2.5 \cdot 10^6$			L-3TC	[b]800 [c]4000	[c]L-FMAUDP	$1.3 \cdot 10^5$

[a] Varga and coworkers [33]
[b] Gallois-Montbrun and coworkers [31]
[c] Krishnan and coworkers [28]

The first nucleoside antiviral drug against herpes viruses, acyclovir [345], was reported to be activated (phosphorylated) mostly by PGK. The phosphorylation of acyclovir was revisited in a recent paper [31] and the catalytic efficiency was found to be $2.5 \cdot 10^4$ $M^{-1}s^{-1}$, which is although 100-times less than for ATP, but it is still a reasonable activity. In fact, this analogue can be incorporated into the viral DNA chain (after its activation) and because of the lack of its 3'-OH group of the ribose (and the ribose ring itself), the DNA chain cannot be elongated further, i.e., terminated by this analogue. This leads to death of the virus. A next generation drug against herpes viruses, gancyclovir (for the chemical formulas cf. Figure 58), was reported to be as active as acyclovir and phosphorylated in the first step to its monophosphate form by the virus thymidylate kinase and to its di- and triphosphate forms by cellular kinases. The last step of its activation is catalysed by PGK [346].

Recently, the activation of tenofovir, an anti-HIV drug was investigated. This drug is also an acyclic analogue, but it is a phosphonate, which means that it has already a "phospho-like" group, where one O atom is missing. (This can be an advantage for its stability, since it cannot be dephosphorylated by 5'-nucleotidases.) This difference, however, makes phosphorylation of tenofovir completely different from either acyclovir or gancyclovir: its main activator is creatine kinase and not PGK [347, 348]. Docking studies have well supported the activity data. Although both tenofovir and acyclovir fit satisfactorily into the active site of the closed (active) conformation of HsPGK (Figure 59 A and C), there are well detectable differences in the binding modes of the two compounds if they are docked into the open (inactive conformation of HsPGK (Figure 59 B and D). If one considers that the active form of PGK is resulted in through a large-scale of hinge-bending domain closure upon binding of the substrates [109, 349], it is reasonable to assume that binding of tenofovir monophosphate in place of ADP is possibly not able to induce the same type of well defined large-scale conformational changes which is required for the catalysis [227, 228]. Indeed, the binding mode of tenofovir monophosphate in the open form of PGK is different from that of the real substrate ADP (Figure 59 B). Not only the interaction with the missing ribose ring is lacking, but the adenine ring occupies a position different from that observed with ADP. Furthermore, the positions and the interactions of the phosphate chain of tenofovir monophosphate with the important catalytic residues, Lys 215 and Lys 219, are greatly different from that of ADP. In the absence of the proper interactions of tenofovir monophosphate with the open PGK, domain closure will possibly not be induced and consequently, catalysis will not occur. As a comparison, acyclovir binds to either conformations of PGK in a very similar way to that of the natural substrate, ADP. Tenofovir and acyclovir are, in fact, different in the nature of base (guanine for acyclovir and adenine for tenofovir) and the number of atoms connecting the base and the phosphate chain. Thus, the distance between the base and the phosphate chain is somewhat shortened in tenofovir compared to acyclovir due to the presence of the phosphonate group in tenofovir in place of the phospho-group in acyclovir. Thus, the active site architecture of PGK, although it is flexible, is not able to tolerate the slight structural differences of tenofovir, relative to the molecule of acyclovir. – In conclusion, conformational flexibility of enzymes, such as large scale domain motions (as required for PGK function) is not always advantageous for nucleoside analogues compared to their corresponding natural substrates. Instead, smaller extent of structural plasticity of the active site cleft (as for CK) seems to be more favourable in respect of handling tenofovir monophosphate, but this requirement is not fulfilled with PGK.

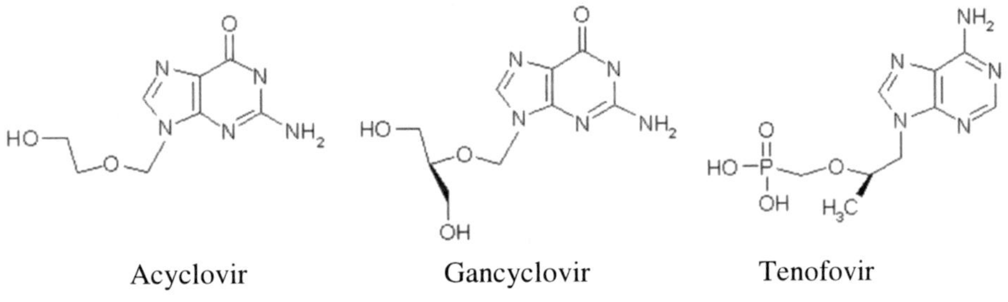

Figure 58. Acyclic analogues used in antiviral therapy.

The possible binding modes of tenofovir monophosphate (A and B) and acyclovir diphosphate (C and D) in comparison with ADP (shown in all figures) to both closed (A, C) and open (B, D) structures of PGK have been determined by docking as described.

In the past decade the importance of L-nucleoside analogues, which are the mirror images of their natural D-counterparts, has been recognized in the inhibition of viral replication and treatment of cancer. From the tremendous work carried out it appears that favourable features of L-nucleoside analogues may include an antiviral activity comparable and sometimes greater than their D-counterparts, more favourable toxicological profiles and a greater metabolic stability. Several L-analogue drugs (e.g., shown in Figure 60) have already approved by the US Food and Drug Administration (FDA) and some others are in clinical trials.

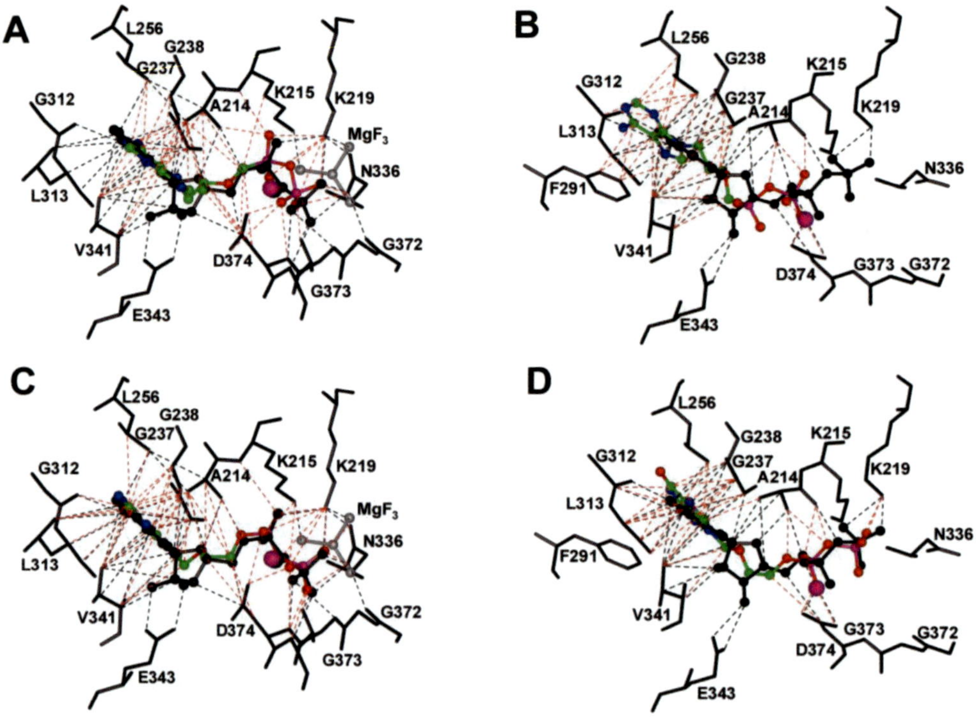

Figure 59. Comparison of the binding modes of tenofovir monophosphate and acyclovir diphosphate to that of ADP in the nucleotide site of PGK possessing both closed and open domain conformations (adapted from [348]): The possible binding modes of tenofovir monophosphate (A and B) and acyclovir diphosphate (C and D) in comparison with ADP (shown in all figures by black ball and stick models) to both closed (A, C) and open (B, D) structures of PGK have been determined by molecular docking. The analogues tenofovir and acyclovir are presented by ball and stick models, coloured according to the nature of their atoms. The positions of Mg^{2+} ion are indicated by purple balls. The residues that form the nucleotide site are indicated by black sticks. The dashed lines (coloured black and red for ADP and the analogues, respectively) indicate the atomic contacts that are formed between the enzymes and their respective ligands.

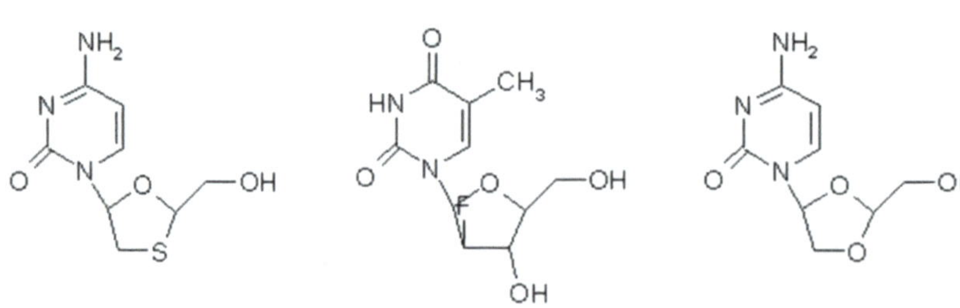

Lamivudine Clevudine Troxacitabine

Figure 60. L-analogues used currently in antiviral therapies.

It was reported by Krishnan and coworkers for the first time that L-nucleoside analogue diphosphates are selectively phosphorylated by 3-phosphoglycerate kinase [28]. They used small interfering RNAs to down-regulate the amounts of PGK in the cell [30] and studied phosphorylation of nucleoside analogue diphosphates applied in anticancer therapies e.g., lamivudine (known as L-3TC or L-SddC), clevudine (L-FMAU), troxacitabine (L-OddC) and gemcitabine (D-dFdC) to monitor the changes in the amount of their di-, or triphosphate forms (for the respective chemical formulas of the L-analogues see Figure 60). When PGK production was down-regulated, the total amount of L-3TC, L-FMAU, L-OddC and its phosphorylated metabolites in the cells also decreased, approving that the phosphorylation of diphosphate metabolites by PGK is really the bottleneck for total uptake of L-nucleoside analogues into the cells. The reverse process, dephosphorylation of the nucleotides catalysed e.g., by 5'-nucleotidases also occurs in the cell and results in the catabolism of nucleoside analogue triphosphates to the respective nucleosides, which can diffuse out of the cells. These processes regulate the total amount of nucleoside and nucleotide analogues in the cell. From the experiment of down-regulation of PGK, it is conceivable that PGK is not only important for initial phosphorylation but also for the maintenance of steady-state levels of L-nucleotides in the cells.

In order to reveal the structural basis of the lack of enantio-selectivity of PGK i.e., to understand the phosphorylation of L-analogues by PGK, two collaborating groups made an attempt to characterise the details of the interactions of PGK with L-ADP and L-CDP as model compounds at atomic level. They found that L-ADP is almost as good substrate for PGK as D-ADP (Table 17) [250, 350]. From the analysis of the structure of PGK·PG·Mg-L-ADP open ternary complex [75] and of PGK·PG·Mg-L-ADP·AlF_4^- closed complex [351] only small differences could be obtained for the interactions of the nucleotide with the enzyme. The base of L-ADP adopting a *syn* conformation, however, tilted by about 40° in the binding site, inserting its amine 1.5 Å deeper into the base-binding pocket than D-ADP (see Figure 61). This adjustment is rendered possible by the malleable water network in the cavity behind the base, and the promiscuous hydrophobic clamp to the adenine. The ribose adopts an O4' endo and a C4' endo conformations with D-ADP and L-ADP, respectively. Nevertheless, hydroxyls of the L-ADP ribose can still form D-ADP-like H-bonds with Glu 343. However, there is an increase in the H-bond length between the 3'-OH group of the ribose and Glu 343. Due to the flexibility of the sugar-phosphate link, the L-ADP α-phosphate is placed in the same position as its D-ADP counterpart, despite differences in sugar orientation. Kinetic studies carried out with E343A mutant PGK showed that differences in catalytic efficiencies between the two analogues stemmed mainly from differences in Glu 343 H-bonding. The k_{cat} and K_m values for D-ADP and L-ADP are different with wild-type PGK and become equal in case of the E343A mutant. This means that different catalytic processing of D- and L-ADP is at least in part due to the differences in Glu 343-ribose hydrogen bond strength. It is notable that this interaction is not essential, but potent enhancer of PGK function [227]. In the case of the D-CDP and L-CDP X-ray structures [75], one can conclude that positioning of the ribose ring can be important for the positioning of the phosphate chain for the catalysis. This can explain the more effective phosphorylation of L-CDP by PGK over D-CDP [33] (Table 17) and this is in line with the existing ribose-Glu 343 interaction in case of L-CDP, which is missing for D-CDP [75].

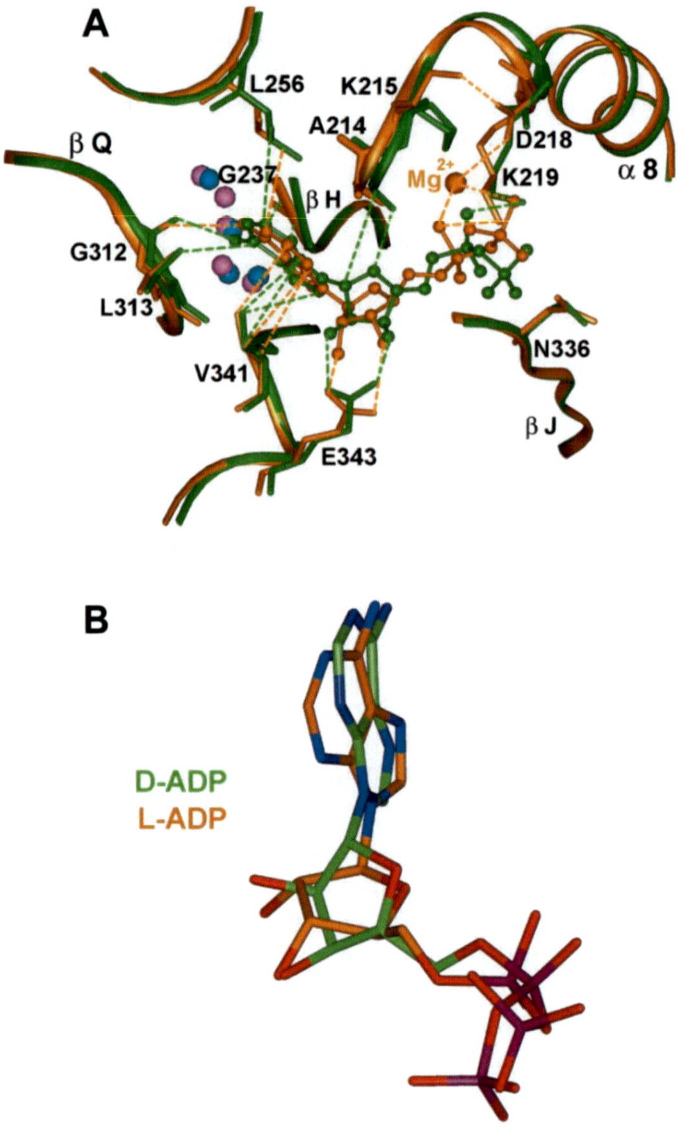

Figure 61. Comparison D- and L-ADP binding in the crystal structures of human PGK (made using the crystal structures [75]): (A) The D- and L-ADP binding structures of human PGK [75] are superimposed according to the inner β-strands of the nucleotide-binding C-domain. Surroundings of the bound nucleotides (ball and stick models) are shown by orange and green ribbons, respectively in cases of D- and L-ADP bound structure. The nucleotide-binding side-chains are represented by stick-models. The bound water molecules around the adenine ring are represented by purple and blue balls in the D- and L-ADP bound structures, respectively. (B) Enlargement of D- and L-ADP molecules (coloured according to their atoms) as they are bound in figure (A).

The broad nucleotide specificity of PGK was further examined by Gallois-Montbrun and her coworkers [31], but instead of phosphorylation they investigated the physiologically less important dephosphorylation process catalysed by PGK. Comparing the dephosphorylation of D- and L-purine analogues, L-analogues can be dephosphorylated almost as fast as D-analogues. Surprisingly, in case of D- and L-pyrimidine analogues PGK dephosphorylates

L-analogues faster than their D-counterparts. The anti-HIV drug, L-3TC (known L-SddC in another publication, which is the drug lamivudine used in anti-HIV treatment, see Figure 60) contains cytosine base without the hydroxyl group of ribose. Gallois [31] and Krishnan [28] tested the phosphorylation of the diphosphates of L-3TC by PGK and obtained the values 800 and 4000 $M^{-1}s^{-1}$, respectively for k_{cat}/K_m. For L-FMAUDP, which is a thymine base containing anti-HIV drug (cf. Figure 60), possessing the 2'-OH group of ribose, Krishnan detected a k_{cat}/K_m $1.3 \cdot 10^6$ $M^{-1}s^{-1}$ for the catalytic efficiency, which decreased to $4.8 \cdot 10^4$ $M^{-1}s^{-1}$ if they replaced Glu 343 with alanine. Thus, these data underline the importance of the interaction(s) of ribose OH group(s) with the side-chain of Glu 343 of PGK in case of pyrimidine base containing drug molecules.

In summary, PGK has an essential physiological function in the metabolism of all types of cells. In the last twenty years, however, it become evident that it has an important role in the activation of antiviral/anticancer drugs, which step enables for the nucleotides to reach their triphoshate forms and thus they can be built either in viral DNA or the DNA of cancer cells. These analogues then can block the DNA chain elongation and lead to virus/cell death. What we learnt investigating this activation step is that if analogues with D-configuration are designed, the purine-base analogues are more effectively phosphorylated than the pyrimidine ones by PGK. Here, one need to take into account that other kinases also participate in the process and for them the purine-nucleotides are also more favourable substrates, thus the total amount of active drug molecules is increased by this way. However, for the L-analogues, PGK is the only kinase that shows reasonable activity. In order to block DNA chain elongation, the analogues (incorporated into the viral DNA) has to miss their 3'-hydroxyl group. However, the absence of this OH-group has a more dramatic drop in the phosphorylation of the pyrimidine rather than the purine analogues by PGK, thus for the L-nucleotides we can conclude that L-purine analogues are activated more efficiently over L-pyrimidines.

Besides the role of PGK in the antiviral/anticancer drug activation it has contradictory roles in the survival of the cancer cells. As discussed above, on one side, PGK promotes tumour formation by facilitating anaerobic ATP production, on the other side, PGK even without having a secretion signal can be secreted by tumour cells and this can counteract with tumour angiogenesis. Unfortunately tumour cells learnt how to regulate PGK secretion and thus they can facilitate their angiogenesis process. It is a future task to uncover this hidden regulatory mechanism in order to be able to block angiogenesis of the tumour cells.

In: Phosphoglycerate Kinase: A Hinge-Bending Enzyme ISBN: 978-1-62808-836-6
Editor: Mária Vas

Chapter V

Kinetic Reaction Pathway for PGK

Corinne Lionne, Laurent Chaloin and Tom Barman
Research Centre on Pathogens and Biotechnologies for Health, CNRS,
University of Montpellier Sud de France, Montpellier, France

There are several reasons why a knowledge of the chemical reaction pathway by which an enzyme catalyses its reaction is important. First, it adds to the general knowledge on how enzymes work. With PGK, the kinetics of the putative hinge bending motion upon the binding of the substrates ADP and bPG is of particular interest. Second, the reactivity of the intermediates of certain enzyme pathways may be of key importance in biological processes. This is exemplified by the mechano-chemical coupling in muscle contraction, which depends on specific interactions of the different myosin ATPase intermediates with actin (for example, [352]). Finally, certain enzymes are potential drug targets for combating diseases. Compounds that are uncompetitive inhibitors (that is to say inhibitors that interact specifically with an enzyme intermediate) may be especially effective as drugs [353, 354]. The design of such drugs requires a detailed knowledge of the chemical properties of the intermediates that make up the reaction pathway (for example [355]). As discussed by several workers, PGK is a potential drug target (e.g., [28, 31, 356, 357]). The exploitation of PGK as a drug target is discussed in detail in Chapter IV. 4.1.

Here our intention is to explain how a reaction pathway for PGK was built and we do not give a full account of the numerous steady-state studies that have been carried out on this enzyme. For these, the reader is referred to the detailed studies of, in particular, [149, 358-361]. Briefly, PGK follows a two-substrate sequential mechanism with random binding of the substrates. However, the detailed kinetic pathway, including the nature and the rates of each elementary step, can only be derived by the aid of the transient kinetic experiments, described below. Here certain steady-state and equilibrium studies have been included but only to aid in the planning and interpretation of the transient kinetic experiments that were so essential in obtaining the pathway. The subject of the detailed transient kinetic studies was mainly the yeast enzyme (YPGK) in the reverse direction, but more recent studies have also been carried out with the recombinant human PGK (HsPGK) in the physiologically relevant forward direction. In the following, unless otherwise stated, "PGK" refers to yeast enzyme. The work on human PGK is at the end of the chapter.

5.1. Type of Mechanism: Dissociative or Associative?

As stated by Knowles [38, 362], the mechanistic pathways of phosphotransferases can be divided into two classes. In the double displacement or dissociative mechanism, the enzyme reacts with the first substrate alone to form a product that is released before the second substrate binds. In the single displacement or associative mechanism, transfer occurs directly between the substrates within a ternary complex.

There has been much discussion concerning the mechanism by which PGK catalyses phospho-transfer. It appears to be difficult to answer this question by X-ray crystallography (e.g., [84]). As judged by its steady-state kinetics at reasonable substrate concentrations, PGK binds its substrates in a random manner [363] and references cited therein [149]. On the other hand, in earlier reports it was suggested that PGK catalyses phospho-group transfer via a phospho-enzyme intermediate involving the phosphorylation of the γ-carboxyl group of a glutamyl residue [364, 365], which would imply a dissociative mechanism and "ping-pong" steady-state kinetics. However, Johnson et al. [363, 366] showed that the "phospho-enzyme" contains stoichiometric amounts of PG rather than a phosphorylated amino-acid residue.

More recent works support an associative mechanism for PGK, that is to say a mechanism that does not involve a phospho-enzyme intermediate. Structural studies show that there are separate sites for PG and ATP and that substrate binding is random (cf. Chapters I. 1.4 and I. 1.6.). Webb & Trentham [367] studied the stereochemistry of the phospho-group transfer by the use O-labelled ATPβS (adenosine 5'-O-[2-thiotriphosphate]) and NMR and showed that it occurs with inversion of the oxygen, a result which is consistent with direct transfer of the phospho-group. The results of transient kinetic studies are in agreement with a reaction scheme (Scheme 2) that involves direct phospho-transfer between the substrates within a ternary complex [368] (discussed further below). Finally, Bernstein & Hol [84] proposed that an important role of the active site of PGK is to stabilise negative charges on the transferring phospho-group, a role that is consistent with an associative mechanism [369].

As discussed by Schlichting & Reinstein [370], knowledge of the structure of the transition state should allow one to distinguish the reaction mechanism of an enzyme but the very transient nature of the state makes this difficult. With enzymes that involve phospho-transfer this problem can be approached by the use of compounds that mimic the transition state such as aluminium fluoride, beryllium fluoride and vanadate. Schlichting and Reinstein applied this method to UMP/CMP kinase which they showed to catalyze phospho-transfer by an associative mechanism. This important method has been also applied to HsPGK (cf. Chapter IX., X. and XI.).

5.2. Steady-State and Equilibrium Studies on Elementary Steps on the PGK Reaction Pathway

We consider here the reaction catalysed by yeast PGK in the reverse direction of bPG formation. To aid the discussion, we give a reaction pathway for PGK in the reverse direction in Scheme 2, as deduced from steady-state and equilibrium studies.

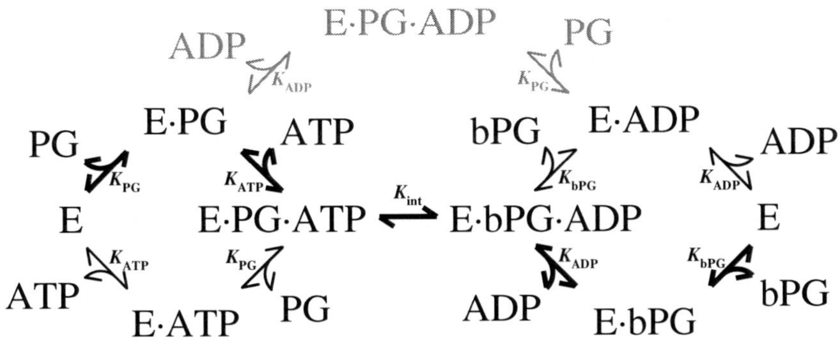

Scheme 2. Reaction pathway for PGK.

In the scheme, it is assumed that both substrate binding and product release are random and that phospho-transfer takes place in one step between the central ternary substrate and product complexes. *K* refers to the dissociation constants for the different ligands. The scheme also includes the abortive E·PG·ADP complex for which there is both structural [74, 75] and kinetic evidence [149, 350]. Estimates for the dissociation constants are given in Table 18.

Table 18. Dissociation Constants for Ligand Binding to Yeast PGK. Effect of 30% Methanol

ligand		complex formed (Scheme 2)	K_d (mM)		
			no cryosolvent		30% methanol
variable	constant		yeast[a]	pig muscle[b]	yeast[c]
ATP	none	E·ATP site 1	0.105	0.27	0.102
		site 2	4	–	–
	PG	E·PG·ATP	–	0.51	–
PG	none	E·PG	0.032	–	0.0015
ADP	none	E·ADP site 1	0.04	0.030	0.028
		site 2	1	–	–
	PG	E·PG·ADP	0.09	0.340	0.3
bPG	none	E·bPG	0.00005	0.00005	–

[a] method: gel filtration [371]; conditions: 20 mM triethanolamine pH 7.5, 1 mM free Mg^{2+}, 1°C [149].

[b] method: TNP-ATP [2'(3')-(2,4,6-trinitrophenyl)adenosine 5'-triphosphate] as a probe; conditions: 50 mM Tris pH 7, 20°C [372].

[c] method: fluorescence spectroscopy; conditions: 30% methanol, 20 mM triethanolamine pH 7.5, 0.1 M K-acetate, 1 mM free Mg^{2+}, 4°C [373].

5.3. The Internal Equilibrium Constant, K_{int} for the Interconversion of the Ternary Complexes PGK·PG·ATP and PGK·bPG·ADP

It has been proposed that systems that operate near equilibrium *in vivo* have K_{int} values that are near unity, even if the corresponding thermodynamic K_{eq} values (the ratio of free products to free substrates at catalytic concentration of the enzyme for the overall reaction) are very different [374], and references cited therein. An extreme example of this is myosin ATPase where K_{int}, [Myosin·ADP·P_i]/[Myosin·ATP], is about 9 whereas K_{eq}, $[ADP]_{free} \times [P_i]_{free} / [ATP]_{free}$, is 1.7×10^7 M [375].

Attempts to measure the value of K_{int} for PGK have been made by two methods. Both are model dependent in that they assume that there is only one each of the ternary complexes, as in Scheme 2. Further, to be sure that the concentrations of free substrates, products and binary complexes were low, the concentration of the PGK had to be high, in the several mM range.

Nageswara-Rao and co-workers [376] carried out a ^{31}P-NMR study of bound substrates (ATP and PG) and products with yeast PGK at equilibrium at pH 7 and 1°C. They obtained an estimate for K_{int} of 0.8 whereas their K_{eq} is about 3×10^{-4} (cf. Chapter IV. 4.1.). Further, they found that the rates of interconversion of substrates and products in the forward and reverse directions are fast and comparable at about 100 s^{-1}. This is in accord with Gutfreund and Trentham [375] who observed that with several enzyme reactions, steps involving chemical catalysis are fast (e.g., the K_{int} step in Scheme 2). Nageswara-Rao and co-workers [376] suggest that the shift in the products to substrates equilibrium in favour of bPG species at stoichiometric enzyme levels results from the very tight binding of bPG. It is noteworthy that with ATPβS as substrate, K_{eq} is an order of magnitude greater than with ATP: 2.5×10^{-3} [377]. Thus, with ATPβS as the donor substrate, time courses for bPG production should be easier to obtain than with ATP (see below).

Schmidt et al. [378] obtained estimates for K_{int} by the "equilibrium-quench" method of [374]. The procedure here is to choose for measurement a substrate – product pair (with PGK, ATP and bPG) and then to determine the dependence of the product to substrate ratio upon the enzyme concentration. The successful outcome of the method depends on having the conditions such that the substrates and products are all in the ternary complexes enzyme·ATP·PG and enzyme·ADP·bPG. Under the conditions used (pH 7.5 and 4°C), K_{int} was 0.11. This is lower than the 0.8 reported by Nageswara-Rao et al., a difference that could be explained by Schmidt et al. having included ADP in their reaction mixtures. Thus, it could be that Nageswara-Rao et al. included the binary enzyme·bPG in their measurements.

These estimates for K_{int} are subject to the caveat that should there be more than one each of the ternary complexes, the methods used merely measure the ratio [*all* PGK·bPG·ADP states]/[*all* PGK·PG·ATP states]. For a discussion of this problem with myosin ATPase, the reader is referred to Lionne et al. [379]. We note that from transient kinetic studies on PGK there is evidence for two each of the ternary complexes (see below). A further difficulty is that the two methods require high concentrations of the enzyme which may present solubility problems. This is also discussed below. Estimates for K_{eq} and K_{int} are given in Table 19.

Table 19. K_{eq} and K_{int} for Yeast PGK: Effect of Cryosolvent

values	Conditions	methods	references
K_{eq} = [ADP]×[bPG]/[ATP]×[PG] (×10^4)			
3 (± 1)	independent of pH	–	[35]
3	0.15 M H-Hepes pH 7.5, 1°C	^{31}P-NMR	[376]
≤8	50 mM 2-methylimidazole pH 7.4, 7.5°C	^{31}P-NMR	[14]
1.5 (± 1)	20 mM triethanolamine pH 7.5, 0.1 M K-acetate, 1 mM Mg^{2+}, 4°C		[378]
0.8 (± 0.2)	ditto + 40% ethylene glycol	[γ-^{32}P]ATP and	[378]
1.2 (± 0.1)	ditto + 30% methanol	[1-^{32}P]bPG measurements	[373]
K_{int} = [E·ADP·bPG]/[E·ATP·PG]			
0.8 (± 0.3)	0.15 M H-Hepes pH 7.5, 1°C	^{31}P-NMR	[376]
			[14]
1.0 (± 0.5)[a]	50 mM 2-methylimidazole pH 7.4, 7.5°C	^{31}P-NMR	[378]
0.11	20 mM triethanolamine pH 7.5, 0.1 M K-acetate, 1 mM Mg^{2+}, 4°C	equilibrium quench method	[378]
0.075	ditto + 40% ethylene glycol	of Burbaum & Knowles [374]	

[a] PGK from halibut muscle

Table 20. Effect of Cryosolvents on the Steady-State Parameters of PGK at 4°C

(adapted with permission from [373]; Copyright 1997 American Chemical Society):

Substrate		k_{cat} (s^{-1})[a]			K_m (μM)[a]			k_{cat}/K_m ($\mu M^{-1} \cdot s^{-1}$)		
variable	constant	water	ethylene glycol	Methanol	water	ethylene glycol	methanol	water	ethylene glycol	methanol
ATP	PG (5 mM)	72 ± 2	9.4 ± 0.7	7.6 ± 0.2	109 ± 11	110 ± 20	62 ± 6	0.66 ± 0.07	0.09 ± 0.02	0.12 ± 0.01
	PG (0.25 mM)	16 ± 1		1.05 ± 0.06	109 ± 18		19 ± 4	0.15 ± 0.03		0.06 ± 0.01
PG	ATP (5 mM)	79 ± 3[a]	8.1 ± 0.4		240 ± 40[a]	225 ± 40		0.33 ± 0.06	0.036 ± 0.007	
	ATP (2 mM)	31 ± 3		2.7 ± 0.1	118 ± 32		16 ± 2	0.26 ± 0.08		0.17 ± 0.02
	ATP (0.25 mM)	17 ± 1		0.71 ± 0.03	79 ± 10		5 ± 1	0.22 ± 0.03		0.14 ± 0.03

[a] By the coupled enzyme method. The concentrations of ethylene glycol and methanol were 40% and 30%, respectively.

5.4. Temperature Dependences of k_{cat}.

The temperature dependence of a rate constant (Arrhenius plot) can lead to mechanistic information when it is non-linear. Thus, when a dependence consists of two straight lines that are joined by a curve with the higher activation energy dominating at low temperatures, then as a first approximation one can conclude that there is a change in rate-limiting step as the temperature is decreased. For a discussion on the significance of "breaks" in Arrhenius plots, the reader is referred to Biosca et al. [380].

Dryden and co-workers [198] obtained the temperature dependence of the k_{cat} of yeast PGK in the reverse direction. At pH 5.6, the dependence was linear with activation energy 37 kJ·mol^{-1}. However, at pH 6.8, there was a break at 19°C (cf. Figure 32B) with activation energy 22 kJ·mol^{-1} above and 46 kJ·mol^{-1} below the break, but with the mutant [H388Q] PGK the dependence was linear with activation energy 34 kJ·mol^{-1}. Dryden et al. interpreted the break by a change in protein dynamics that is associated with a destabilisation of the waist region of PGK. Another explanation for the break is that there is a change in the rate-limiting step as the temperature is decreased. The dependences by Dryden et al. were obtained at high concentrations of the substrates: 15 mM PG and 5 mM ATP; which is under substrate activating conditions (for substrate activation cf. Chapter VI.). Schmidt et al. [378] showed that at lower substrate concentrations, when there was little substrate activation (2 mM PG and 1 mM ATP), the dependence was linear in the temperature range 4 to 30°C with $\Delta H^{\ddagger}$ = 30.5 kJ·mol^{-1} (Figure 62). In the same buffer but with 40% ethylene glycol or 30% methanol the dependence was also linear in the range –15 to 30°C, with the same slope; the significance of this is discussed below.

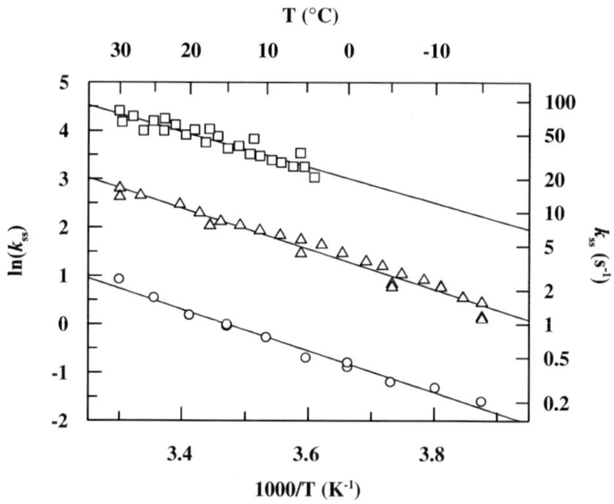

Figure 62. Arrhenius plots of the steady-state rates of PGK in the direction of bPG formation: Effects of cryosolvent: (□) without cryosolvent, (△) 40% ethylene glycol and (○) 30% methanol. The dependences were linear with $\Delta H^{\ddagger}$ = 30.5 ± 2.4 kJ·mol^{-1} in water, 35.0 ± 1.3 kJ·mol^{-1} in 40% ethylene glycol [378], and 36.3 ± 2.0 kJ·mol^{-1} in 30% methanol [373].

Thomas & Scopes [381] give Arrhenius plots for the k_{cat} of PGK from certain mesophilic and thermophilic bacteria. The plots are linear up to about 60°C at which temperature the enzymes were denatured.

Therefore, at low, non-activating substrate concentrations, because of their linearity, the Arrhenius plots of the steady-state rates of PGK give little mechanistic information.

5.5. Building a Reaction Pathway for PGK in the Reverse Reaction

Ideally, to build an enzyme reaction pathway one should identify chemically and have the concentration of each intermediate under a given set of conditions. To allow such measurements, one must work at high concentrations of the enzyme and because of the resulting rapidity of the kinetics, this requires transient kinetics equipment, e.g., Rapid Quench Flow (QF) and Stopped-Flow [382].

An initial strategy to build a reaction pathway is to obtain the time course of total product formation by QF. This is a chemical sampling method: enzyme and substrate are mixed, the reaction mixture allowed to age whilst samples a few msec to seconds old are quenched in, for example, acid and a product assayed [383]. It is assumed that any enzyme intermediate with bound product decomposes in the acid so the measurement reports *total* rather than just free product. If one obtains a rapid rise of product that is followed by the steady-state phase ("burst" experiment), an intermediate containing product accumulates in the steady-state because the product release kinetics are rate-limiting. If there is a transient "lag" phase, a step involved with substrate binding is rate-limiting and substrate containing intermediates accumulate. On the other hand, if there is no transient phase, then the chemical step (phospho-transfer step with kinases) is rate-limiting and there is little information. In this event one must in some way induce one's system to give a transient phase, for example by temperature or solvent perturbation.

The QF method was first applied to yeast PGK in the reverse direction and using [γ-^{32}P]ATP, because the product [1-^{32}P]bPG is convenient to measure. Thus, by quenching in a suitable acid, the bPG decomposes quantitatively to PG and [^{32}P]P_i which can then be measured [378]. This method is based upon the assumption that the PGK reaction pathway does not include an acid-labile phosphoenzyme intermediate (see above).

A typical time course for bPG formation at 4°C is illustrated in Figure 63 A: there is a transient burst phase of amplitude about 0.25 mol bPG/mol PGK that is followed by a steady-state phase of 25 s^{-1} until equilibrium is reached. The amplitude of the burst phase increased linearly with the PGK concentration, which is evidence that intermediates involving bPG accumulate in the steady-state. If one assumes that the steady-state phase represents free bPG formation and that there is an equivalent amount of free ADP, the K_{eq} thus calculated agrees reasonably well with that expected: about 2×10^{-4} compared with 1.5×10^{-4}. The *kinetics* of the burst phase however, were too fast to be measured with the apparatus used, even at 4°C, which confirms the conclusions of Nageswara-Rao et al. [376] that the formation of bPG on PGK (i.e., as E·bPG·ADP in Scheme 2) is rapid. Furthermore, the linear steady-state phase was over at about 25 msec and it was difficult to measure since it was squeezed between the transient phase and the final equilibrium. In any event, to confirm a transient phase, one must have its *kinetics*.

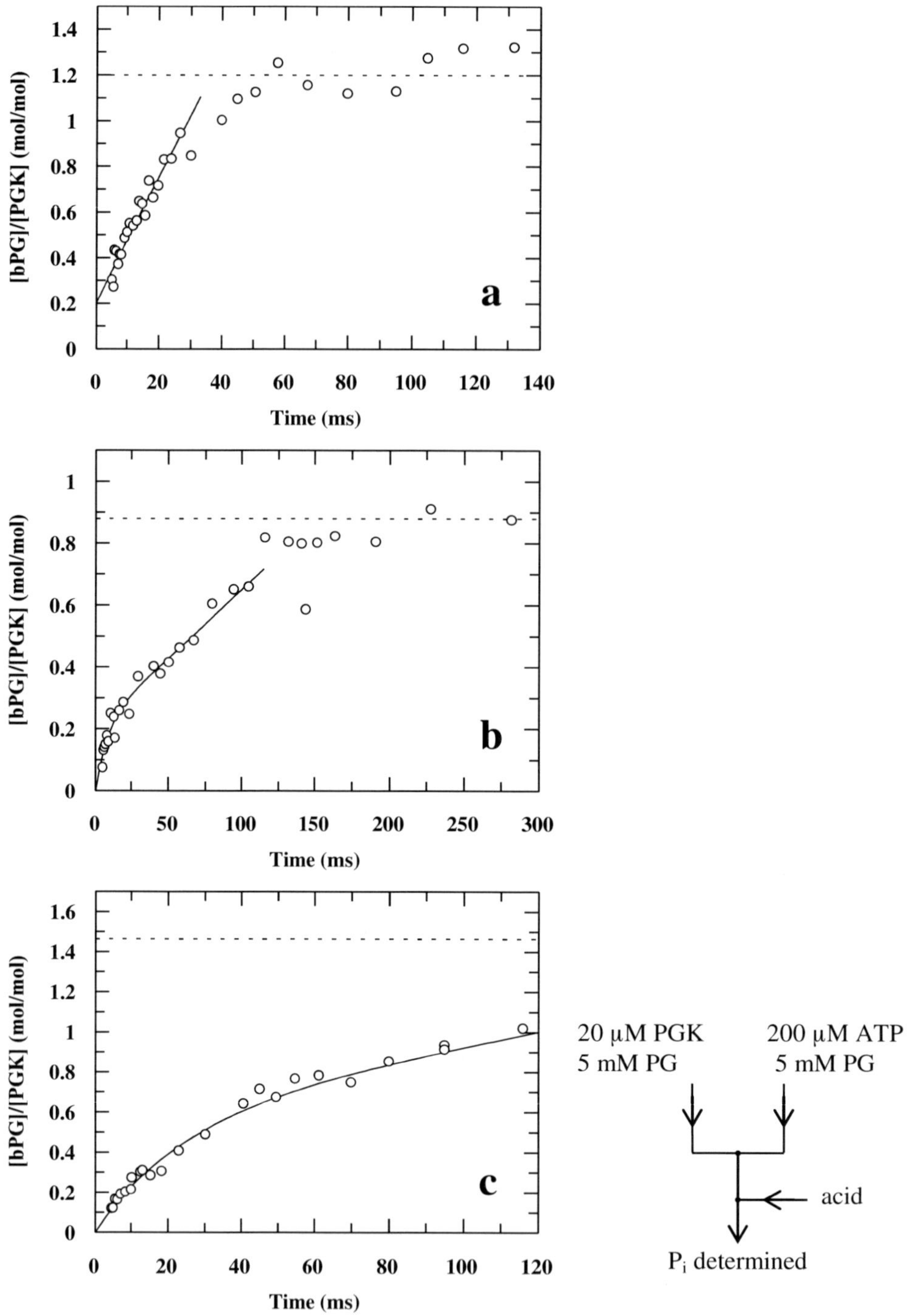

Figure 63. Effects of solvent composition on the transient kinetics of bPG formation by PGK at 4°C (adapted with permission from [373] and [378]; Copyright 1995 American Chemical Society): (A) water, (B) 40% ethylene glycol [378], (C) 30% methanol [373]. The concentrations in the reaction mixtures were ½ those in the flow diagram and they were the same in the three experiments. The buffer was 20 mM triethanolamine pH 7.5, 0.1 M potassium acetate and 1 mM free Mg^{2+}. The final equilibrium of the reaction mixture is indicated by the dotted line.

Perturbation of the PGK Reaction Pathway by Cryoenzymology

A powerful way to obtain information on an enzyme reaction pathway is to work under cryoenzymic conditions. This technique involves two perturbants: temperature and an antifreeze, usually an organic solvent. It may permit the accumulation of reaction intermediates that cannot be obtained under conventional conditions, by slowing down the kinetics of their formation, by a change in a rate-limiting step, or by shifts in equilibria [384].

40% Ethylene glycol has been used as a cryosolvent with ATP handling systems such as myosin ATPase (e.g., [385-387]) and creatine kinase [388]. A time course of bPG formation with the PGK system in 40% ethylene glycol is illustrated in Figure 63B. As with the experiment in water, the time course consists of a transient burst phase of bPG that is followed by the steady-state phase. The amplitude of the burst was similar to that in water but its kinetics was slowed down enough to allow an estimate: 120 s^{-1}. The steady-state rate was also slowed down to 4.5 s^{-1} [378].

The effects of ethylene glycol on the PGK system are interesting but not enough to allow a more detailed study: the kinetics of the transient burst phase remained fast and its amplitude low. To make these transient kinetic parameters more favourable for measurement, other cryosolvents were tried. Finally settled on 30% methanol [373]. A typical bPG progress curve is illustrated in Figure 63C: as can be seen, 30% methanol had the effect of increasing the burst amplitude to 0.56 mol bPG/mol PGK and decreasing its kinetics to 42 s^{-1}.

Before continuing the study with 30% methanol in the buffer, we checked on the overall effects of this solvent on the PGK system [373]. First, its effect on the stability of PGK was tested by the linked assay system (by the use of glyceraldehyde-3-phosphate dehydrogenase). At 22°C and a concentration of 0.4 μM PGK, there was little loss of activity after 5 hours incubation in 30% methanol. At 30°C, the enzyme lost 35% of its activity. However, PGK is less soluble in 30% methanol – up to 0.2 mM compared with several mM in the absence of methanol.

As can be seen in Table 19, neither 30% methanol nor 40% ethylene glycol had a significant effect on K_{eq}. 40% ethylene glycol had little effect on K_{int}. However, because experiments in methanol with PGK are limited to 0.2 mM, K_{int} could not be estimated in this solvent by the equilibrium-quench method, a method that requires mM concentrations of PGK. The effects of the two solvents on the steady-state parameters of PGK are summarized in Table 20. Both solvents reduced k_{cat} by about an order of magnitude, an effect that was also observed with other ATP handling systems ([373] and references cited therein). It is noteworthy that methanol reduced the K_m values for both ATP and, in particular, PG: thus in this solvent PGK can *be saturated with PG without anion activation* (cf. Chapter VI.). Finally, as seen in Figure 62, neither cryosolvent had any significant effect on the activation energy of the k_{cat} of PGK. The results of these checks allowed to continue the transient kinetic study on PGK in 30% methanol.

Burst Experiments at Low PG Concentrations

The transient burst experiments in Figure 63 had been carried out at a high concentration of PG (5 mM), i.e., at a concentration that activates strongly the overall PGK reaction (Table

20). Furthermore, at high PG concentrations the system is pushed towards free bPG production (as determined by K_{eq}, Table 19) so even with the unfavourable K_{eq} there was a steady-state phase, although of a very short duration, until the final equilibrium for the reaction was reached (Figure 63A).

The effect of lowering the PG concentration to 20 μM on the time course of total bPG formation at 4°C and in 30% methanol is illustrated in Figure 64. As can be seen, under this condition the bPG progress curve could be fitted to a single exponential – there was little sign of a steady-state phase. Thus, at low PG concentrations the PGK system barely turns over and the experiments are essentially single turnovers. It is noteworthy that neither the amplitude nor the kinetics of the transient is affected by the PG concentration. This supports the ideas of Scopes [361] and Szilágyi & Vas [129], that substrate activation is the result of the acceleration of a late step on the PGK reaction pathway (cf. Chapter VI.).

To facilitate the analysis of the bPG time courses and to avoid complications introduced by substrate activation, unless otherwise stated, the experiments below were carried out at low PG concentrations.

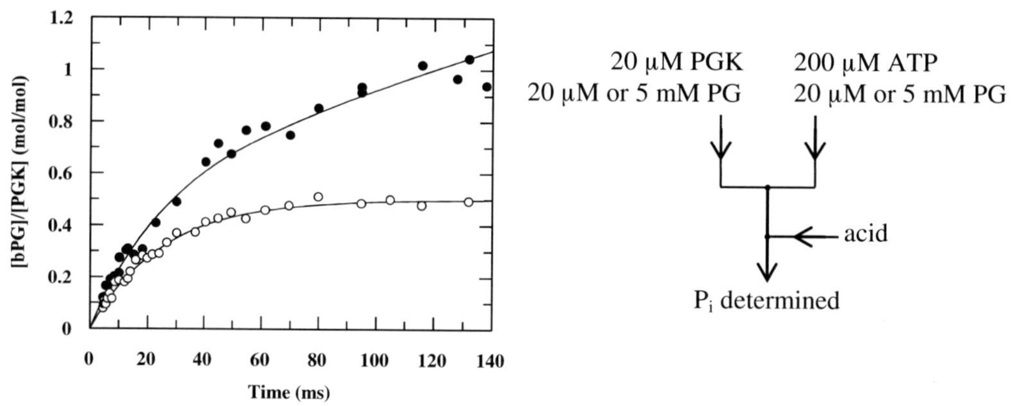

Figure 64. Effect of the PG concentration on the time course of bPG formation (adapted with permission from [373]; Copyright 1997 American Chemical Society): In the upper curve with 5 mM PG (•), the data were fitted to a transient burst phase of kinetics k_{obs} = 42 s^{-1} and amplitude B = 0.56 mol bPG/mol PGK, followed by a steady-state rate, k_{ss} = 3.7 s^{-1}. In the lower curve with 20 μM PG (o), the data were fitted to a single exponential of k_{obs} = 40 s^{-1} and plateau B = 0.50 mol bPG/mol PGK [373]. The buffer (legend to Figure 63) contained methanol (30%) and the temperature was 4°C.

Substrate Binding: Kinetics of Formation of the Binary E·ATP and E·PG Complexes (Scheme 2)

There is general agreement that PGK binds its substrates in a random manner (e.g., [361, 372]) and there is evidence that the binding is antagonistic that is to say, the affinity of one substrate is reduced by the presence of the other [110, 372] (cf. Chapter VII., also see section on human PGK, below). There seems to be only one site for PG but there are two for ATP, one strong the other weak [149, 376]. Each substrate on its own induces conformational changes (e.g., [36, 70] but both are needed for the full hinge bending motion (cf. Chapter II. 2.9.) As pointed out by the Vas and co-workers [7], the variation in the literature for the K_d

values for nucleotide binding may be explained by the presence of two sites (cf. also Chapter II. 2.1.). There is uncertainty as to the role of the weak site; it is probably not involved in the catalytic process (e.g., [7, 149]; and references cited therein). Table 18 summarises the dissociation constants for the interaction of certain ligands with PGK and the effect of 30% methanol on the constants. Data for pig PGK is included for comparative purposes.

There is little information on the *kinetics* of formation of the binary substrates complexes. Scopes [149] attempted to measure the kinetics of formation of the E·ATP by a stopped-flow method involving the competitive inhibitor cibacron blue, but they were too fast to be measured. Geerlof et al. [373] investigated the kinetics of E·ATP and E·PG formation by a fluorescence stopped-flow technique. These were unsuccessful, even at low reagent concentrations (10 μM ATP or 20 μM PG) and in 30% methanol at 4°C. Thus, on mixing PGK with either substrate, there was a sharp decrease in fluorescence that occurred completely within the dead time of the stopped-flow apparatus (2 msec).

The rapidity of the binding of PG to PGK, even in 30% methanol, is illustrated in Figure 65. As can be seen, the kinetics of the transient burst phases of bPG are identical whether or not the enzyme had been preincubated with PG.

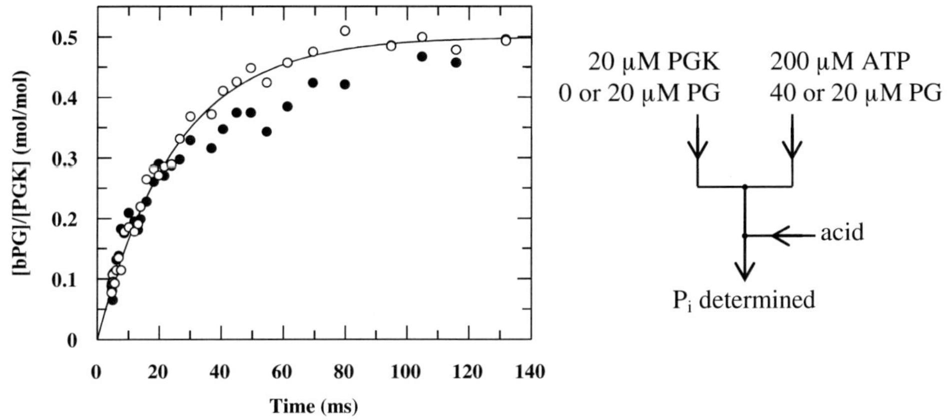

Figure 65. Time courses for bPG production (adapted with permission from [373] and [378]; Copyright 1995 American Chemical Society): The starting material was either PGK alone (o) or PGK·PG (•). The conditions used are in the legend to Figure 64.

Kinetics of the Binding of ATP to E·PG: Formation and Interconversion of the Ternary Complexes E·PG·ATP and E·bPG·ADP

We now return to the transient burst experiments as illustrated in Figures 63C, 64 and 65. Unless otherwise stated, in these and the following, the PGK had been preincubated with PG so the starting materials were E·PG and ATP. As a first approximation, a burst situation can be fitted to the simple Scheme 3 in which the "burst" is the manifestation of the accumulation of enzyme bound bPG:

$$\mathrm{E{\cdot}PG + ATP \rightleftharpoons E{\cdot}PG{\cdot}ATP \rightleftharpoons E{\cdot}bPG{\cdot}ADP \rightleftharpoons E + bPG + ADP}$$

Scheme 3. The kinetic "burst" can be attributed to the accumulation of enzyme bound bPG.

Consider first the ATP binding process: the question is whether this is merely diffusion controlled, implied in Scheme 3, or whether it also involves a protein isomerisation step (an induced-fit process [389] as in Scheme 4:

$$\mathrm{E^{\circ}{\cdot}PG + ATP \rightleftharpoons E^{\circ}{\cdot}PG{\cdot}ATP \rightleftharpoons E^{*}{\cdot}PG{\cdot}ATP}$$

Scheme 4. ATP binding includes a protein isomerization step.

where * indicates an ATP induced conformational change of PGK.

This question can be answered by determining the dependence of the kinetics of the burst phase, k_{obs}, upon the ATP concentration: if k_{obs} becomes independent of the ATP concentration at high ATP concentrations ("saturation kinetics"), then a two step binding process for ATP is indicated [382]. We assume that on the time scale of our experiments, the phospho-transfer step is a rapid equilibrium. The dependences of k_{obs} and also the amplitude of the burst phase, B, are illustrated in Figure 66.

The k_{obs} dependence fitted well to a straight line up to 200 μM with a slope of 0.37 $\mu M^{-1}{\cdot}s^{-1}$ and an intercept on the k_{obs} axis of 8 s^{-1}. The slope represents the second order binding constant for ATP; it is at least an order of magnitude too low to represent a diffusion controlled process (for instance, [382]) which suggests that ATP binds to E·PG in two steps. Unfortunately, because of the rapidity of the binding, this could not be confirmed by carrying out experiments at higher than 150 μM ATP, in order to attain saturation kinetics. Nevertheless, because of the low slope, it is reasonable to fit the dependence data to a two-step binding scheme:

$$\mathrm{E^{\circ}{\cdot}PG + ATP \overset{K_1}{\longleftrightarrow} E^{\circ}{\cdot}PG{\cdot}ATP \underset{k_{-2}}{\overset{k_2}{\rightleftharpoons}} E^{*}{\cdot}PG{\cdot}ATP \overset{K_3}{\longleftrightarrow} E^{*}{\cdot}bPG{\cdot}ADP \overset{k_4}{\longrightarrow} E^{\circ} + bPG + ADP}$$

Scheme 5. A four step kinetic pathway of bPG formation.

In this pathway, there are 4 types of step: formation of the collision complex E°·PG·ATP, a protein conformational change (defined by k_2 and k_{-2}), phospho-transfer (a rapid equilibrium, $K_3 = k_{-3}/k_3$; [376]) and the release of ADP and bPG in one step of kinetics k_4 (Scheme 5). Since there is a significant transient burst phase of bPG, the ternary products complex accumulates in the steady-state which means that k_4 is the rate-limiting step on the pathway.

The dependence of B on the ATP concentration fitted well to a hyperbola with a limiting B^{max} of 0.56 mol bPG/mol PGK and $ATP_{0.5}$ of 8 μM (Figure 66 B). Since B_{max} represents the concentration of an intermediate, the PGK titrated at least 0.56 equivalent active site.

To build a more complete scheme we now addressed the problem of the order of release of the products – are ADP and bPG released together as in Scheme 5 or are they released sequentially?

Kinetics and Order of Release of bPG and ADP

In solution, bPG is relatively unstable as it is hydrolyzed to PG and P_i with a half-life of 27 min at pH 7.2 and 38°C [390]. When bound to PGK, however, bPG is protected from hydrolysis "over a period of weeks" [14]. This stability can be explained by the binding of bPG being able to stabilize the closed conformation of PGK in which the active site is protected from the bulk solvent

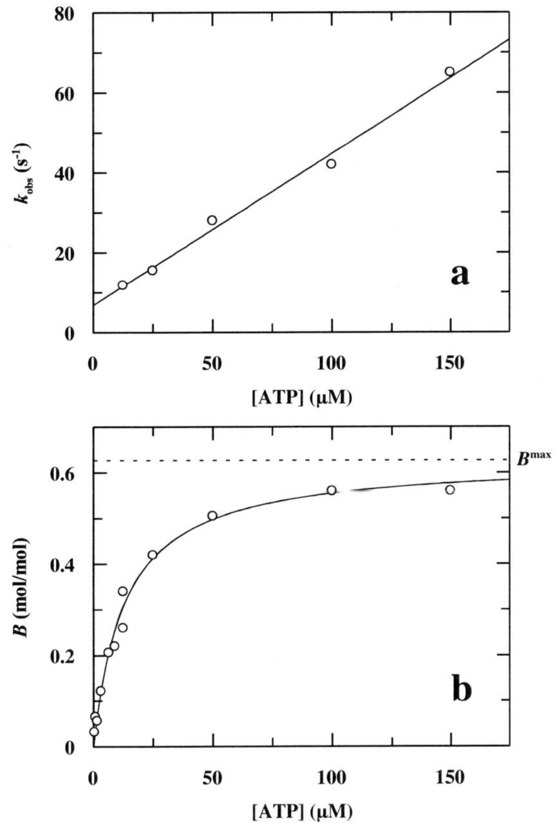

Figure 66. ATP dependences of (A) the kinetics, k_{obs}, and (B) the amplitude, B = [bPG]/[PGK], of the transient burst phases (adapted with permission from [373]; Copyright 1997 American Chemical Society): The time courses of bPG formation was measured at 20 μM PG (Figure 64). The parameters obtained are in Table 21.

There is little information on the details of the kinetics of the release of bPG from PGK in the reverse reaction. In 1989, Kvassman & Pettersson [391] wrote "no kinetic evidence has been suggested which would indicate that bPG is the last-leaving product in the PGK reaction". On the other hand, Scopes [149] proposed that because of its tight binding (K_d about 50 nM), bPG is released after ADP (K_d = 40 μM), that is to say, that the products dissociation is ordered. Furthermore, Scopes argued that bPG dissociates from PGK in two steps, an argument that arises from a contradiction: how is it that high concentration of PG activate PGK when there is only one binding site for both bPG and PG? Scopes proposed that bPG binds first by either one of its phosphate groups (1- or 3-) to form two types of E·bPG complex in which the bPG is bound relatively loosely. This is presumably followed by a conformational change by which the second phosphate (3- or 1-) also binds which results in the tight binding of the bPG. This tight binding of bPG may induce the closed conformation of PGK. Scopes terms the initial type of bPG complex "half dissociated forms". To continue his argument: "in these forms, one can imagine that PG (or other anion) inserts itself into the vacant half-site on the enzyme, *so preventing the formation of the final tight E·bPG complex*". In this way, the kinetics of the bPG release would be accelerated as it would be from a "loose" rather than a "tight" type of bPG complex.

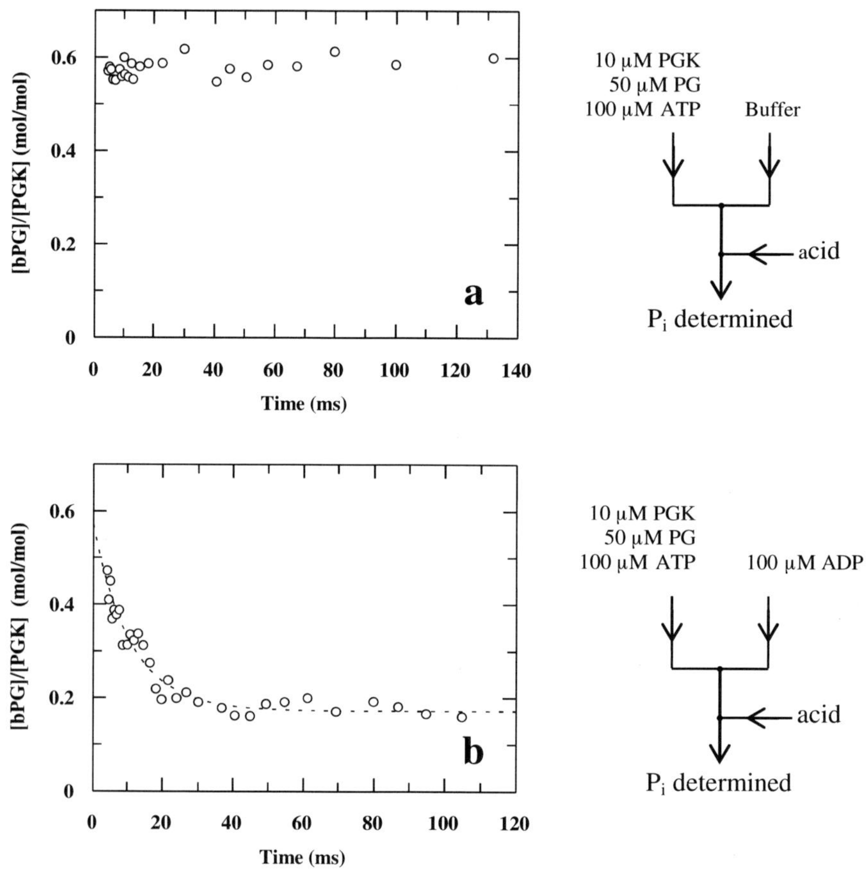

Figure 67. Perturbation of PGK reaction mixtures by dilution (A) or by 50 μM ADP (B): (adapted with permission from [368]; Copyright 2005 American Chemical Society): The reaction mixtures (initially 10 μM PGK, 50 μM PG, 100 μM ATP and 25 μM Ap5A), were mixed with buffer (a) or 100 μM ADP (b) in the rapid quench flow apparatus and the final mixtures aged for the time shown. Ap5A (cf. Abbreviations) was included to inhibit any myokinase activity in the PGK preparation. In (b), the data were fitted to an exponential decay of kinetics $k_{per} = 92\ s^{-1}$ and low plateau $B_{per} = 0.17$ mol bPG/mol PGK. The conditions used are in legend to Figure 66.

The proposition of Scopes is supported by Vas and co-workers [7] who recently hypothesized that with PGK in its open conformation, the two negatively charged phosphates of bPG span the 12–13 Å gap between Lys 215 in the C-domain and Arg 65 in the N-domain. These initial interactions would facilitate the movement of the side-chain of the lysine to its right place in the closed conformation of the enzyme. The proposition is also in agreement with the finding that the early steps on the PGK reaction pathway are unaffected by PG activation (Figure 62). It also implies that *bPG release is the rate-limiting step on the PGK reaction pathway.* Finally, it is supported further by the results of the following perturbation experiments and by more recent transient kinetic studies of the forward reaction with HsPGK [126] (see also below).

Perturbation of PGK Reaction Mixtures: Formation of ATP from a Binary E·bPG Complex and ADP

Since bPG is relatively stable when bound to PGK, it is possible to carry out experiments on PGK reaction mixtures at low PG concentrations, that is to say, when there is little bPG released. For example, when 200 μM [γ-^{32}P]ATP + 20 μM PG is mixed with an equal volume of a solution containing 20 μM PGK + 20 μM PG, within less than 100 msec the concentration of [1-^{32}P]bPG in the reaction mixture is 5.5 μM (for example, Figure 65). We assume that the bPG is enzyme-bound (0.55 mol bPG/mol PGK) because the PG concentration was low. Two experiments were carried out with this mixture, both in the QF apparatus.

First, the equilibrium mixture was diluted with an equal volume of buffer and the diluted mixture was aged for different times, the reaction quenched in acid, and the [1-^{32}P]bPG measured. As shown in Figure 67A, there was little change in the bPG concentration in the time range 4-140 msec. Thus, the equilibrium of enzyme reaction remained unaffected and little bPG is released from the active site.

In the second experiment, the reaction mixture was perturbed by the addition of ADP and the change in the [1-^{32}P]bPG concentration followed with time. An experiment at 50 μM ADP is shown in Figure 67B: following the addition of ADP, the concentration of bPG decreased rapidly to a low plateau that represents the new equilibrium situation. We interpret this experiment as follows. The ADP binds to a binary E·[1-^{32}P]bPG complex in the original reaction mixture to give the ternary E·[1-^{32}P]bPG·ADP which is then converted to E·PG·[γ-^{32}P]ATP and finally free [γ-^{32}P]ATP, neither of which is measured, as ATP is stable in the acid quench. Thus it appears that, under the conditions used, E·bPG is an important intermediate on the PGK reaction pathway.

The experimental points in Figure 67B fitted well to a single exponential of kinetics $k_{per} = 92\ s^{-1}$ and a final plateau of 0.17 mol [1-^{32}P]bPG/mol PGK that is a manifestation of the new equilibrium, B_{per}.

The dependences of k_{per} and B_{per} on the ADP concentration are shown in Figure 68. These experiments were limited to a narrow ADP concentration range: at low ADP concentrations the amplitude of the transient was low and a k_{per} difficult to obtain whereas at high ADP k_{per} became too fast for our equipment.

k_{per} appears to increase linearly with the ADP concentration up to 50 μM; a linearity that was supported by an experiment at 150 μM ADP. Thus, at this concentration, k_{per} was about 240 s^{-1} (first sampling at 4.5 ms at about 80% completion of the transient). With the data up to 50 μM ADP, the slope of the dependence is 1.3 $\mu M^{-1}\cdot s^{-1}$ and the intercept on the k_{per} axis, 44 s^{-1}.

Because of the limited concentration range and the rapidity of the kinetics, these estimates must be regarded as very approximate. Nevertheless, the slope appears too low to represent a diffusion controlled process [382] which suggests that ADP binds to an E·bPG complex in two steps.

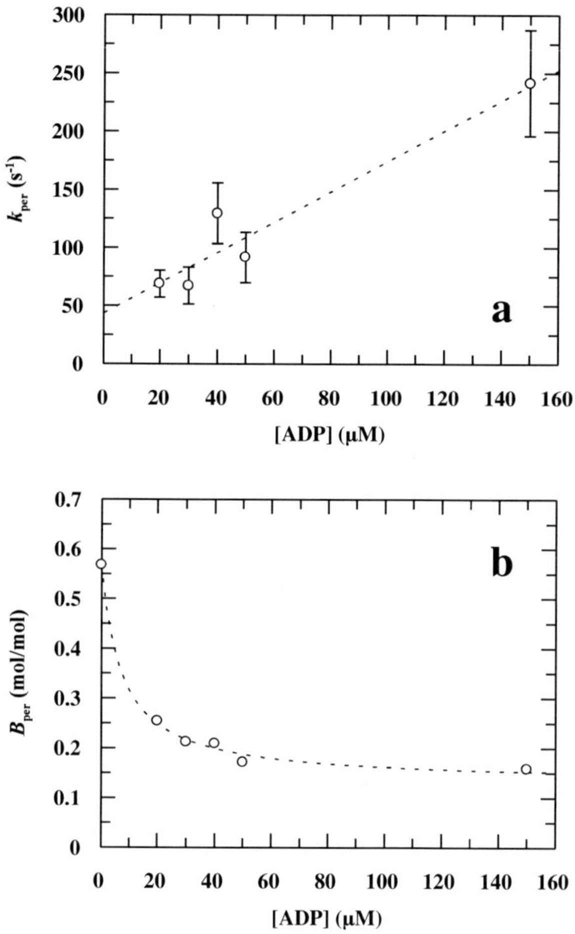

Figure 68. ADP dependences of the kinetics, k_{per} (A), and low plateau, B_{per} (B), in ADP perturbation experiments (adapted with permission from [368]; Copyright 2005 American Chemical Society): The parameters obtained are in Table 22.

Table 21. Estimates of the Transient Kinetic Parameters Obtained in bPG Burst and ADP Perturbation Experiments (Scheme 5)

bPG bursts: ATP dependence of k_{obs} (Figure 66)		
slope	$= k_2 / K_1$	$= 0.37\ \mu M^{-1} \cdot s^{-1}$
intercept	$= k_{-2} \times K_3 / (1+K_3)$	$= 8\ s^{-1}$
ADP perturbations: ADP dependence of k_{per} (Figure 68)		
slope	$= k_{-4} / K_5$	$= 1.3\ \mu M^{-1} \cdot s^{-1}$
intercept	$= k_4 \times (1+K_3) / K_3$	$= 44\ s^{-1}$

To conclude, the results of these ADP perturbation experiments suggest strongly that a binary E·bPG complex is an important intermediate on the PGK reaction pathway because the bPG release process is rate-limiting. Accordingly, we extend Scheme 5 to include ADP release:

$$E^{\circ}\cdot PG \xrightleftharpoons[ATP]{K_1} E^{\circ}\cdot PG\cdot ATP \xrightleftharpoons[k_{-2}]{k_2} E^{\bullet\bullet}\cdot PG\cdot ATP \xrightleftharpoons{K_3} E^{\bullet\bullet}\cdot bPG\cdot ADP \xrightleftharpoons[k_{-4}]{k_4} E^{\bullet}\cdot bPG\cdot ADP \xrightleftharpoons[ADP]{K_5} E^{\bullet}\cdot bPG \xrightleftharpoons[k_{-6}]{k_6} E\cdot bPG \xrightleftharpoons[bPG]{K_7} E$$

Scheme 6. Complete kinetic pathway of the PGK reaction including ADP release.

where E and E° are open as well as $E^{\bullet}$ and $E^{\bullet\bullet}$ are closed PGK conformations, respectively.

The scheme includes bPG release, a process that occurs in at least two steps ([149], see above). As the experiments were carried out at low substrate concentrations, there was little release of bPG; the steps are included merely to complete the scheme. Release of bPG is not included in the following arguments.

In Scheme 6 there are various open (E and E°) and closed ($E^{\bullet}$ and $E^{\bullet\bullet}$) enzyme conformations (see Chapter I. 1.8. of this book and [116]). There are three types of step: formation of collision complexes (rapid equilibria defined by K_1 and K_5), phospho-transfer (a rapid equilibrium, K_3; [376]) and protein conformational changes (k_2, k_{-2}, k_4 and k_{-4},). It is assumed that on the time scale of the experiments, the kinetics are manifestations of the conformational changes, modulated by the rapid equilibria: k_2 and k_{-2} in the bPG bursts and k_4 and k_{-4} in the ADP perturbations.

In Scheme 6, the bPG measurements represent the sum of three intermediates: $E^{\bullet\bullet}$·bPG·ADP, $E^{\bullet}$·bPG·ADP and $E^{\bullet}$·bPG so as mentioned above they do not give directly K_{int} (K_3 in the scheme).

Estimates of the Kinetic Constants in the Proposed Reaction Scheme for PGK

The difficulty in attaching values to the individual rate constants of a multi-step enzyme reaction pathway is well known (for example, [382, 392]). Here, as a first step to obtain such estimates with PGK, the transient kinetic parameters obtained in the bPG burst and ADP perturbation experiments were expressed in terms of the constants of Scheme 6 and these approximations are given in Table 22. The expressions are approximations only because each constant is more or less modulated by all of the other constants in the scheme, depending upon their relative values. However, their use was important in enabling the exploitation of a global fitting procedure to obtain plausible estimates for the constants.

Briefly, the global fitting procedure involved fitting the experimental data obtained at different ATP (in the direction of bPG formation, for example Figure 64) or ADP concentrations (ATP formation, Figure 67B) by the use of Scheme 6 (excluding the bPG release steps) and the program Scientist. The following constraints were made: 1) on the time scale of the experiments, steps 1, 3 and 5 are rapid equilibria; 2) all the constants are positive; 3) the active site titration, equivalent active site per mole PGK, is < 1 but ≥ 0.56 (B_{max} in Figure 66 B and Figure 65) the approximate values of the second order constants k_2/K_1 and

k_{-4}/K_5 in Table 22. For full details of the global fitting procedure used, the reader is referred to [368].

The estimates for the constants in Scheme 6 are given in Table 22. The errors were given by the computer analysis and they do not include inadvertently introduced experimental errors. The problem of errors has been discussed [393].

Table 22. Estimates of the Kinetic Constants Describing Scheme 6 (bPG release excluded)[a]

constant[b]	estimate from experimental data		estimate from global fit
	how obtained	value	
From kinetics of bPG formation (Figures 64 and 65)			
K_1 (mM)	linearity of k_{obs} dependence	>> 0.2	2.8 ± 0.5
k_2 (s^{-1})	linearity of k_{obs} dependence	>> 80	1000 ± 200
k_2/K_1 (μM^{-1} s^{-1})	slope of k_{obs} dependence	0.36 ± 0.02	0.36[c]
k_{-2} (s^{-1})	intercept of k_{obs} dependence	28 ± 7	28 ± 6
K_2	k_{-2}/k_2	<< 0.35	0.028 ± 0.012
From kinetics of ATP formation (Figures 67 and 68)			
K_5 (mM)	linearity of k_{per} dependence	>> 0.15	7.5 ± 7.5
k_{-4} (s^{-1})	linearity of k_{per} dependence	>> 250	1 ± 0.9 × 10^4
k_{-4}/K_5 (μM^{-1} s^{-1})	slope of k_{per} dependence	1.3 ± 1.2	1.3[c]
k_4 (s^{-1})	intercept of k_{per} dependence	13 ± 11	25 ± 5
K_4	k_4/k_{-4}	<< 5.2 × 10^{-2}	2.5 ± 2.5 × 10^{-3}
Constant not obtained experimentally			
K_3	–	–	0.4 ± 0.1[d]

[a] In 20 mM triethanolamine, 0.1 M potassium acetate, 1 mM free Mg^{2+}, 30% methanol, pH 7.5 at 4°C
[b] K_1, K_2 and K_3 are $K_i = k_{-i}/k_{+i}$; K_4 and K_5 are $K_i = k_{+i}/k_{-i}$
[c] Imposed
[d] In [373] the value is 0.25

It is noteworthy that the experimental and global fit estimates agree well, especially k_{-2} and k_4. This is surprising because in the experimental estimates, steps far removed from those observed were not considered whereas in the global fits *all* of the steps were taken into account. The agreement could be due to the imposition of the second order constants for ATP and ADP binding (Table 21) in the global fits. It could also be due to the model used (Scheme 6) in which steps involving relatively slow, and therefore observable kinetics, are separated by rapid equilibria. Further, it cannot be excluded that the data can be fitted to other sets of constants in addition to that in Table 22. Finally, even if the progress curves, whether describing bPG formation or disappearance, fitted well to single exponentials, it cannot be excluded that each curve consists of several exponentials with similar rate constants and, therefore, that there are intermediates in addition to those in Scheme 6.

A Seven-Step Pathway for PGK: Structural Implications

It is generally agreed that substrates bind to enzymes in two steps – initial formation of a collision complex which is followed by a conformational change – and that product release is the reverse of this [382]. There are three such processes in Scheme 6: ATP binding (steps 1 and 2), ADP release (steps 4 and 5) and finally, bPG release.

Consider first the binding of PG and ATP. PG alone induces a conformational change in PGK that leads to the E° conformation (e.g., [199, 200]). This is illustrated by the perturbation of tryptophan upon the binding [128, 194, 302, 373]. This structural change of PGK is not sufficient for the hinge binding motion, a motion that involves the closure of the cleft between the N- and C- domains. As stated by LaDine & Cross ([216], and references cited therein), this closure serves to exclude water from the active site which provides a low dielectric environment in which nucleophilic attack of the γ-phosphate of ATP by the phospho-group accepting substrate is facilitated and futile hydrolysis of ATP prevented. For the closure to occur *both* PG and ATP are needed, as shown by SAXS-measurements (cf. Chapter II. 2.9.). Thus, when ATP binds to E°·PG, the full conformational change takes place which leads to the ternary $E^{\bullet\bullet}$·PG·ATP *via* the collision complex E°·PG·ATP. *We propose that k_2 and K_1 are a reflection of the hinge-bending motion in PGK.*

Consider next the ADP release. We propose that the $E^{\bullet\bullet}$·bPG·ADP complex undergoes a small conformational change to $E^{\bullet}$·bPG·ADP in which the enzyme remains in its mostly closed conformation; with the bPG remaining bound by its two phosphate groups to both the domains of PGK. This is a key feature of the scheme – we propose that it allows the ADP to escape (or to enter in the direction of ATP formation in the ADP perturbation experiments) *from a bPG-bound closed conformation of PGK* (cf. also *Kinetics and Order of Release of bPG and ADP*).

Finally, there is the bPG release. The transient kinetics of this process has not yet been studied; as proposed by Scopes [149] (discussed above) it almost certainly takes place in at least two steps (steps 6 and 7 in Scheme 6). From the results of the transient kinetic experiments outlined above, it appears that a binary E·bPG complex accumulates in the steady state. *This implies that the overall k_{cat} is limited by the bPG release kinetics* as already suggested by Scopes [149]. The elementary step that limits k_{cat} is probably k_6.

5.6. Transient Kinetics of Human PGK in the Forward Reaction: Implications

Because of its possible involvement in the activation of pro-drugs (see Chapter IV. 4.2.), it seemed important to study the reaction pathway of HsPGK in the direction of ATP formation. Thus, we investigated its transient kinetics with bPG and ADP as substrates by the rapid quench flow and stopped-flow methods. We summarize the salient points of the results of these experiments. First, we show that an intermediate that contains ATP accumulate in the steady state. Therefore, a product release step is rate limiting [126].

Second, HsPGK is subject to substrate antagonism in the forward reaction as determined by fluorescence stopped-flow [351]. As with YPGK, the tryptophan fluorescence of HsPGK is perturbed by the addition of substrates. Typical time courses are illustrated in Figure 69.

From the dependences of ΔF upon bPG concentrations, K_d values for bPG in the presence or not of ADP were determined. The association rate constant values, k_{on}, were estimated from the slope of the dependences of k_{per} upon bPG concentrations. The dissociation rate constants correspond to the intercept on the Y axis. The values estimated for the kinetics of bPG interaction with HsPGK are in Table 23. The effects of bPG or PG on ADP interaction kinetics with HsPGK were also studied and are summarized in Table 23.

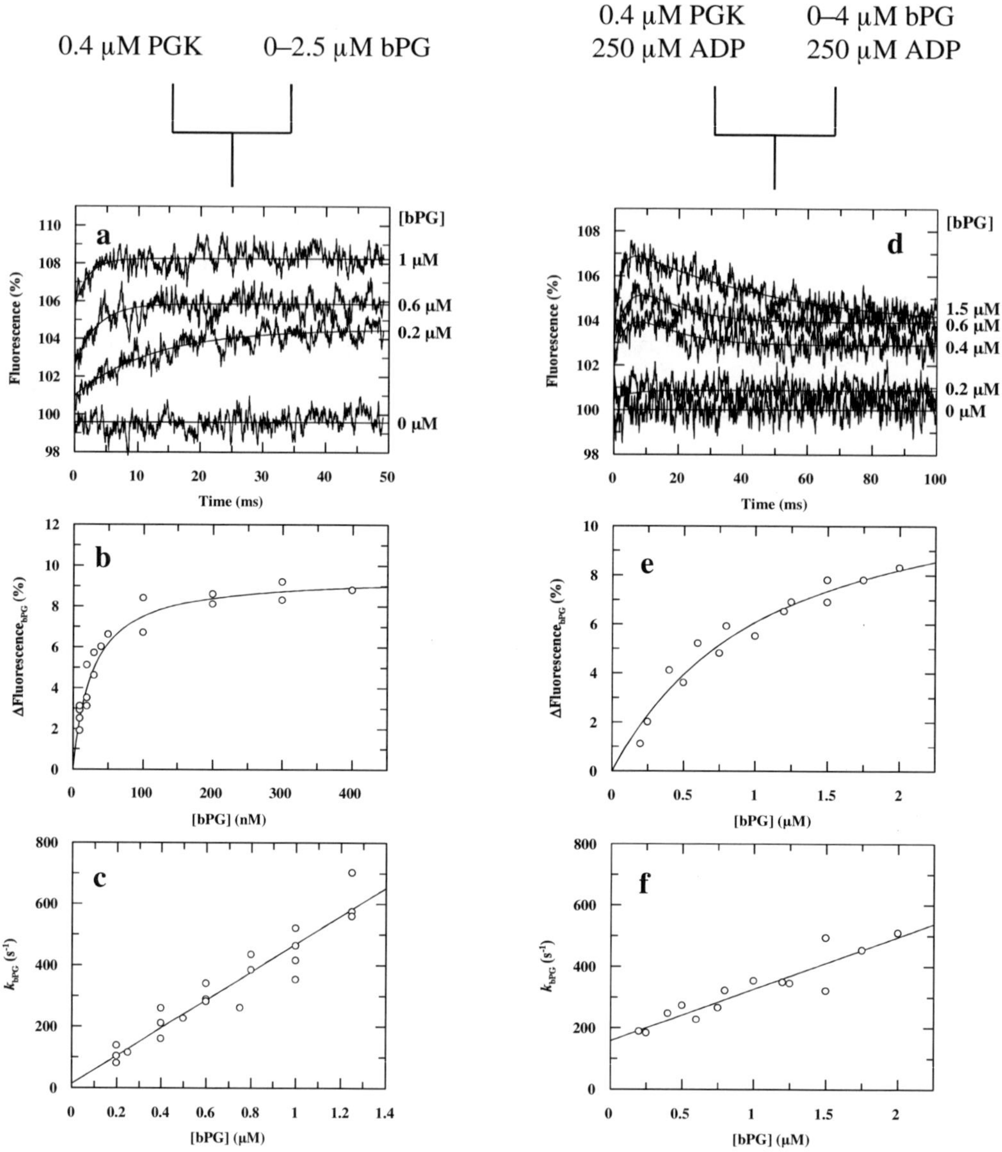

Figure 69. PGK Trp fluorescence changes induced by the binding of bPG in the presence or not of ADP (adapted from [351]): The parameters obtained, amplitude (Δ Fluorescence$_{bPG}$) and the rate constant (k_{bPG}) of the fluorescence change, and plotted against the bPG concentrations.

Table 23. Fluorescence Stopped-Flow Experiments: Comparison of the Transient Kinetics of the Interactions of Substrates with PGK (adapted from [351])

substrate						
variable	constant	k_{on} (μM^{-1} s^{-1})	k_{off} (s^{-1})	k_{off}/k_{on} (μM)	K_d (μM)	with ref. to Scheme 7
bPG	E	450 ± 40	14 ± 26	0.03 ± 0.06	0.03 ± 0.01	K_{bPG1}
bPG	E·ADP	170 ± 30	160 ± 30	0.94 ± 0.24	0.92 ± 0.23	K_{bPG2}
PG	E				0.7 ± 0.2	K_{PG1}
PG	E·ADP				6.5 ± 0.7	K_{PG3}
ADP	E	6.1 ± 0.3	38 ± 10	6.2 ± 1.7	7.7 ± 0.6	K_{ADP1}
ADP	E·bPG	4.1 ± 0.5	270 ± 30	66 ± 11	100 ± 20	K_{ADP2}
ADP	E·PG	4.6 ± 0.6	313 ± 40	68 ± 12	50 ± 3	K_{ADP3}

The data from these experiments were interpreted by scheme 7 in which open conformations of HsPGK are represented by ° and closed conformations by *. There are three ternary complexes: the catalytic intermediates E*·bPG·ADP and E*·PG·ATP, and the abortive dead end complex E*·PG·ADP.

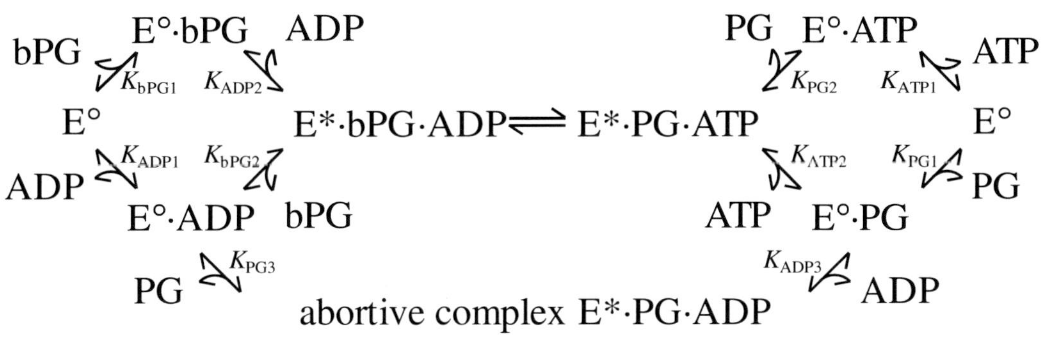

Scheme 7. Complete kinetic pathway for PGK.

It is noteworthy that the K_d value determined for one ligand in the presence of the other is always larger than that obtained in its absence. In other words, the presence of a ligand in one domain of PGK weakens the affinity of the protein for the second ligand in the other domain. As proposed for another kinase, 6-phosphofructokinase [394], there are several possible causes for substrate antagonism between two similarly charged substrates: electrostatic repulsion between the substrates, steric hindrance or a conformational change of the kinase. Due to the small effect observed on k_{on} (because PGK is still open) and the large effect on k_{off} (when the two substrates are approximated), it could be that the substrate antagonism observed with HsPGK is due to a mutual repulsion of the charged substrates when the PGK closes up (cf. also discussion in Chapter VII.).

Also included in Table 23 are estimates for K_{ADP1} and K_{ADP3} and for K_{PG1} and K_{PG3}. These confirm that under our experimental conditions, substrate antagonism also occurs in the back reaction with HsPGK [350]. It is noteworthy that the kinetics of formation and the structures of the abortive E*·PG·ADP and catalytic complex E*·PG·ATP are similar [351].

Finally, steady state and equilibrium studies have lead to the suggestion that the highly conserved Lys 215 is catalytically important (cf. Chapter II. 2.7.). We have confirmed this suggestion by a detailed transient kinetic study on the K215A mutant of HsPGK, showing that this lysine is indeed involved in the chemical phospho-transfer step itself [126].

In: Phosphoglycerate Kinase: A Hinge-Bending Enzyme
Editor: Mária Vas
ISBN: 978-1-62808-836-6

Chapter VI

Anomalous Kinetic Behaviour: Activation by Anions and Excess Substrates

Mária Vas
Institute of Enzymology, Research Centre for Natural Sciences,
Hungarian Academy of Sciences, Budapest, Hungary

PGK exhibits a rather interesting regulatory behaviour: *activation* by *low* concentrations of multivalent anions, including the non-substrate Mg^{2+}-free ATP and *inhibition* at their *high* concentrations [361, 395, 396]. In addition, *high* concentration of substrates, 3-PG or MgATP, also causes *activation* [359, 361, 397]. These properties were observed long before structural information became available, but such information itself could not lead to understanding of the mechanism of activation at molecular level.

While *inhibition* by the non-substrate anions can reasonably be attributed to direct competition with the anionic substrates, to explain *activation*, the existence of a separate anion binding site, outside the catalytic centre, has been assumed [395]. The location of this regulatory site has been suggested either close to the terminal phosphate of the bound substrate, MgATP [149] or in the 'basic patch' region on the inner surface of the N-terminal domain [36, 136, 140, 148]. Scopes [149] assumed that occupancy of this site by substrates or anions activated the enzyme by accelerating the release of the product, 1,3-bisphosphoglycerate. Release of this tightly bound product from the enzyme active site has indeed been shown to be rate-limiting by transient kinetic studies (cf. Chapter V. 5.6.).

Confirmation of the nature of the rate limiting step, however, has not been accompanied by evidence on the existence and the location of a separate regulatory site. Site directed mutagenesis apparently underlined the importance of the basic patch residues [156, 223], but has not led to definitive conclusions about the activating site [132, 224, 225]. As for the existence of any possible secondary substrate site, ligand binding studies could only detect a single site for PG [361], whereas for the nucleotide substrates the data are equivocal. Numerous publications claimed the existence of a secondary nucleotide site, although sometimes without presenting completely satisfactory binding curves [139, 140, 142, 149, 215, 376, 398], while others report about only a single nucleotide site [130, 154, 399-402]. In any case, high resolution X-ray structural data cannot yet give evidence on the location of a separate regulatory site for substrates or activators, since the data have only indicated a single site for each substrate, e.g., [6, 7, 15, 72-74].

Thus, even two decades after the original discovery of the activation phenomenon of PGK by anions and excess of substrates, neither the nature nor the spatial location of a possible activating site is established. At that time, however, Szilágyi and Vas [129] have conducted detailed kinetic studies on pig muscle PGK using systematically varied analogues of PG of different sizes, bearing one, two or three negatively charged groups. Based on these studies a kinetic model was proposed which indeed described quantitatively all the data (cf. the curves with wild type PGK on Figure 70). Accordingly, a common secondary site for regulation (activation) of all effectors, separate from the primary catalytic site of PG, and a common site for inhibition, identical with the catalytic site of PG, do exist, in spite of the fact that crystallographic and solution binding data apparently supported the existence of only a single site for each of the two substrates. In order to resolve this discrepancy a model was derived by an assumption that the regulatory anionic site is formed only temporarily during domain closure, i.e., in the coarse of the catalytic cycle. As outlined below, this hypothesis, indeed, satisfactorily explains activation by either the excess substrates or multivalent physiological anions [129].

Quantitative Aspects of Activation and Inhibition by Multivalent Anions

The effects of different analogues (for structural formulas see Table 24) of the substrate, PG, on the activity of PGK were investigated by Szilágyi and Vas [129] at relatively low, constant concentrations of the substrates PG and MgATP (Figure 70). All of the analogues, namely 2-phosphoglycolate (Figure 70A), 2,3-bisphosphoglycerate (Figure 70B), glycerol 2-phosphate (Figure 70C) and pentane-1,5-bisphosphonate (Figure 70D), exhibit activation at low concentrations and inhibition at high concentrations. As a comparison, activation and inhibition by the metal-free ATP (Figure 70G), as well as by simple anions, such as phosphate (Figure 70F) and sulphate (Figure 70E), were also determined under identical conditions. The extent of the observed activation varies in the same range (about 1.5-2 fold), and is apparently independent of the size of the molecule and the number of the charged groups. Both activation and inhibition could also be detected at much higher concentrations of substrates. At higher PG concentration the extent of activation is slightly increased, while the extent of inhibition is greatly decreased (Table 24).

In general, the maximum activation occurs at millimolar concentrations of the analogues, at 15-20 mM for sulphate or phosphate and about 1 mM ATP. As most of the effectors are present *in vivo*, similarity of these values to the physiological concentrations, at least for metal-free ATP (about 0.4 mM [403]) and for inorganic phosphate (about 8 mM [404]), indicates that activation may occur naturally.

These unusual kinetic effects of anions remained the same even after 10-20 minutes preincubation of the enzyme with the ligands; thus activation cannot be attributed to slow equilibration with the ligands. In case of rapid equilibration, however, the activation and inhibition by any ligand, L, cannot be described by a simple model of 1:1 binding stoichiometry. In a model of 1:1 stoichiometry raising the concentration of ligand cannot first activate and later inactivate, since the ratio between species with the activating and inactivating (sub)sites filled is independent of ligand concentration. Thus, in spite of the absence of independent evidence in favour, a model of 2:1 binding stoichiometry must be assumed. According to this model, activation, caused by the occupation of the first site turns into inhibition as soon as the second site becomes saturated. Thus, the net dependence of the enzyme activity on ligand concentration is given by the curve that has a maximum. The experimental points in Figure 70A-G can indeed be satisfactorily fitted by theoretical curves calculated on this basis:

$$v = v_u + v_u \cdot (a-1) \cdot \frac{[L]}{K_{d(act)} + [L]} - \left\{ v_u + v_u \cdot (a-1) \cdot \frac{[L]}{K_{d(act)} + [L]} \right\} \cdot \frac{[L]}{K_{d(inh)} + [L]} \qquad \text{Eq. 3.}$$

where the first two terms stand for activation, while the third term with negative sign stands for inhibition. The term v_u is the activity of the unligated form of the enzyme (at fixed substrate concentration), whereas a is the factor by which activity is increased, i.e., at infinite ligand concentration, [L] the activity $v = v_u \cdot a$. Accordingly by definition the relation of $a > 1$ should be always valid. The physical meaning of the coefficient $v_u \cdot (a\text{-}1)$ is the net activity increase at saturation of the activating site by the ligand L. The terms $K_{d(act)}$ and $K_{d(inh)}$ are the dissociation constants for the sites of activation and inhibition, respectively.

Fitting Eq. 3 to the activity values (v), determined experimentally as a function of [L] (e.g., Figs 70A-H), the parameters $K_{d(act)}$, $K_{d(inh)}$ and the factor a, characteristic for the investigated effector (L), could be obtained and listed in Table 24. The fitted parameters were obtained by keeping the initial estimates of the activation factor, a, close to each other for the anionic ligands with various chemical structures. This assumption was based on their closely similar extent of activating effects (Figure 70), indicating no particular specificity.

Figure 70 also shows activation and inhibition (dotted and dashed lines, respectively), the difference between them yields the actual dependence of activity on [L]. The *activation* component, given by the first two terms of Eq. 3, represents a simple hyperbolic dependence on [L], whereas the *inhibition* component, given by the third negative term of Eq. 3, represents a more complex sigmoidal dependence on [L]. It has to be noted that the values obtained for K are not the real, but *apparent* dissociation constants of the activating and inhibiting ligand complexes, because the substrate(s), present in the assay mixtures, may compete with the anions. This explains the variation of K with the substrate concentration: at higher 3-PG concentration the values of the apparent K constants are increased (Table 24).

Activations by ATP and by the other anions are mutually exclusive as demonstrated by separate experiments. Figure 70H for example shows that the extent of activation caused by ATP becomes smaller and smaller as the concentration of sulphate is increased, i.e., ATP and sulphate antagonise each other's activating action. In agreement, $K_{d(inh)}$ constant for the first anion apparently increases as the concentration of the second anion is increased. Thus, the mechanisms of activation by ATP and the other anions are very similar to each other, and they all may bind to the same regulatory site. On the other hand, as for the inhibiting site, double inhibition experiments have revealed additive inhibition of PG-analogue anions, such

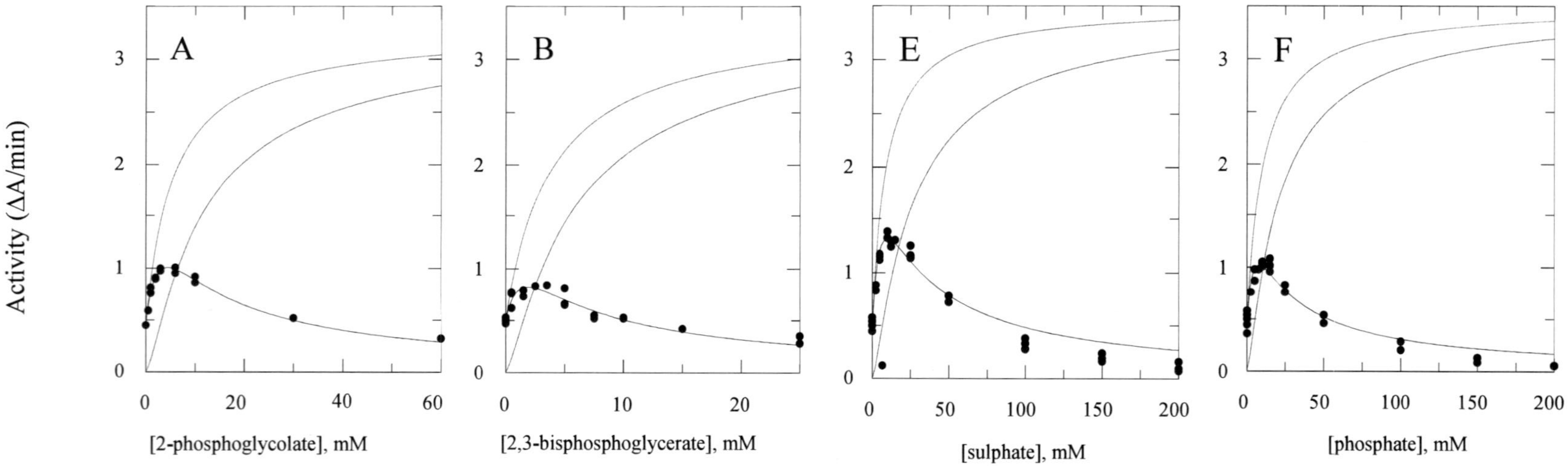

Figure 70. Activation and inhibition by various anions (adapted with permission from [129]; Copyright 2014 American Chemical Society).

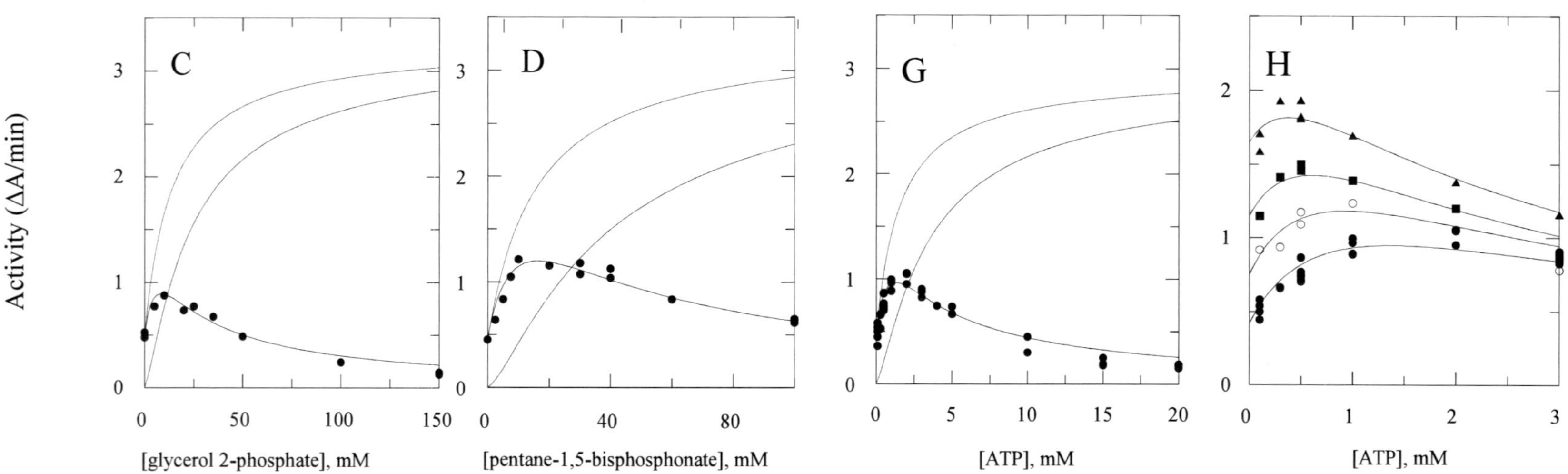

Table 24. Summary of the kinetic parameters for activation by excess substrate as well as for activation and inhibition by various anions

Substrate or Effector		[MgCl$_2$] mM	[3-PG] mM	v_S or v_u min^{-1}	a	$K_{S(act)}$ or $K_{d(act)}$,mM	$K_{S(cat)}$ or $K_{d(inh)}$,mM
3-phospho-glycerate	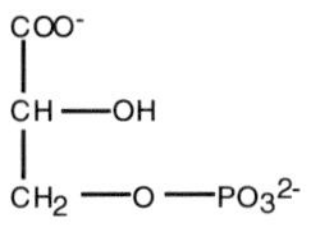	1	various	80 ± 7	21.7±3.6	9.5±3.2	0.019±0.008
		10	various	231 ± 15	6.8±0.9	12.6±4.5	0.027±0.015
2-hydroxy-4-arsonobutyrate	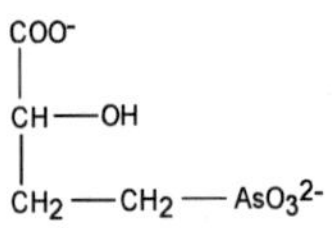	1	no	0.15 ± 0.05	n.d.	n.d.	n.d.
			0.05	n.d.	15.1±4	70±30	6.5±2.5
			0.2	n.d.	13.5±3.5	50±25	6.1±2.3
glycerol 3-phosphate	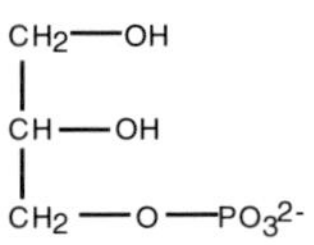	1	0.2	251 ± 14	5.96±1.8	15.0±6.6	1.4±0.4
			2	869 ± 80	5.76±2.1	19.8±5.2	4.8±1.1
2-phospho-glycolate	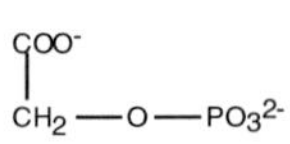	1	0.2	241 ± 20	6.75±1.5	5.4±0.3	6.4±2.8
2,3-bis-phospho-glycerate	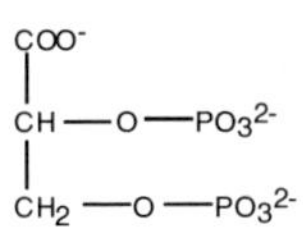	1	0.2	248 ± 12	6.72±1.2	3.86±0.7	2.43±0.5
			2	850 ± 76	6.28±1.3	4.2±0.8	4.3±0.9
glycerol 2-phosphate	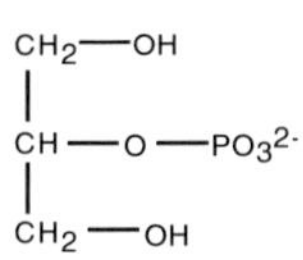	1	0.2	245 ± 22	6.67±1.6	13.8±3.5	11.6±3.9
			2	818 ± 75	6.62±1.8	26.8±6.7	38.2±9.2
		5	0.2	170 ± 10	5.2±0.8	16.2±5.8	17.9±5.2
pentane 1,5-bis-phosphonate	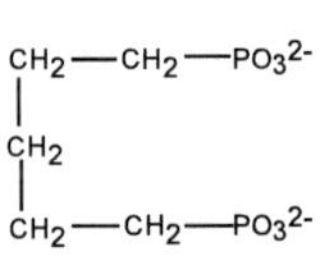	1	0.2	230 ± 12	7.4±1.3	15.8±3.8	27.4±5.4
sulphate		1	0.2	230 ± 12	7.77±1.3	8.9±2.1	17.4±3.2
phosphate		1	0.2	240 ± 20	7.47±1.4	10.3±2.6	10.6±2.5
ATP		1	0.2	210 ± 15	6.27±0.9	1.6±0.3	2.04±0.4

The data are from the papers [107, 129]

as glycerol-3-phosphate and ATP. Thus, two different type of inhibiting sites exist for ATP: one of these sites is a relatively weak anionic site (where multivalent anions are also bound); the other site may be the non-ionic adenosine site.

Kinetic Equations Describing Activation by the Excess of Substrates

The other anomaly in the kinetic behaviour of PGK is the non-hyperbolic character of both PG- and MgATP saturation curves, as shown in Fig 71A and B. Thus, the curves of velocity *versus* substrate concentration cannot be fitted by a single hyperbola. This non-Michaelian character of the substrate saturation has been interpreted long before any structural information was available about the enzyme, by assuming the existence of two independently functioning catalytic sites. This assumption, however, cannot be held in the light of the X-ray structural data, showing the presence of only one catalytic site [69, 73, 84] (cf. also Chapters I. 1.4. and 1.6.).

An anomalous substrate saturation curve might also be a consequence of slow formation of the active enzyme-substrate complex, relative to the rate of the catalysed reaction, often coupled with a protein conformational change. For PGK, indeed, a large scale hinge-bending domain closure was assumed a long time ago, as an essential requirement for occurrence of the enzyme reaction (Chapter I. 1.5.), supported by small angle X-ray scattering studies (Chapter I. 2.9.). It was found however that 10-20 minutes preincubation of PGK with both of its substrates (3-PG and MgATP) had no influence on the measured rate of the catalysed reaction. This result clearly shows fast equilibration with the substrates, as with the anionic effectors. Even if domain motion occurs before the catalytic reaction, this protein isomerisation step, which probably accompanies substrate binding, must be also faster than the enzymatic reaction itself. In agreement with this, Geerlof and coworkers [373] had already proposed that the putative hinge-bending domain motion occurred before establishment of the steady state. Thus, the anomalous substrate saturation curve cannot be due to slow equilibration with the substrates.

Szilágyi and Vas [129] have shown, that activation by the excess of the substrate and activation by various anionic effectors are related phenomena and both can be described by a similar two-site model. Accordingly, excess of the substrate, which is itself an anion, can bind not only to the catalytic site, but also to the same regulatory site where the other activating anions bind. The experimental data in Figure 71, indeed, can be satisfactorily described by the following equation:

$$v = v_S \cdot \frac{[S]}{K_{S(cat)} + [S]} + v_S \cdot (a-1) \cdot \frac{[S]}{K_{S(cat)} + [S]} \cdot \frac{[S]}{K_{S(act)} + [S]} \qquad \text{Eq. 4}$$

The first term represents a simple hyperbolic saturation of the catalytic site with the substrate (S), i.e., by analogy to v_u in Eq. 3, it expresses the activity in the absence of activation. Thus, v_S stands for the hypothetical activity at saturation of the catalytic site by the substrate, if activation were not taking place. Similarly to Eq. 3, the second term is a representative of saturation of a regulatory site, which is here a non-hyperbolic function. $K_{S(cat)}$ and $K_{S(act)}$ are the dissociation constants for the catalytic and the regulating (activating) sites, respectively. The coefficient *a*, as above, is the factor by which activity increased by binding of substrate to the regulatory site, i.e., $a \cdot v_S$ is the activity of the enzyme when both catalytic and activating sites are filled by the substrate. Fitting Eq. 4 to the activity values (v), measured at various [S] (e.g., Figure 71), the parameters $K_{S(cat)}$ and $K_{S(act)}$ and the activation factor a can be obtained. The parameters that gave the best fit are shown in Table 24. Thus, the enzyme molecules, in which only the catalytic site is filled by the substrate, may function only with a lower activity. With a large excess of substrate, the weaker regulatory sites may also be filled, with concomitant enhancement of the activity.

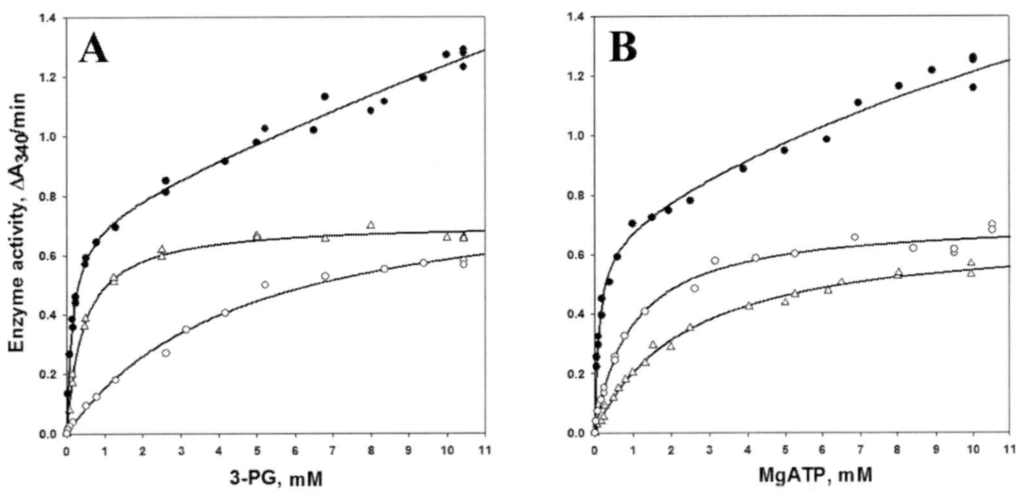

Figure 71. Substrate saturation kinetic curves with 3-PG (A) and with MgATP (B) of the wild type and the active site mutants of human PGK (adapted with permission from [129]; Copyright 1998 American Chemical Society): Activity of 9 nM wild type (•), 4 μM K215A (△) and 4 μM R38A (o) mutant PGKs were measured as a function of 3-PG concentration in the presence of 10 mM ATP and 12 mM $MgCl_2$ (A) and as a function of MgATP concentration in the presence of 20 mM 3-PG (B). The continuous lines represent the best fits of the experimental points to Eq. 4.

Activation of the Reaction with PG by a Weak Substrate

The above model assumes simultaneous binding of two PG molecules to the enzyme (one to the catalytic, and one to the regulatory site), for which direct evidence is still lacking. Neither structural nor binding studies indicate binding of a second substrate molecule to PGK. The hypothesis that a separate regulatory site exists is strongly supported by the following experiment.

Szilágyi and Vas [129] also investigated whether a weak substrate, namely the arsonomethyl-analogue of PG, i.e., 2-hydroxy-4-arsonobutyrate, can activate PGK in the reaction with its physiological good substrate, PG. Keeping the concentration of PG at a constant, non-saturating level, the concentration of the weak substrate was progressively increased. A large acceleration of the enzyme reaction was detected, in spite of the fact that the activity with the weak substrate itself is clearly negligible under these conditions (Figure 72). This activation can only be due to binding at another site, since any binding at the catalytic site would have caused inhibition by gradual replacement of the good substrate, PG, by the increasing concentration of the poor substrate. The experimental data on Figure 72 can, indeed, be described by Eq. 5 that assumes activation, and also takes into account competition of the two substrates for the catalytic site:

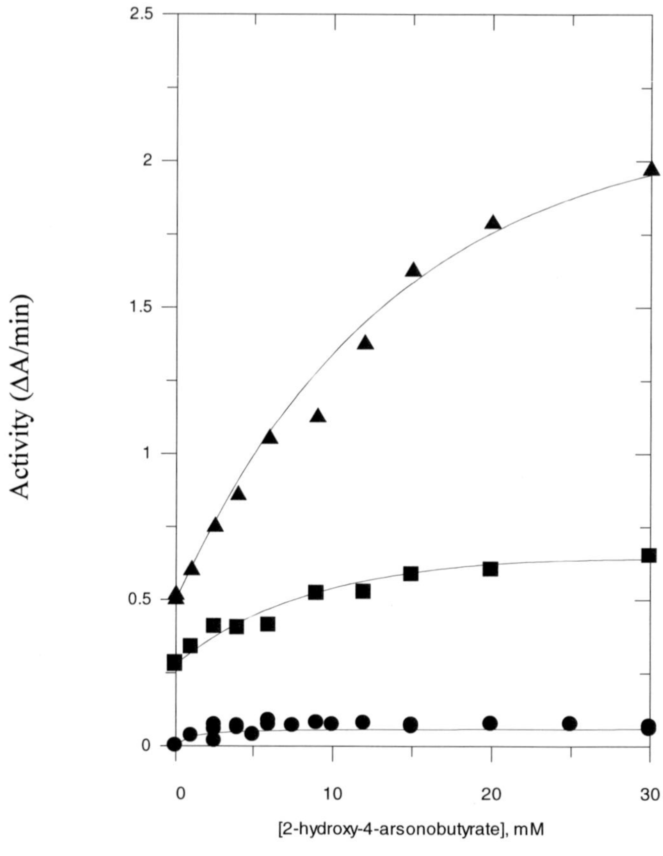

Figure 72. Effect of 2-hydroxy-4-arsonobutyrate on the reaction of PGK with 3-PG (adapted with permission from [129]; Copyright 1998 American Chemical Society)**:** The activity of 40 nM enzyme was measured in the presence of 0.5 mM MgATP (provided by 1 mM $MgCl_2$ and 0.6 mM ATP) in the absence (o), and in the presence of 0.05 mM (□) and 0.2 mM (Δ) 3-PG as a function of 2-hydroxy-4-arsonobutyrate concentration. Curve fitting was carried out by using Eq. 5.

$$v = v_S \cdot \frac{[S] \cdot K_{S'(cat)}}{[S] \cdot K_{S'(cat)} + K_{S(cat)} \cdot K_{S'(cat)} + [S'] \cdot K_{S(cat)}} + \\ + v_S \cdot (a_{S'} - 1) \cdot \frac{[S] \cdot K_{S'(cat)}}{[S] \cdot K_{S'(cat)} + K_{S(cat)} K_{S'(cat)} + [S'] \cdot K_{S(cat)}} \cdot \frac{[S']}{K_{S'(act)} + [S']} \qquad \text{Eq. 5}$$

This equation describes the activation caused by the weak substrate (S’) on the reaction with the good substrate (S); it gives the increase of activity as a function of [S’] if [S] is kept at a constant value. When [S']=0 the Eq. 5 will be simplified into the Michaelis-Menten equation:

$$v_0 = v_S \cdot \frac{[S]}{K_{S(cat)} + [S]} \qquad \text{Eq. 6}$$

The kinetic experiment illustrated by Figure 72, thus, provides direct evidence in favour of the existence of a separate regulatory site.

Modelling of the Regulatory Anionic Site

Since the direct experimental detection of a secondary (possibly weak) anionic site has met difficulties, other, molecular graphical approaches have been applied. First of all the available X-ray structural data of PGK have been inspected. The highly conserved structure of this enzyme allows comparison of data of different origin. Several structures of various complexes with substrates exist at high resolution, but none of them showed binding of a second substrate molecule. Thus, by this way no information could be provided about the location of the postulated regulatory anionic site. Even from structural data obtained in the presence of high concentration of sulphate [4], the anionic site could not be clearly identified although in this structure all of the possible anionic sites, including the activating one, should have been occupied by sulphate. Furthermore, the anionic site(s) observed from NMR data in the basic patch [140] (cf. also Chapter II. 2.1.) was/were not clearly separated from the substrate sites or from the active centre area.

Therefore the suggestion could be made that the regulatory site may not exist originally in the structure of the substrate-free enzyme, but may only be formed as consequence of the conformational change, i.e., domain closure that occurs upon binding of both substrates. The kinetic equations, quoted above, are already based on this assumption. Moreover, further evidence have been obtained by computer modelling (docking) of anion binding to the closed PGK conformation present in the crystal structure of TbPGK [84]. A new anion-binding site could indeed be seen in the structure of this closed conformation (Figure 73). As a control, docking was also carried out to the open conformation of pig muscle PGK [69], but with a less definitive results, since in the open conformation only part of this site exists. Figure 73B illustrates the binding of an activating phosphate ion outside of the closed active site. The anionic site is constituted by participation of Arg 65 (N-domain) and of Lys 215 (C-domain), respectively. It is remarkable, that the docking suggested interaction of the anion with the same Lys 215, which may also periodically interact with the nucleotide phosphate (cf. Chapter I. 1.4.). This bound anion may regulate domain closure and opening. Site-directed mutagenesis of Lys 215 either to Ala or to Arg confirmed this assumption. The substrate saturation curves are changed into hyperbolic ones (Figures 71A and B), which lack activation at high excess of the substrates; and the mutants cannot be activated, just inhibited by anions (Figure 73A) [107]. Mutagenesis of other side-chains responsible for domain closure, such as Arg38 (cf. Figures 71 and 73A) or Lys 219, Asn 336 and Glu 343 into Ala (not shown) also lead to loss of the activation phenomena. Thus, modelling and mutagenesis studies have approved the assumed relationship between anion-activation and domain closure.

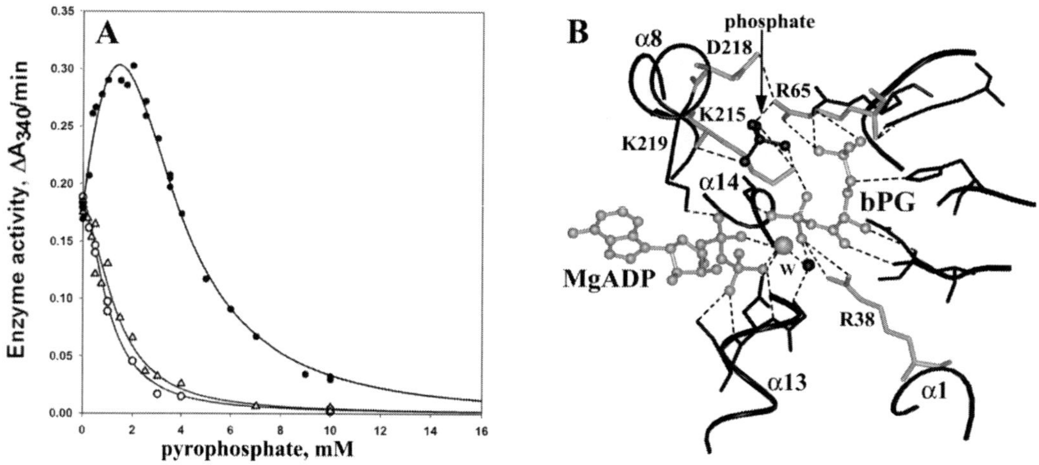

Figure 73. Anion activation of PGK (A) and its possible structural origin (B) (adapted from [349]): Activity of 9 nM wild type (•), 4 μM K215A (△) and 4 μM R38A (o) mutant PGKs were measured as a function of pyrophosphate concentration in the presence of 0.5 mM PG and 0.5mM MgATP (A). The continuous lines represent the best fits of the experimental points to the Eq. 3. The possible binding mode of an activating phosphate ion (black ball and stick model) in the active site of the closed structure of Tb PGK is shown by molecular docking in (B). The important secondary structural elements and side-chains are highlighted by black ribbons and by grey stick models. The side-chains are numbered according to the human PGK sequence numbering. The substrates MgADP (as bound in *Tb* PGK crystal structure) and bPG (as modeled into the place of bound PG) are shown as grey ball and stick models.

In summary, taking into account anion activation, the catalytic cycle of PGK briefly consists of the following events. Upon binding of both substrates, under their concerted action closure of the two domains is taken place. Simultaneously, an anion-binding site is formed by contributions of Lys 215 (C-domain) and Arg 65 (N-domain) (Figure 71B). The activating anion can bind here and stabilizes the closed active site, in which the reactive groups are precisely oriented and the phospho-group is transferred between the two substrates. Anions may also accelerate the consequent domain opening and dissociation of the products. Up to now, however, the precise relationship between anion activation and domain closure could not yet be uncovered by direct structural identification of the activating anionic site(s).

In: Phosphoglycerate Kinase: A Hinge-Bending Enzyme
Editor: Mária Vas ISBN: 978-1-62808-836-6

Chapter VII

Antagonistic Binding of the Two Substrates

Mária Vas
Institute of Enzymology, Research Centre for Natural Sciences,
Hungarian Academy of Sciences, Budapest, Hungary

It has been reported that, in solution, binding of the nucleotide substrate is weakened by the bound substrate, 3-phosphoglycerate, PG [372]. The reciprocal effect, i.e., weakening of PG-binding in the presence of the nucleotide substrates, has also been observed, even earlier [154]. It was also shown that this mutual weakening of substrate binding is somehow due to the presence of the charged groups, namely, the carboxylate of PG and the phosphate-chain of the nucleotides. In fact, glycerol-3-phosphate, the carboxyl-lacking analogue of PG does not affect nucleotide binding and as expected, it is a non-competitive inhibitor towards MgATP [93]. Furthermore, adenosine that lacks phosphate-chain does not weaken at all binding of PG [154]. Thus, at a face value, antagonism could be attributed to an assumed mutual repulsion of the negatively charged groups of the two substrates that should approach each other during the enzyme reaction. Very soon it became evident that this is probably not the case. Equilibrium dialysis binding studies have revealed that PG has no any weakening effect on binding of Mg^{2+}-free ADP, although it possesses the full phosphate chain [110]. These substrate binding data in solution are summarised in Table 25. It is seen that besides PG, the product 1,3-bisphosphoglycerate, bPG also largely weaken binding of MgATP. The 'antagonistic binding' of the substrates could be best interpreted by assuming a conformational transition upon formation of the ternary complexes [372]. However, the question still remained open whether this conformational transition can be identified as the domain closure itself that, in fact, has only been observed in the crystal structure of the ternary enzyme-substrate complexes and not (or only to a very small extent) in the respective binary complexes (cf. Chapters I. 1.8. and II. 2.9.). It is also conceivable that the conformational change, occurring during the transition from the binary to the ternary complex, may affect the binding characteristics of the simultaneously bound ligands relative to their individual binding. Naturally, the binding properties of the substrates in the catalytically competent ternary complex are important for the catalysed phospho-transfer.

Table 25. Dissociation constants (mM) of substrate binding to PGK

Ligand	Second ligand	
	—	Mg^{2+} and nucleotide
PG	0.032±0.003 [372] 0.03±0.01 [154]	0.25±0.08 (MgATP) [154] 0.20±0.10 (MgADP) [154] 0.12±0.05 (MgAMP) [154] 0.03±0.01 (adenosine) [154]
bPG	0.00005±0.00002 [149] 0.00005±0.00001 [154]	—
Glycerate-3-P	0.40±0.10 [154]	—

Ligand	Second ligand		
	—	Mg^{2+}	Mg^{2+} and PG/bPG/ Glycerate-3-P
ATP	0.21±0.03 [402] 0.33±0.15 [372]	0,15±0,03 [93] 0.23±0.03 [402] 0.27±0.09 [372]	0.51±0.22 (PG) [372] 0.38±0.06 (PG) [110] 2.39±0.68 (bPG) [372]
ADP	0.27±0.04 [402] 0.41±0.12 [110] 0.34±0.05 [110]	0.04±0.01 [93] 0.06±0.01 [402] 0.048±0.006 [372]	0.34±0.13 (PG) [372] 0.38±0.05 (PG) [110] 0.058± 0.001(Glycerate-3-P) [372]
AMP	—	0.29±0.05 [93] 0.36±0.05 [402]	—
Adenosine	1.05±0.20 [93] 0.65±0.05 [402]	—	—

Tirations of the enzyme with the substrates have been carried out with various methods: by ANS-labelling [154], by the extent of protection against chemical modification [93], equilibrium dialysis [110, 402] or competitive titration using a fluorescent nucleotide analogue [110, 372].

In order to test the hypothesis of a possible relationship between substrate antagonism and domain closure a rather unusual approach has been applied by Merli and coworkers [110]. Using the technique of single crystal microspectrophotometry, these authors determined the binding constants of MgADP and MgATP to the crystalline state of the binary complex of the pig muscle enzyme and PG, possessing an open domain conformation as known from the X-ray study [15]. It was also known that the large scale conformational transition of domain closure is prevented by crystal lattice forces in this binary complex with PG; namely, it maintains the open conformation even after diffusing the nucleotide analogue 5'-adenylylimidodiphosphate (AMP-PNP) through the crystal channels [69]. Therefore, substrate antagonism in the crystals of PG binary complex is expected not to take place, if this feature really depends on domain closure.

In order to characterise nucleotide binding in the crystals of PGK, these authors have exploited the chromophoric properties of the nucleotide analogue 2'3'-O-(2,4,6-trinitophenyl)ATP (TNP-ATP), which has been successfully used with PGK in solution, to substitute for either MgADP or MgATP [372] as well as an affinity label for other ATP-dependent enzymes [405-407]. Using single crystal microspectrophotometry, competitive titrations of the nucleotide substrates were performed, similar to those previously done in solution [372], to determine the dissociation constants of the nucleotide substrates in the various available crystal forms of the pig muscle enzyme [110]. The dissociation constants of MgADP and MgATP binding to the crystal of PG-binary complex of pig muscle PGK was obtained to be 0.096±0.021 and 0.21±0.02 mM, respectively. MgADP binding was also investigated in the ternary complex obtained by crystallising pig muscle PGK in the presence of both PG and the ATP analogue MgAMP-PCP possessing also open conformation [72]. For this purpose, in advance, the bound AMP-PCP has been replaced by TNP-ATP. In this latter case the binding constant of MgADP was determined to be 0.048±0.016 mM. Thus, in comparison with the solution data (Table 25) significantly stronger binding of the nucleotides could be detected in both crystals in the presence of PG. This result indicates that the antagonistic substrate binding, characteristic of the enzyme in solution, is not retained in the investigated open crystal forms.

The absence of substrate antagonism in the crystals containing bound PG and having open domain conformations is consistent with the assumption that substrate antagonism is related to domain closure. Furthermore, in solution there are two notable findings: *i.* 3-PG binding practically abolishes the strengthening effect of the Mg^{2+} ion on the binding of ADP; and *ii.* no antagonism in detected, when PG is replaced by glycerol-3-phosphate (cf. Table 25), in line with the possible importance of the PG carboxylate in the interdomain communication leading to domain closure. Taken into account these observations, Merli and coworkers [110] provided a possible structural explanation of the substrate antagonism by comparison of the crystallographic data of different PGK binary (with either 3-PG or MgADP) and ternary (with both 3-PG and MgADP) complexes, having open and closed conformations, respectively (Figure 74). They noted that the specific interaction between the 3-PG carboxylic group and a conserved arginine side-chain (Arg 38) (Figure 74A) is changed during domain closure, and through interdomain communication, this change may be transmitted to the site in which Mg^{2+} binds the ADP phosphates (Figure 74B). One basic element of this communication network is the H-bond between Gly 372(N) of helix 13 and Ser 392(O) of the β-strand L in this PG-binary complex structure (Figure 74C), which have been shown to be part of the network essential for operation of the main hinge point at this region (cf. also Chapter X.). In Figures 74A, B and C the high resolution crystallographic data are shown for the binary complexes of pig muscle PGK with PG [15] and of BsPGK with MgADP [71] in the open conformations and the PG·MgADP ternary complex of TbPGK in a closed form [84]. Because of the highly conserved structure of PGK, such a comparison seemed to be reasonable. Furthermore, considerations referring to the unproductive ternary complex containing bound PG (substrate) and MgADP (product) could possibly be extended towards the productive ternary complex. The facts that PG exerted an antagonistic effect towards the ATP-analogue MgAMP-PCP (Kovári, Flachner, Náray-Szabó & Vas, unpublished) as well as towards ATP (Table 25) support this assumption.

This H-bond network represents a continuous link from the 3-PG carboxyl group through Arg Tb39/38/Bs36 to the most important hinge at β strand L and thereby to the N-terminus of helix 13, as can be visualised in Figure 74C. This network largely contributes to the interdomain communication in the ternary complex and may perturb the Mg^{2+}-ion linked interaction of the nucleotide phosphates with the N- terminus of helix 13. In fact, the position of Mg^{2+}-ions in the two conformations differs by about 0.8-1.0 Å when the respective PGK molecules are superimposed (Figure 74D).

These changes around the metal-ion may be brought about not only by the immediate changes in the interactions between 3-PG and the contacting conserved side-chains, as described above. A more complex, highly co-ordinated conformational change may be propagated from the PG site to the nucleotide site and *vice versa*, concomitant with domain closure (as also discussed in Chapter X.). By this way both bound substrates can mutually weaken each other's binding, as it is actually found, not only by the above discussed authors, but in some more recent systematic studies, too [350, 351].

In these latter works transient kinetic studies have been explored with HsPGK, using stopped flow (cf. Chapter V. 5.6.) by the aid of changes in protein fluorescence occurring upon interaction with either of the two substrates. By this way both k_{on} and k_{off}, i.e., the rates of association and dissociation of the substrates, both in the absence and presence of each other have been systematically determined. By this way Lallemand and coworkers provided, by the first time, a kinetic explanation of the substrate antagonism. Namely, PG only slightly decreases k_{on}, the association rate of MgADP, but largely increases (about 8-fold) k_{off}, i.e., its dissociation rate from the enzyme [350]. Later on the work has been extended towards the functioning enzyme by the same group [351] and determined the effect of MgADP on bPG binding and *vice versa,* the effect of bPG on binding of MgADP. The results are summarized in Table 26. The derived K_d data are not comparable with the previously determined values in Table 25, since in order to be able to measure kinetics, these workers had to apply cryo-conditions (i.e., 30% methanol and 4° C). By this method, they concluded that the substrate binding kinetics that lead to the formation of the catalytic PGK·bPG·MgADP complex is mutually antagonistic with MgADP. The presence of MgADP induces a large decrease (about 30-times) in the affinity of PGK for bPG. This antagonistic effect of MgADP on the interaction of bPG with PGK is mainly due to a large increase (about 11-times) in the k_{off} of bPG and, to a lesser extent, to a moderate decrease (2.6-times) in $k_{on.}$

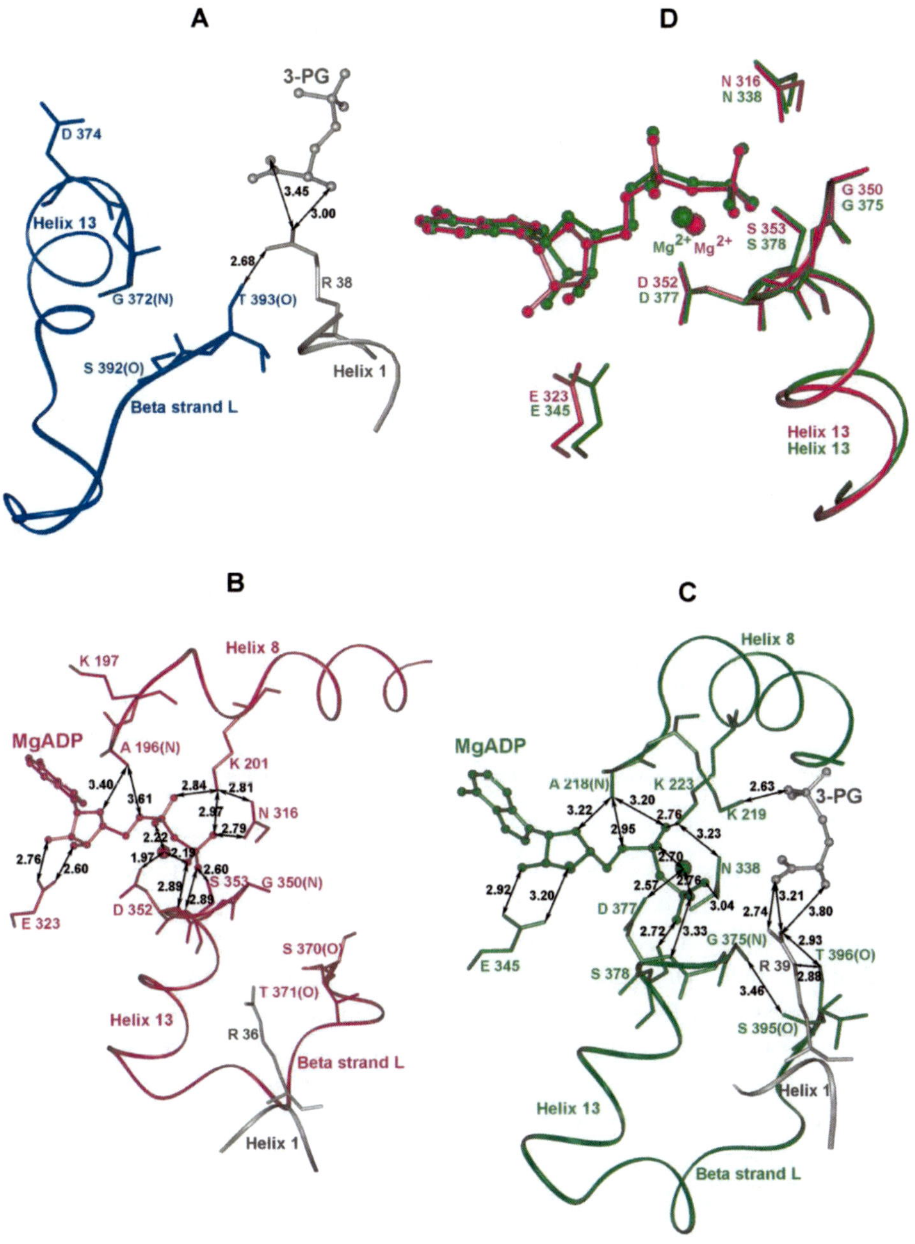

Figure 74. Surroundings of the bound substrates in the X-ray crystallographic structures of PGKs of various origin (adapted with permission from [110]; Copyright 2002 American Chemical Society): Part of the structures, represented by ribbon diagrams with the stick model of few critical side-chains around the bound substrates (ball and stick models) are shown: the binary complex of pig muscle PGK with PG (A); the binary complex of BsPGK with MgADP (B) and the ternary complex of TbPGK with MgADP and PG (C). Superimposition of the two MgADP-bound structures, optimising the alignment of all atoms of the conserved N-terminal residues of helix 13, is also shown (D). The bound nucleotide and the part of the C-terminal domain are coloured according to the structures of different origins: blue (pig muscle), red (Bs) and green (Tb) PGK. The bound PG and the respective part from the N-terminal domain are shown in grey. The indicated distances are in Å. The conserved amino-acid residues equivalent structural positions possess different numbering in the PGK sequences of various origins.

Table 26. Comparison of the transient kinetics of the interactions of substrates with HsPGK (adapted from [351])

Variable ligand	Constant	k_{on} (μM^{-1} s^{-1})[a]	k_{off} (s^{-1})[a]	k_{off}/k_{on} (μM)	K_d (μM)[b]
bPG	E	450 ± 40	14 ± 26	0.03 ± 0.06	0.03 ± 0.01
bPG	E·ADP	170 ± 30	160 ± 30	0.94 ± 0.24	0.92 ± 0.23
D-ADP	E	6.1 ± 0.3[c]	38 ± 10[c]	6.2 ± 1.7[c]	7.7 ± 0.6[c]
D-ADP	E·PG	4.6 ± 0.6[c]	313 ± 40[c]	68 ± 12[c]	50 ± 3[c]
D-ADP	E·bPG	4.1 ± 0.5	270 ± 30	66 ± 11	100 ± 20
L-ADP	E	3.1 ± 0.2[c]	55 ± 6[c]	18 ± 2[c]	17 ± 1[c]
L-ADP	E·PG	2.4 ± 0.2[c]	79 ± 10[c]	33 ± 5[c]	28 ± 2[c]
L-ADP	E·bPG	3.2 ± 0.5	75 ± 30	23 ± 10	27 ± 7

Using the signal of the changes in protein fluorescence either the pure protein (E) or its ligand complexes (as indicated) were titrated with the variable ligands. The time courses of ligand binding were followed by stopped flow apparatus [351].

[a] The values of k_{on} and k_{off} were obtained from the dependences of the transient kinetics on the substrate concentration

[b] From the dependences of the fluorescence amplitude on the substrate concentration

[c] Data adapted from Gondeau et al. [350]

It is notable that these transient kinetic works, besides the natural D-ADP substrate have been also extended towards to its mirror image counterpart, L-ADP. As referred in Chapter IV. 4.2., PGK is able to act also on L-ADP, although with somewhat reduced activity [250]. It was found interestingly that, in contrast to the D-form of MgADP, no any antagonistic effect of PG on binding of the L-form of MgADP could be observed (Table 26). Gondeau and coworkers [350] explained this difference by a molecular modelling (docking) study showing that the β-phosphates of D- and L-ADP indeed have different orientations when bound to the active site of human PGK. As indicated by the contact distances in Table 27 the interaction with D-ADP (namely through Glu 343 and Asp 374) is weakened in the presence of PG in the closed ternary complex, while no such weakening is detectable in the interaction with L-ADP. The importance of the interaction with Glu 343 in relation with the different catalytic efficiencies of PGK towards D- and L-ADPs has already been discussed in Chapter IV. 4.2.

When the modelled structures of E·PG·D-ADP and E·PG·L-ADP are superimposed, the phosphates can be seen to occupy different positions, which leads to a small change in the position of helix 13. As a consequence, interdomain communication may be altered. Moreover, with D-ADP in the presence of PG, Asp 374 is displaced; therefore Mg^{2+} ion is less connected to this residue, which leads to a weaker binding of Mg^{2+} ion to D-ADP. This observation is in good agreement with the suggestion of Merli et al. that PG weakens the binding of D-ADP via Mg^{2+} ion. Indeed, from the docking results, the distance between Mg^{2+} ion and the carboxylic group of Asp 374 is increased upon PG binding with D-ADP but not with·L-ADP (Table 27).

Table 27. Contact distances (Å) of the nucleotide substrates with selected protein residues as determined by docking of the nucleotides into the open structure model used for the binary complexes and the closed structure model for the ternary complexes
(adapted with permission from [350]; Copyright 2008 American Chemical Society)

Nucleotide groups	N6 (adenine)	2'-OH (ribose)	3'-OH (ribose)	Mg^{2+}
HsPGK residues	Gly312	Glu343	Glu343	Asp374
in absence of PG				
D-ADP	2.98	3.17	2.70	2.03/2.12*
L-ADP	2.87	2.73	2.70	2.00/2.03
in presence of PG				
D-ADP	2.87	2.71	**4.74**	**1.98/3.49**
L-ADP	3.44	2.68	**2.73**	**2.05/2.09**

*The two values indicate distances of Mg^{2+} from the two O-atoms of Asp 374 carboxylate.

Meanwhile, in addition to the docking studies, crystal structures of the ternary complexes of HsPGK with bound PG and either D-ADP or L-ADP have been determined in both the open [75] and in the closed [351] conformations. In the latter case the presence of the transition state analogue, AlF_4^-, has facilitated the appearance of the closed state in the crystal. Besides noting the changes in the binding mode of L-ADP compared to the D-form that places the adenine ring in a tighter position within the hydrophobic pocket, the X-ray data of Lallemand and coworkers also indicated a different position of the β-phosphate of·L-ADP bound to the open conformation, in agreement with the above docking results. Lallemand et al. pointed out that a possible reason of the absence of antagonism between L-ADP and PG (or bPG) may be a lower repulsion between the two negatively charged substrates that are further apart in the PGK·PG·L-ADP structure than in the PGK·PG·D-ADP structure. However, in the closed conformations of the complexes with D- and L -ADP no such differences could be observed by crystallography. In the transition state analogue complexes PGK, naturally, stabilizes the phosphate-chain of the nucleotides in a position that is competent for catalysis (i.e., oriented towards AlF_4^-). Namely, the interactions between the protein residues and the ligands were similar. Thus, L-ADP has been confirmed as a good substrate for PGK (cf. [250]), as its β-phosphate and the metal-fluoride (which mimics the transferring γ-phosphate) are perfectly positioned for catalysis in the same orientation as the D-enantiomer.

Concerning the substrate antagonism the question arises what would be its significance in respect of the enzyme function. As pointed out by Merli et al. [110] weakening of the metal-mediated co-ordination of the nucleotide phosphates caused by the bound PG or bPG, indeed, may have functional relevance. In general, it is assumed that the co-ordination pattern of the metal ion to the nucleotide changes significantly during the enzyme-catalysed phospho-group transfer, concomitant with the ADP-ATP transformation. There are still controversial results concerning this pattern, as well as about the role of the metal-ion in catalysis (cf. also Chapter VIII.) but, in most cases, Mg^{2+} is found to be co-ordinated to both the α- and the β-phosphates of ADP and to both the β- and the γ-phosphates of ATP. This implies the movement of Mg^{2+}-ion during catalysis, which seems to be consistent with the accumulating structural observations in various kinases, including PGK [69, 84, 146]. Based on these considerations, Merli and coworkers [110] suggested that the 'substrate antagonism' reflects the initial weakening of the Mg^{2+}-mediated link between the bound nucleotide phosphates and PGK, which may be a prerequisite for the occurrence of phospho-transfer. – Alternatively, Lallemand and coworkers have put forward a more general view about the advantage of substrate antagonism in respect of PGK kinetics. Namely, substrate antagonism results in an increase in the dissociation rate constant for the second substrate, and therefore an increase in the overall kinetics of the interaction of E·S1 with S2 (cf. Schemes 2 and 5 in Chapter V. 5.5.). Even though the ratio [E·S1·S2]/[E·S1]×[S2] is decreased, which is not a kinetic advantage for the enzyme reaction, enhancing the exchange rate of substrates could have advantages. Because substrate antagonism also occurs in the backward reaction, it could be that it aids the release of PG and ATP, which is rate-limiting in the forward reaction. It has been demonstrated that the enzyme remains mostly in an open conformation, with short bursts of closure for catalysis (see Chapter XI.). In this situation, an enhanced mechanism for substrate and product exchange has clear advantages.

In: Phosphoglycerate Kinase: A Hinge-Bending Enzyme ISBN: 978-1-62808-836-6
Editor: Mária Vas

Chapter VIII

Role of the Metal-Ion in Nucleotide Binding and in the Catalysis

Mária Vas
Institute of Enzymology, Research Centre for Natural Sciences,
Hungarian Academy of Sciences, Budapest, Hungary

PGK, like other kinases, can only use the metal complex of ATP or ADP as substrates. This conclusion was already deduced from early kinetic studies. It was shown that the protein displayed only a weak, non-specific interaction with the metal-ion (Mg^{2+} or Mn^{2+}), but rather binds the nucleotide-metal complex [408]. Thus, the metal-complex appeared to be the active form of the nucleotide substrate; the free nucleotide and free metal-ion both seemed to have inhibitory effects only [358].

The various possible roles of the metal-ion, such as shielding of the negative charges on the attached phosphates (and thereby, facilitating the attack of the nucleophile in an associative reaction as well as assisting in development of the negative charge and departure of the leaving group), stabilizing the transition state, or optimally orienting the reacting substrates cannot be generalized for kinases and has to be investigated in individual cases, such as the PGK-catalyzed reaction [38, 186, 409].

In an early kinetic study the metal-ion specificity of PGK has been characterized and the following order of the bivalent metal-ions has been established: $Mg^{2+} > Mn^{2+} > Co^{2+} > Ca^{2+} > Zn^{2+} > Ni^{2+} > Cd^{2+}$ with decreased specificity ranging from 90 % to 15 % [3]. In an attempt to correlate this order with any properties of the metal-ions, however, could not give any clear answer. Correlation between the crystal ionic radius and the efficiency of divalent metal ions as activators, as found e.g., for creatine kinase [410] did not hold for PGK. Thus, the size of the metal-ion cannot be the sole factor responsible for effectiveness of the metal-ion in question. An important function of the metal-ion might be to polarize the bound, which is being cleaved during the reaction. The polarization capacity roughly parallels the acidity of the metal-ion and increases in the following order: $Ca^{2+} < Mg^{2+} < Mn^{2+} = Ni^{2+} = Co^{2+} = Zn^{2+} = Cd^{2+} < Be^{2+} < Fe^{2+}$. However, this order does not give an explanation why Ni^{2+}, Co^{2+} and Zn^{2+} are not as good cofactors as Mg^{2+} or why no activity of PGK was noticed with Be^{2+} and

Fe^{2+}. It seems obvious from these results that the metal-ion specificity and the effectiveness of metal-ions are determined by a rather complicated function of kinetic, thermodynamic and structural properties of the metal-ion, which of course all are reflections of the electronic structure of this ion.

Better insight into the role of the metal-ion in nucleotide binding to PGK as well as in the catalysis might be obtained from the ligation pattern of the metal, i.e., its interactions with both the nucleotide and the protein. The available crystallographic data on the ADP and ATP-complexes of PGK (cf. Chapter I. 1.4.) have shown ligation of Mg^{2+} to both ADP (α- and β-phosphates) and the protein, while in complex with ATP ligation of Mg^{2+} to only the nucleotide β- and γ-phosphates, but not to any protein side-chains, has been observed. These observations, in one side, are in accordance with the facts that Mg^{2+} strengthens ADP binding, while does not affect ATP binding to PGK [402]. On the other hand, comparison of the two nucleotide-bound structures suggests some movement of the metal-ion during catalysis. This movement that possibly occurs during ADP to ATP transformation, has also been supported by NMR solution studies [146] as described in Chapter II. 2.1. Other NMR-studies using CoADP and the CoATP complexes [141] have shown that the cation is directly coordinated to all the phosphate groups of the nucleotide in both ADP and ATP complexes. The coordination of the metal to both α- and β-phosphates of ADP is in good agreement with the crystallographic data obtained with various PGKs [5, 6, 84]. However, for ATP bound to PGK, the direct coordination of the cation to all three phosphate groups does not agree with the X-ray results with bound ATP [4, 7], although, coordination to all three phosphates has been observed in the crystal structures of Tm [73], yeast [70] and pig muscle [69] PGK-s where the analogue AMP-PNP substituted the bound ATP (cf. Chapter I. 1.4.).

Thus, in the light of these early X-ray and NMR structural data the coordination pattern of the metal either to all three α-, β- and γ-phosphates or only the β- and γ-phosphates of PGK-bound ATP cannot be clearly decided. However, the movement of the coordinating metal ion during the catalytic turnover, i.e., the ADP-ATP transformation, looks to be reliable.

Moreover, movement of the metal complexed with the nucleotide phosphates has an additional aspect. Namely, the metal-ion may move together with the phosphates of the enzyme bound ATP. As discussed above (Chapter I. 1.4.), crystal structures of PGK with various nucleotide triphosphates have revealed at least two alternative positions of the bound nucleotide phosphates. Thus, we may assume that the ATP phosphates can fluctuate between two alternative positions in the open enzyme conformation, corresponding to the N-termini of helices H8 and H13, where Asp 374 and Asp 218, respectively, are located. This fluctuation as also confirmed by independent solution kinetic, stability and thiol-reactivity studies (cf. Chapters II. 2.5 and II. 2.8). These aspartates are the potential metal-binding sites that may mediate fluctuation of the ATP phosphates between H8 and H13 and thereby these helices are moved closer to each other, i.e., formation of the active closed conformation thereby can be promoted. Indeed, a very recent mutagenesis study, namely preparing single and double Ala-mutants of the potential metal-binding residues, D374A and D218A, clearly confirmed the importance of these aspartates as well as the metal-ion in the catalysis [411]. This is demonstrated by the kinetic results with these mutants (Figure 75).

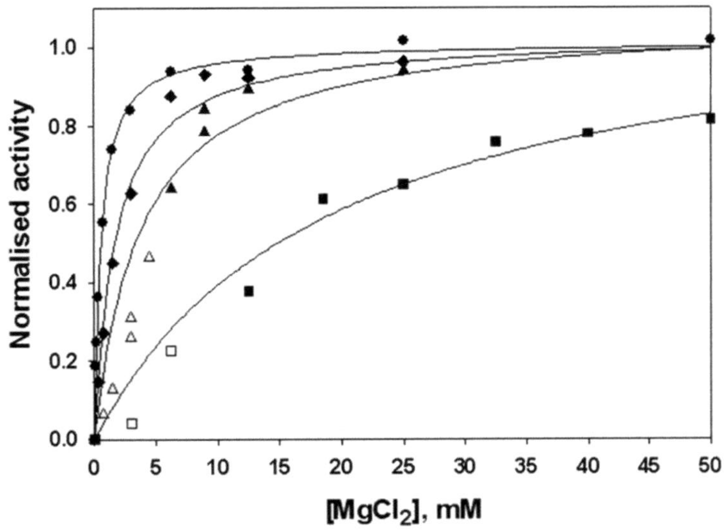

Figure 75. Dependence of enzyme activities of the investigated mutant PGKs on the concentration of Mg^{2+}-ion (adapted with permission from [411]; Copyright 2012 American Chemical Society): Activities of 9nM wild-type (●), 25.5 nM D218A (◆) 19 nM D374A (▲) and 206 nM D218AD374A (■) enzyme were measured at 4 mM ATP and 10 mM PG concentration with increasing concentration of the added $MgCl_2$. In cases of D374A and D218AD374A mutants the first few points (open symbols) at low concentrations of $MgCl_2$ are omitted from the hyperbolic fitting due to the possible inhibitory effect of ATP uncomplexed with Mg^{2+}, concentration of which is not negligible at this concentration range of $MgCl_2$.

On the other hand, as discussed above (cf. Chapter II. 2.3.), binding data with exchange-inert nucleotide-metal complexes (e.g., [160]) are in line with metal-coordination to all three phosphates of ATP. The hydrodynamic measurements on the dead-end combination of substrates support this concept [167] (cf. Chapter II. 2.9.). Experiments with anions, such as vanadate or arsenate were initiated in an attempt to find a substitute for the missing phospho-group in the dead-end complex. Vanadate is an inhibitor of several enzymes that catalyse phospho-transfer reactions [168], among them, of PGK [169]. The inhibitory action of vanadate is attributed to the ability of the pentavalent vanadium to adopt a stable trigonal-bipyramidal geometry that is analogous to the transition state in phospho-transfer reactions [170]. The EPR spectra for the dead-end complex of PGK with vanadate exhibit inhomogenous brodening both from α- and β-^{17}O-labeled forms of ADP. These observations demonstrate that Mn^{2+} remains bound to both phosphate groups of ADP even in this complex, in good agreement with complexation of the metal with all phosphates of the nucleotides.

Furthermore, using another exchange-inert complex of $Rh(H_2O)_4ATP$, entry of the other substrate PG or bPG into the coordination sphere of the metal complexed, at the same time, with ATP has been concluded [185] (cf. Chapter II. 2.3.), in agreement with the concept of optimal alignment of the two reacting substrates by the metal-ion. In a coupled analytical work conserved glutamates, either Glu Y398/400, Glu Y401/403 (in helix 13/14) or Glu Y190/192 (in helix 5/7) may be localised as possible metal-binding sites [186]. Metal-ion binding to yeast PGK was also detected by independent thiol-reactivity studies [412]. However, the existence of an independent metal-ion binding site has been never confirmed by any crystallographic work with PGK.

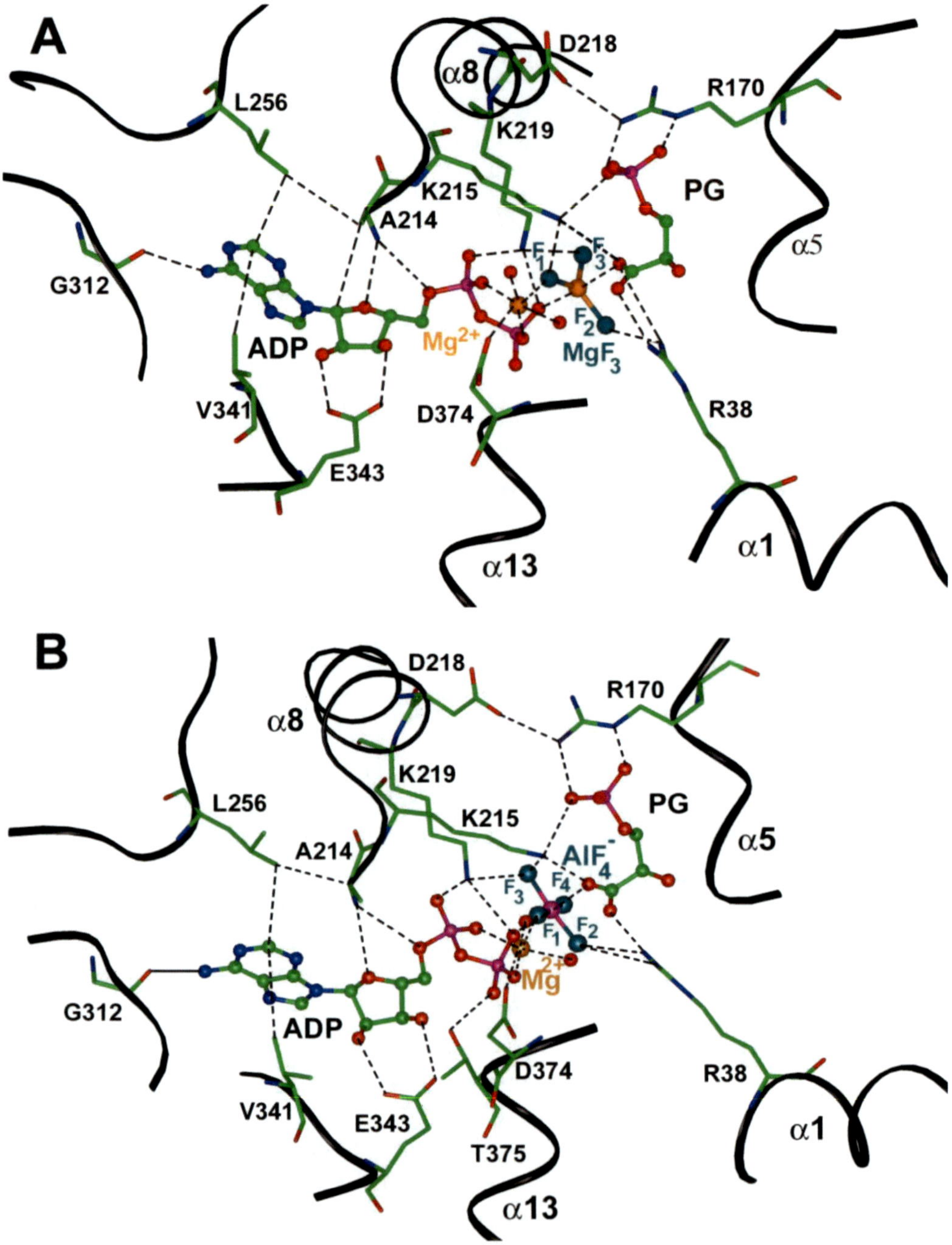

Figure 76. The closed active site structures of HsPGK·PG·MgF_3^-·ADP (A) and the HsPGK·PG·AlF_4^-·ADP (B) TSA complexes: The figures were prepared by using the X-ray coordinates 2WZB (A) and 2WZC (B). The bound substrates (ball and stick models) and the interacting side-chains (stick models) are coloured according to atoms. The atomic interactions are represented by black dashed lines. Black ribbons represent the secondary structural elements that surround the active site.

The recently published high resolution crystal structures of the transition state analogue (TSA) complexes of human PGK (HsPGK) (cf. next Chapter) provided evidences in favour of the complexation of Mg^{2+} to all three phosphates of ATP [76]. The catalytically relevant coordination of the essential Mg^{2+} in the structures of either the HsPGK·PG·MgF_3^-·ADP or the HsPGK·PG·AlF_4^-·ADP TSA complex (Figures 76A and 76B, respectively) is an octahedral one coordinated by the pro*R* oxygen atoms of both α- and β-phosphates of ADP as well as the metal fluoride that mimics the transferring γ-phospho-group. The remaining ligands comprise D374:OD1 and two water molecules. It is the first example of an anticlockwise facial coordination of the catalytic Mg^{2+} in any nucleotide phospho-transfer enzyme (Figure 77). Up to now four possible geometries of the catalytic magnesium coordination have been observed in crystal structures of nucleotide TSA-complexes of various ATP-handling enzymes (Figures 77A, B, C and D). The priorities of the phosphate groups involved in coordinating the catalytic Mg^{2+} are β > α > γ. The anticlockwise facial coordination arrangement, determined for PGK, seems to be a unique one (Figure 77C). The β and γ phosphate only geometry, as seen in F_1-ATPase (Figure 77D), are also found in the majority of nucleotide TSA complexes.

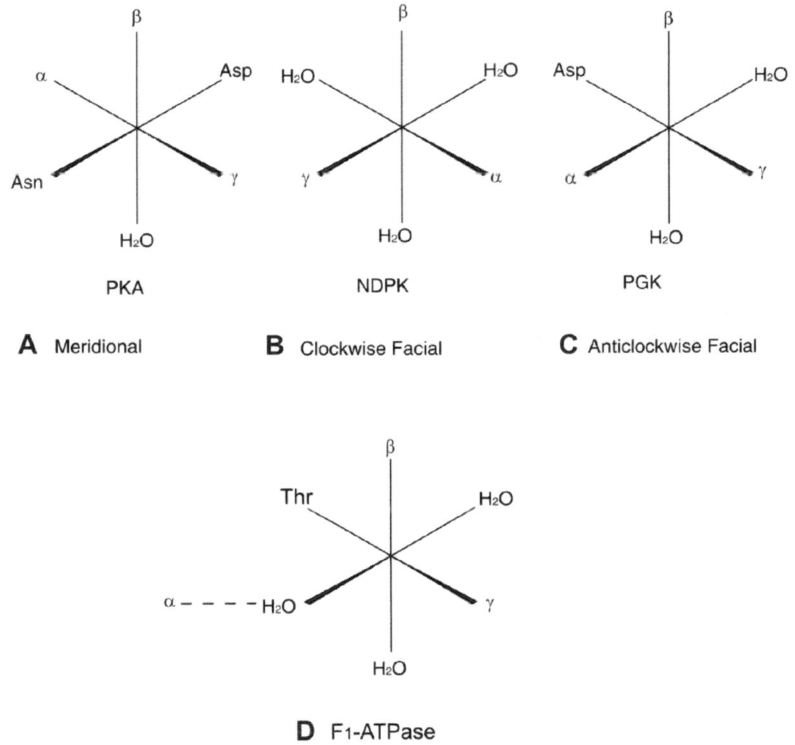

Figure 77. The four observed geometries of catalytic Mg-coordination in crystal structures of nucleotide TSA-complexes (adapted with permission from the Supplement of [76]; Copyright 2010 American Chemical Society)**:** The priorities of the phosphate groups involved in coordinating the catalytic Mg^{2+} are β>α>γ. (A) The meridional arrangement as found in protein kinase A (PKA; 1l3r). (B) The clockwise facial arrangement as found in nucleotide diphosphate kinase (NDPK; 1kdn), where the α to γ-rotation is clockwise around the β-phosphate-Mg axis. (C) The anticlockwise facial arrangement, as seen in the HsPGK TSA structures. (D) The β and γ-phosphate only geometry, as seen in F1-ATPase (1h8e) and the majority of nucleotideTSA complexes.

It is notable that in this crystal structure the catalytic Mg^{2+} does not connect directly PG and the nucleotide, probably because this is a non-functioning complex. However, alignment of the two reacting substrates by the metal-ion, as suggested by the earlier work with $Rh(H_2O)_4ATP$ [185], has been indeed supported by crystallography, simply by washing out the bound TSA (MgF_3^-) from the crystal (i.e., soaking it with only MgADP and PG), followed by soaking (3 hours) with MgATP and PG. This procedure allowed occurrence of a single turnover and finally resulted in a potentional functioning complex of MgADP and bPG in the crystalline state as approved by its crystal structure at 2.1 Å resolution (M. Bowler, 2011, personal communication). The atomic interactions within the active site of this complex are depicted in Figure 78. It illustrates clearly that the catalytic Mg^{2+} interacts simultaneously with both substrates: the α- and β-phosphates of ADP as well as with the transferring, otherwise very labile, acil-phosphate of bPG, in agreement with the solution studies by Pappu et al. [185], thus, supporting their conclusion of the proper alignment of the reacting substrates for catalysis. Besides, interaction of Mg^{2+} with the transferring TSA analogues (cf. Figures 76A and 76B) implies interaction with all α-, β- and γ-phosphates of ATP (in the reverse direction of the catalysis) that would serve to make the coordinated ADP a better leaving group. Furthermore, metal ion coordination to the terminal phospho-group of ATP (or to the 1-phosphate of bPG) may also serve to neutralise charge on the reacting phospho-centre facilitating the attack of the nucleophile (the carboxylate of PG or the β-phosphate of ADP) in an associative reaction.

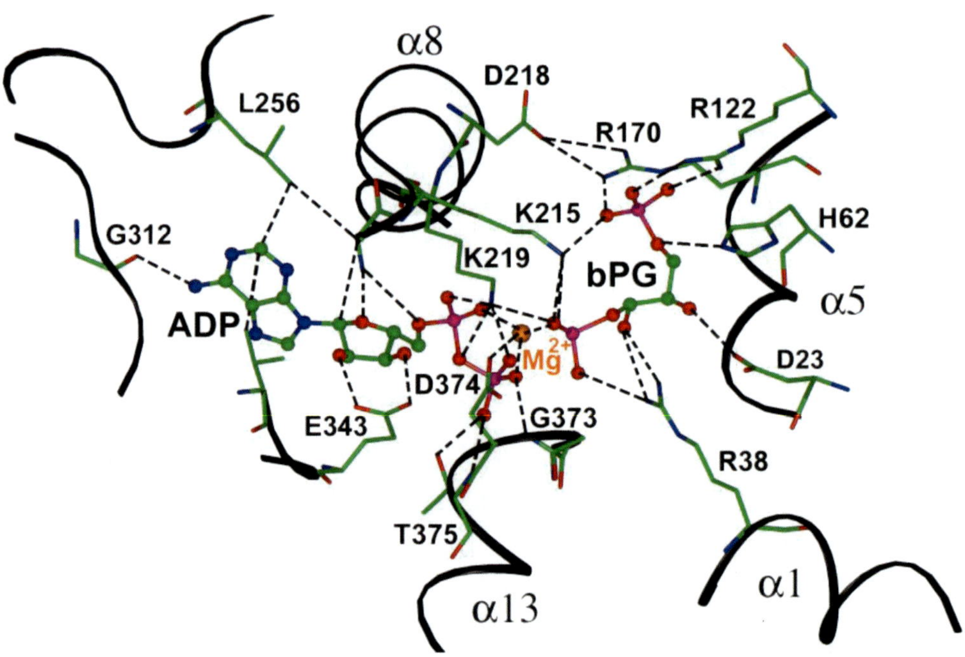

Figure 78. The closed active site structure of the complex of HsPGK·bPG·MgADP (based on unpublished data by M. Bowler, pdb code 2X15): The bound substrates and the interacting side-chains are represented in the same way as shown in Figure 76.

Part III

Relationship Between Structure and Function

In: Phosphoglycerate Kinase: A Hinge-Bending Enzyme
Editor: Mária Vas
ISBN: 978-1-62808-836-6

Chapter IX

Architecture of the Active Site and the Importance of Charge Balance in Catalysis

Jonathan P. Waltho[1], Matthew W. Bowler[2,3], Matthew J. Cliff[1] and Mária Vas[3]

[1]The Krebs Institute & Department of Molecular Biology and Biotechnology, The University of Sheffield, Sheffield, UK
[2]EMBL Grenoble, Grenoble, France
[3]Unit for Virus Host Cell Interactions, UMI 3265 (CNRS-EMBL-UJF), Grenoble, France
[4]Institute of Enzymology, Research Centre for Natural Sciences, Hungarian Academy of Sciences, Budapest, Hungary

Early crystal structures, even of the binary enzyme-substrate complexes (cf. Chapters I. 1.4. and 1.6.), together with various structural and mutagenesis studies of the solubilised enzyme (cf. Chapters II. 2.1.-2.9.) have already provided preliminary information on the steric position and the nature of the substrate binding sites, including identification of some important side-chains interacting with the substrates (see also Figures 4., 5. and 7.). Later on, crystal structures of various PGK ternary complexes have been also determined, but only two of them (Tm and TbPGKs) have appeared approximately in the possible active form that contains the two domains in the so-called closed conformation (cf. Chapter I. 1.8.). Comparing the open and closed active sites, the formation of several new side-chain contacts and new interactions with the bound substrates have been revealed for the closed form (cf. Figure 9). Furthermore, modelling of either bPG or the transferring phospho-group into the active site of TbPGK, representing the most closed structure at that time [84], have suggested interactions exclusively characteristic of the functioning closed active site of PGK (cf. Figure 14).

This picture of the architecture of PGK active site, however, not only confirmed, but further refined by the recently published high resolution crystal structures of two different transition state analogue (TSA) complexes of HsPGK [76]. Besides, domain closure for the

HsPGK has been already demonstrated by SAXS experiments in solution (Chapter II. 2.9.), in the presence of PG and MgATP [109].

TSA-s (such as aluminium and magnesium fluoride species) of the transferring planar PO_3^- group are often used in studying the fundamental mechanism of phospho-group transfer by enzymes [413, 414]. The metal-fluoride TSA complexes are particularly informative because ^{19}F NMR can be used to validate the interpretation of X-ray diffraction data and to report on the local electrostatic and protonic environments in the vicinity of the transferring phospho-group [415-417]. Indeed, the trigonal bipyramidal trifluoromagnesate (MgF_3^-) complexes readily assemble in the active site of a range of phospho-transfer enzymes [414, 415, 418, 419], where MgF_3^- is axially coordinated by donor and acceptor O-atoms. MgF_3^- is isoelectronic and almost isosteric with the transferring PO_3^- group. However, in the presence of Al^{3+}, MgF_3^- was shown in a number of enzymes to be readily out-competed by tetrafluoroaluminate (AlF_4^-) [415], which, although having the same net charge as PO_3^-, has non-isosteric octahedral geometry. This indicates that these enzymes are prioritizing retention of charge rather than geometry in their selection of a surrogate for PO_3^-. Moreover, accommodation of a single, negatively charged moiety, such as MgF_3^- or AlF_4^-, ensures that the number of positively and negatively charged groups contributing to the active site is exactly balanced [415]. A dominant role for such charge balance in transition state stabilization by phospho-transfer enzymes (the so-called charge-balance hypothesis) was suggested earlier [155]. The crystallographic structures of HsPGK·PG·MgF_3^-·MgADP and HsPGK·PG·AlF_4^-·MgADP TSA complexes (1.47 Å and 1.50 Å resolution, respectively) satisfactorily supported the charge-balance hypothesis as well as reported for the first time the catalytically relevant, fully closed conformation of any PGK [76]. Thereby, these structures resolved the ambiguities associated with previous interpretations of PGK-catalysed phospho-transfer.

The crystal structure of HsPGK·PG·MgF_3^-·MgADP TSA complex has revealed an even better closed conformation than any of the previously observed ones (Figure 79A). The domains are rotated 33.5° towards each other from their positions in the open conformation of the pig muscle PGK·PG·MgADP complex and a further 5.5° from their positions in the structure previously thought to be fully closed (TbPGK·PG·MgADP). The electron density maps were of excellent quality for almost all of the residues and the ligands were clearly located in the active site (Figures 80A and 80B); with an MgF_3 moiety occupying the position of the transferring PO_3^- group. The MgF_3^- has TBP geometry with the equatorial positions occupied by the three fluorine atoms and the two axial positions occupied by donor and acceptor oxygen atoms from ADP and PG. Figure 80A also details the coordination of the three equatorial fluorides. The ^{19}F NMR spectrum of this complex (Figure 80C, spectrum (1)) allowed assignment of the resonance at 161 ppm as corresponding to F_1 (cf. Figure 76A) owing to its small deuterium sum isotope shifts 17 and relatively narrow line width, but the environments of F_2 and F_3 are too similar to permit an assignment of resonances F_A and F_B.

The details of the active site in the other, HsPGK·PG·AlF_4^-·ADP TSA complex structure (Figure 76B) are very similar in respect of substrate and side-chain interactions, with one exception. Namely, instead of AlF_3^0 the species of AlF_4^- has been identified by both crystallography based on the electron density map (Figure 80B) and by NMR (Figure 80C, spectrum (2)). In common with the HsPGK·3PG·MgF_3^-·MgADP TSA complex, the presence

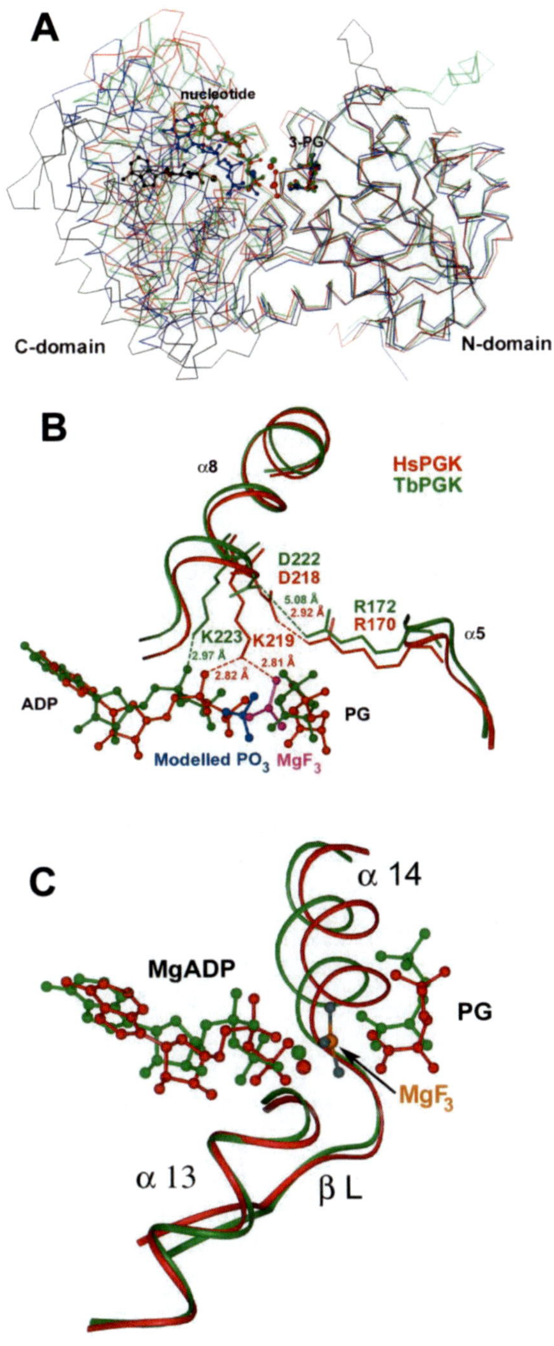

Figure 79. Comparison of the conformation of HsPGK·PG·MgF_3^-·ADP complex with the previously determined PGK structures: (A) The Cα traces of HsPGK (red), Tb·PG·MgADP (green), TmPGK·PG·MgAMP-PNP (blue) and pig muscle PGK·PG·MgADP (black) are superimposed according to the inner β-strands of the N-terminal domain; (B) New interactions of D218 and K219 (stick models) of HsPGK in comparison with TbPGK; (C) Movement of helix 14 relative to helix 13 in HsPGK in comparison with TbPGK. The secondary structural elements in (B) and (C) are illustrated by ribbon diagrams of the respective colours. The figures were prepared by using the respective X-ray coordinates.

of four well-resolved peaks of the NMR spectrum (2) indicates that the fluorine atoms are each bound in a distinct site and do not exchange environments on the NMR timescale. These observations clearly argue in favour of the charge-balance hypothesis, i.e., that the charge of TSA is more important than its geometry. The square planar geometry of the AlF_4^- moiety, indeed, has little effect on the active site. Figure 80D shows that, while the geometries of the MgF_3^- and the AlF_4^- moieties are different, the residues responsible for binding the fluorine atoms are identical. Fluorine F_1 coordinated by the catalytic magnesium and Lys 215:Nζ is located in the same position as F_1 in the HsPGK·PG·MgF_3·ADP TSA complex (positional difference = 0.02 Å). In particular, the side-chain of Lys 219 is also in the same position in the two structures. - The accommodation of F_4 results only in a small perturbation of the backbone geometry involving residues Thr 393–Gly 396, giving an associated main chain heavy atom RMSD of 0.15 Å. Overall, the active site is robust and insensitive to the steric demands of switching from a TBP to an octahedral metal fluoride TSA complex. Otherwise, the structures of the two TSA complexes are almost identical: overall RMSD is only 0.21 Å for all main-chain heavy atoms, which improves to 0.12 Å at the domain level. The latter improvement arises from the domain orientations of the two complexes differ slightly, the HsPGK·PG·AlF_4^-·ADP TSA complex being closed by only a further 0.5°.

Domain closure occurs primarily through alterations in the backbone dihedral angles in the two previously identified hinges, at the C-terminus of helix 7 and one at the N-terminus of the β-strand L [74, 80] (cf. Chapter I. 1.11.), which reduces the distance between the donor and acceptor oxygen atoms to 4.3 and 3.9 Å, respectively, in the MgF_3^- and the AlF_4^- containing structures (compared to 5.0 Å in the TbPGK·PG·MgADP complex). The closure is associated with the formation of three salt bridges: two of them (Arg 65-Asp 218 and Lys 219-Glu 293; HsPGK numbering) identified previously in the TbPGK·PG·MgADP complex and Arg 170-Asp 218. This latter new salt bridge is formed exclusively in the TSA complexes and acts in concert with the relative movement of the α-helix 8 and with the newly demonstrated interaction of Lys 219 with the TSA (Figure 81B).

The position and this new interaction of Lys 219 is remarkable, since it represents an accurate identification the direct role of this key basic side-chain of HsPGK in the chemical phospho-transfer. Previous site-directed mutagenesis studies (cf. Chapter II. 2.7.) have already identified Arg 38 [107, 125] and Lys 215 [107, 126] as essential components of the catalytic machinery, both of which coordinate fluorides in the HsPGK·PG·MgF_3·MgADP TSA complex. An important role for Lys 219 was also inferred from mutagenesis [227], but in all previous PGK structures, this residue has a common position too distant from the putative position of the transferring phospho-group to interact with it. Uniquely, in the HsPGK·PG·MgF_3·MgADP TSA complex the Lys 219 side-chain is rotated 24° around the axis of helix 8 so that the ε-amino group forms a hydrogen bond with F_3 of the MgF_3– moiety (Figure 81B). Lys 219 has previously been highlighted as a critical residue in the hydrogen bonding networks that alter upon domain closure, through its interaction with Asn 336:Oδ [349]. Thus, the movement of Lys 219:Nζ induced by the negative charge at the transferring phosphate position is likely to be transmitted throughout the interdomain region; triggering and/or stabilizing the closure mechanism (cf. Chapter X.).

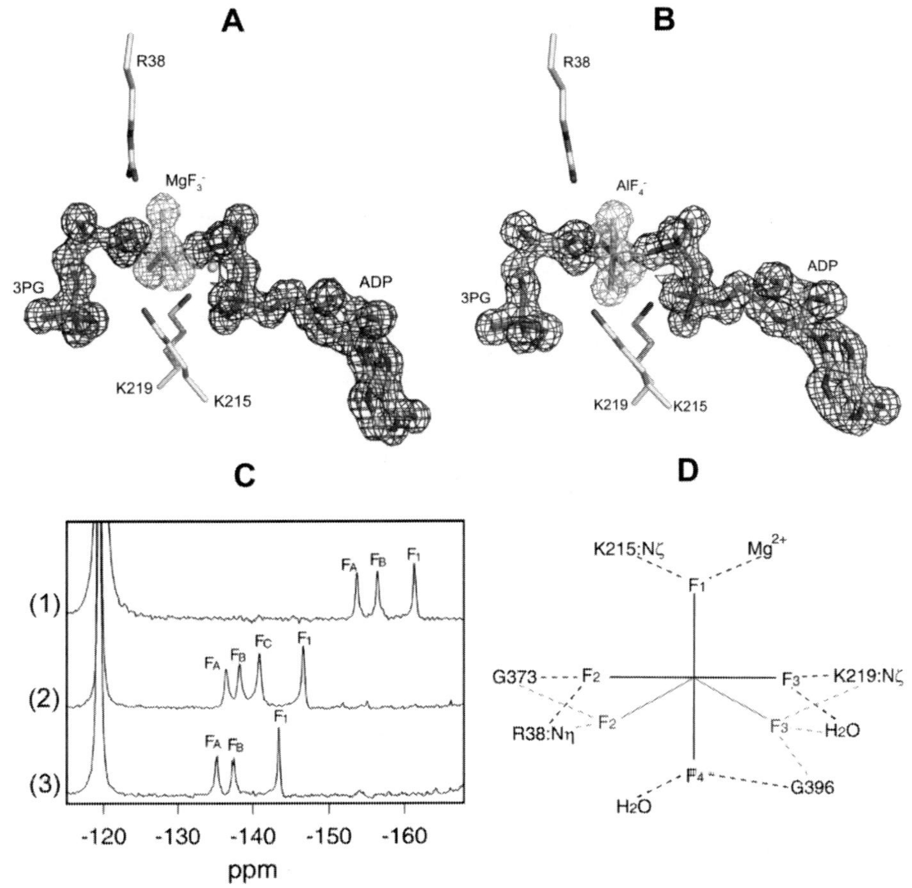

Figure 80. Comparison of the TSA complexes of HsPGK with MgF_3^- and AlF_4^- (adapted with permission from [76]; Copyright 2010 American Chemical Society): (A) and (B) are the electron densities (1.47 and 1.50 Å resolutions, respectively). (C) represents the ^{19}F NMR spectra of the complexes of the wild type enzyme with MgF_3^- (1) and AlF_4^- (2), respectively. Spectrum (3) refers to the K219A mutant complex with AlF_3. (D) is a schematic illustration of the equivalence of the fluorine atoms between the HsPGK·PG·AlF_4^-·ADP (black) and the HsPGK·PG·MgF_3^- (grey) TSA complexes.

As for the better closure of domains, the state of the main hinge at βL that connects the helices α13 and α14 is compared to the less closed TbPGK (Figure 79C). Indeed, α14 moved more relative to α13 in case of the TSA structure of HsPGK, due to the changes in the shape of βL. The details of these changes are seen at the level of atomic interactions in cases of the open HsPGK·PG (Figure 81A), the closed TbPGK·PG·MgADP (Figure 81B) and the completely closed HsPGK·PG·MgF_3·MgADP TSA (Figure 81C) complexes. It is evident that the N-termini of helices α13 and α14 are approach better and better each other and with their partial positive charges contribute to binding of the negatively charged substrates and TSA.

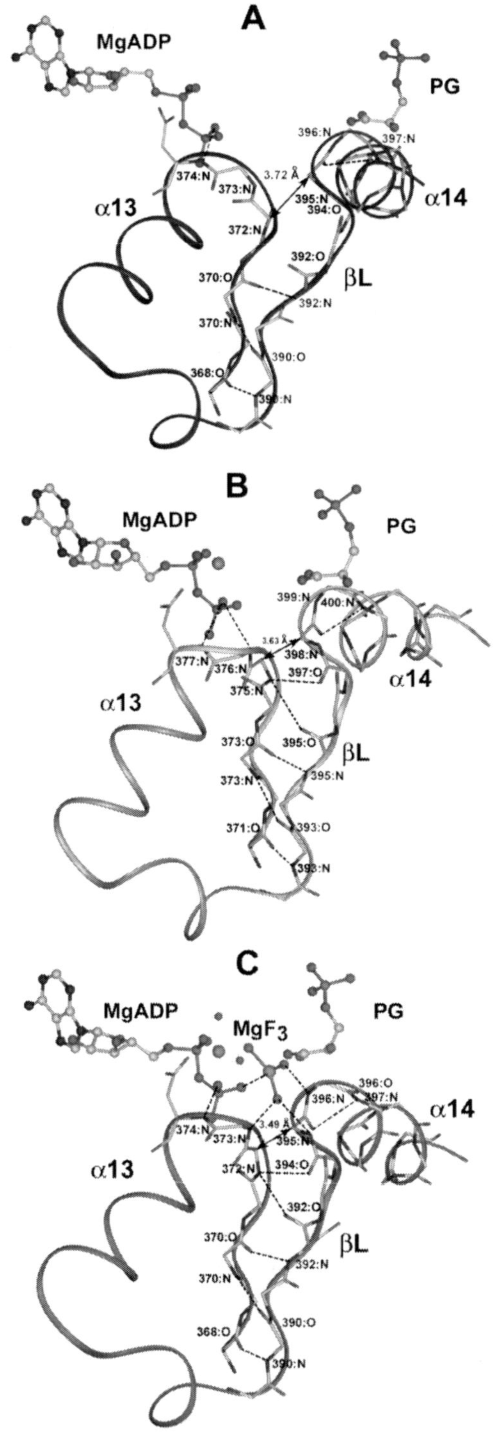

Figure 81. Atomic interactions in the main hinge regions in the open conformation of pig muscle PGK (A) and the closed conformations of TbPGK (B) and HsPGK (C): The figures were prepared by using the respective X-ray coordinates. The ribbon diagrams of the helices α13 and α14 and the connecting βL as well as the bound substrates (ball and stick models) are illustrated in all three cases. The atomic interactions are shown by dashed lines.

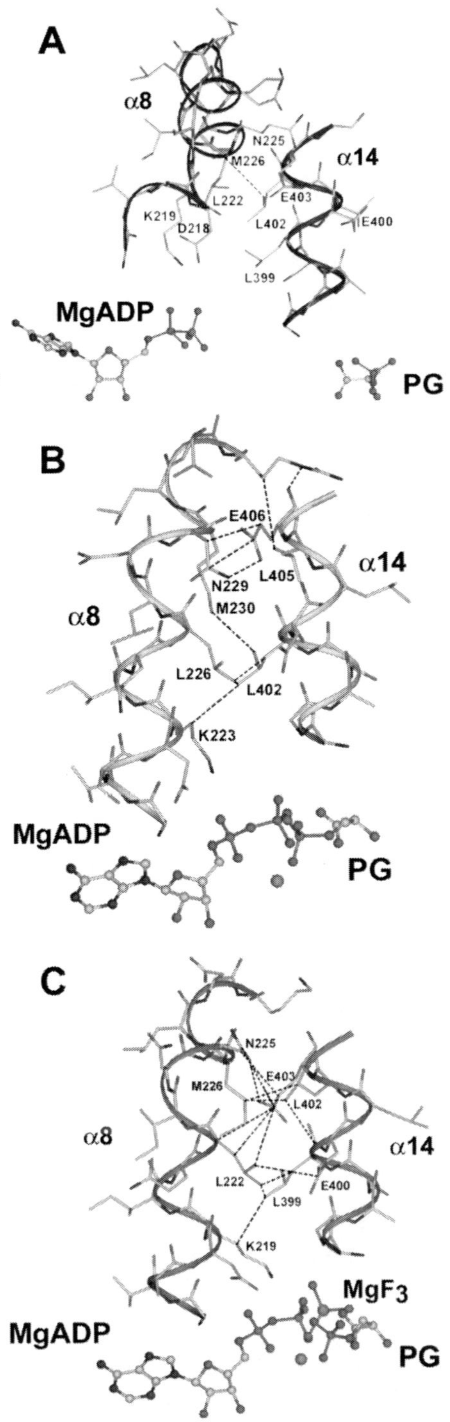

Figure 82. Atomic interactions between helices α8 and α14 in the open conformation of pig muscle PGK (A) and the closed conformations of TbPGK (B) and HsPGK (C): The figures were prepared by using the respective X-ray coordinates. The ribbon diagrams of the helices α8 and α14 and the connecting βL as well as the bound substrates (ball and stick models) are illustrated in all three cases. The atomic interactions are shown by dashed lines.

Another interesting feature of the domain closure is a characteristic conformational change of α8 upon nucleotide binding and its movement relative to α14, as being constituents of the C- and N-terminal domains, respectively. By this mechanism the open channel between them, i.e., between the two domains, is prepared for closure (Chapter I. 1.10). The closed state is stabilised by the interactions formed between the complementary side-chains of the two helices, as demonstrated with differently closed stages of PGK (Figures 82A, B and C).

In addition to revealing the full extent of domain closure, these TSA complexes have also shown, for the first time, the catalytically relevant coordination of the essential Mg^{2+} in the enzyme active site (Figures 76A and 76B), as also discussed previously (Chapter VIII). This coordination to TSA indicates a distinct role for γ-phosphate coordination by Mg^{2+} during catalysis and, for example, explains the specificity of PGK towards chiral thiophosphate analogues (the α-S_p and β-S_p diastereomers) of ATP, which was observed long time ago [159] (cf. Chapter II. 2.3).

The importance of the local charge balance in the active site of the functioning PGK is also further supported by NMR and X-ray investigation of the mutant K219A of HsPGK in complex with fluoroaluminate (but not with fluoromagnesate) and with PG and MgADP. NMR measurements (not shown) revealed three ^{19}F resonances with 1:1:1 ratio of integrals characteristic of trifluoroaluminate, AlF_3^0, TSA species. Moreover, HsPGK(K219A) does not measurably bind an AlF_4^- species, even at elevated fluoride concentrations (up to 70 mM). Thus, these data establish that removal of the positive charge contributed by Lys 219 in the coordination of TSA in the wild-type enzyme induces a corresponding unit reduction in negative charge through the loss of one fluoride from the aluminium fluoride moiety in the mutant complex. Thus, the chemical mutation arises as a consequence of the protein mutation, and the dominance of AlF_3^0 over AlF_4^- in the HsPGK(K219A)·PG·AlF_3^0·MgADP TSA complex is exactly that predicted by the charge-balance hypothesis. The crystal structure of this complex was also solved to a resolution 1.56 Å. The differences are minor between the mutant and the wild-type Al-fluoride complexes: the positions previously occupied by the side-chain Lys 219, including the ε-amino group, are occupied by water molecules. A further surprising result from the HsPGK(K219A)·PG·AlF_3^0·MgADP TSA structure is that, unlike all other phospho-transfer enzyme TSA complexes previously reported to contain an AlF_3^0 moiety, the aluminium ion has octahedral rather than the TBP geometry. From the four equatorial coordinating atoms of AlF_3^0 only three can be identified as fluorines, but the fourth one is a species with less electron density, namely a water molecule. The fact that the trigonal entity does not form, even in circumstances where an AlF_3^0 moiety is the overwhelmingly preferred metal-fluoride, supports the proposal that many previously reported AlF_3^0 TSA complexes of other phospho-transfer enzymes may, in reality, be MgF_3^- complexes [415].

In summary, the behaviour of HsPGK in the three different metal-fluoride complexes provided experimental support for the charge-balance hypothesis. Namely, in HsPGK TSA complexes the charges from seven cationic groups of the enzyme (Arg 38, Arg 65, Arg 122 and Arg 170 from the N-domain and Lys 215, Lys 219 and Asp 374-Mg^+ from the C-domain) balance the charges from the seven anionic groups of the TSA ligands (3 on ADP, 3 on PG and 1 on MgF_3^- or AlF_4^-). Analysis of the charge distributions within the TSA complexes of several other phospho-transfer enzymes have shown that the local zero net charge is their common feature [76].

In: Phosphoglycerate Kinase: A Hinge-Bending Enzyme
Editor: Mária Vas
ISBN: 978-1-62808-836-6

Chapter X

Substrate-Assisted Cooperative Interaction of the Two Domains: Mechanism of Doman Closure at the Level of Atomic Interactions

Mária Vas
Institute of Enzymology, Research Centre for Natural Sciences,
Hungarian Academy of Sciences, Budapest, Hungary

Considering the above details of the functionally relevant active site of the TSA complexes of PGK, the question arises about details of the machinery of its formation through domain closure, i.e., about operation of the defined molecular hinges (cf. Chapter I. 1.11.).

X-ray crystallographic data on PGK from various sources together with various type of solution experiments (chemical modification, DSC and SAXS measurements) suggested high level of cooperativity between the two domains and that domain closure, required by the catalysis, will take place only in the presence of the simultaneously bound substrates, i.e., only in the enzyme–substrate ternary complexes (cf. Chapters II. 2.5., 2.8., 2.9.). Molecular contacts have been also identified upon separate binding of each substrate which describe the possible route of transmission of the substrate-induced structural effects from one domain to the other (cf. Figure 39) [108]. Yet, the operation of the main molecular hinge still had to be elucidated in terms of atomic details and of substrate binding.

In order to compare the consorted effect of both substrates with their separate effects, molecular contacts of the main hinge region (β-strand L) of PGK, as determined in 10 different crystal structures, including the substrate-free enzyme and its binary and ternary complexes with substrates (or analogues) were analyzed [109]. Although each static structure reflects time and space average of numerous dynamic structures, the crystal structures of the various enzyme–substrate complexes can be considered as representatives of various stages of substrate-caused conformational changes on the route from open to closed domain conformations.

Table 28. Lists of the atomic contacts in the main hinge region at β-stand L (adapted from [109])

Structural elements involved	Pig muscle PGK (open)				*B. stearotherm.* PGK (open)			*T. maritima* PGK (closed)			*T. brucei* PGK (closed)		
	Atom1	Atom2	Substrate free	PG binary	Atom1	Atom2	MgADP binary	Atom1	Atom2	ternary	Atom1	Atom2	ternary
H1-βL	R38:NE	T393:O	5.81	4.04	R36:NE	T371:O	5.57	R36:NE	T374:O	4.40	R39:NE	**T396**:O	**2.88**
	R38:NH1	T393:O	6.12	**3.53**	R36:NH1	T371:O	6.73	R36:NH1	T374:O	5.19	R39:NH1	T396:O	4.64
	R38:NH2	T393:O	5.16	**2.68**	R36:NH2	T371:O	4.76	R36:NH2	T374:O	**2.90**	R39:NH2	T396:O	**2.93**
	R38:NE	T393:OG1	8.21	5.70	R36:NE	T371:OG1	7.52	R36:NE	T374:OG1	**3.30**	R39:NE	T396:OG2	5.33
PG-H1	3-PG:O1	R38:NH1		3.52				PG:O1	R36:NH1	2.71	3-PG:O1	R39:NH1	2.74
	3-PG:O2	R38:NH1		4.06				PG:O1	R36:NH2	3.07	3-PG:O1	R39:NH2	3.21
Nucleotide –H13 and βJ		T375:N			ADP:βO1	S353:N	2.89	[a]A-NP:βO1	S356:N	2.92	ADP:βO1	S378:N	2.95
		T375:OG1			ADP:βO1	S353:OG	2.60	[a]A-NP:βO1	S356:OG	2.42	ADP:βO1	S378:OG	2.72
		N336:OD1			ADP:βO3	N316:ND2	2.79	[a]A-NP:N	N317:OD1	3.02	ADP:βO3	N338:ND2	3.04
		G373:N			ADP:βO2	G351:N	3.31	[a]A-NP:βO2	G354:N	3.25	ADP:βO2	G376:N	3.42
		G373:N						[a]A-NP:γO2	G354:N	2.96			
βK-H14	G371:O	S398:OG	5.45	**2.76**	G349:O	S376:OG	7.01	G352:O	S379:OG	5.45	G374:O	S401:OG	6.41
βK-βJ	G371:O	N336:ND2	6.55	4.21	G349:O	N316:ND2	**3.04**	G352:O	N317:OD1	**3.39**	G374:O	N338:ND2	**2.72**
βK-βL	I370:O	S392:N	2.97	2.90	I348:O	S370:N	2.90	V351:O	S373:N	2.65	I373:O	S395:N	2.70
	G372:N	S392:O	4.62	4.16	G350:N	S370:O	5.36	G353:N	S373:O	**2.73**	G375:N	S395:O	**3.46**
	G372:N	G394:O	4.47	4.90	G350:N	G372:O	7.20	G353:N	G375:O	4.84	G375:N	G397:O	**3.24**
within βL	T393:OG1	G394:N	3.30	3.37	T371:OG1	G372:N	3.31	T374:OG1	G375:N	3.32	T396:OG1	G397:N	3.42
	S392:OG	G394:N	4.03	**2.97**	S370:OG	G372:N	**3.42**	S373:OG	G375:N	**3.23**	S395:OG	G397:N	**2.84**
H13-βL	G373:N	T393:O	6.40	6.54	G351:N	T371:O	7.75	G354:N	T374:O	**3.31**	G376:N	T396:O	5.10
βL-H14	S392:OG	S398:OG	4.38	4.58	S370:OG	S376:OG	**2.72**	S373:OG	S379:OG	5.04	S395:OG	S401:OG	**3.00**
	G394:O	S398:OG	3.04	2.63	G372:N	S376:OG	3.49	G375:O	S379:OG	2.99	G397:O	S401:OG	2.59

Atomic distances are given in Å-s for the most characteristic open and closed crystal structures, selected from the total investigated 10 ones. The H-bonding and ionic interactions that are formed upon substrate binding or upon domain closure are indicated in bold.

[a]A-NP stands for AMP-PNP (see Abbreviations)

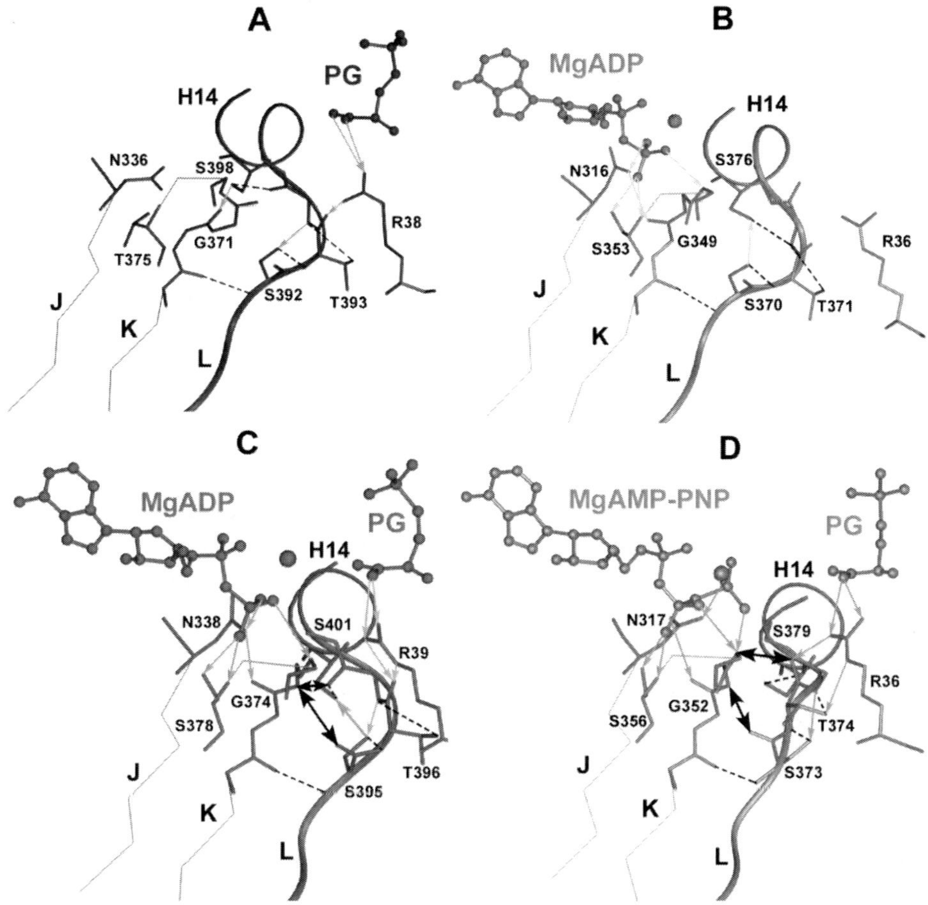

Figure 83. Transmission of the substrate caused conformational changes from the substrate site towards the main hinge at βL (adapted from [109]): In (A) and (B) the PG and MgADP caused effects are shown separately in cases of the open conformations of pig muscle and Bs PGKs, respectively. In (C) and (D) the simultaneous effects of PG and the nucleotides (MgADP or MgAMP-PNP) are shown in cases of the closed conformations of Tb and Tm PGKs, respectively. The substrates are represented by ball and stick models, while the side-chains are shown by stick models. The elements βL and α14, constituting the main hinge, are represented by ribbon diagrams.

The inter-atomic distances in the most relevant structures are listed in Table 28. As expected, the crucial C-terminal part of β-strand L and the connecting loop towards helix 14, i.e., the part, which conformation is characteristically changed between the open and domain-closed structures (cf. Figure 16), is a highly conserved region. It includes the sequence Ser392-Thr393-Gly394-Gly395-Gly396-Ala397-Ser398 (mammalian PGK numbering). Among these residues even the less conserved Ser 398 is identical at least up to 74% among the PGK sequences available in ExPASy Molecular Biology Server. From Table 28 it is evident, that independent of the presence of substrates, the functional OH-groups of Ser 392/Bs370/Tm373/Tb395, Thr 393/Bs371/Tm374/Tb396 and Ser 398/Bs376/Tm379/Tb401 form three invariable H-bonding interactions with the peptide N- and O-atoms of this region, existing in all the known PGK crystal structures (also in those structures, including HsPGK, not listed in Table 28). These H-bonds, therefore, contribute to the basic folding feature of this region, which is independent of substrate binding or the conformational states of the protein molecule, but which may serve as a prerequisite for the operation of the hinge. The usual H-bonds between the peptide atoms of parallel parts of the β-strands K and L (e.g., between Ile 370/Bs348/Tm351/Tb373:O and Ser 392/Bs370/Tm373/Tb395:N) may have similar roles. A part of the invariant H-bonds (dashed black lines) are shown in Figure 83.

For the present analysis, however, the most interesting atomic contacts are formed exclusively in the specific enzyme–substrate complexes, especially in the closed ternary complex structures, in addition to the contacts already present in the respective binary complexes. Several PGK ternary complex crystal structures are available from various sources [69, 72-74, 80], but some of them are open structures [69, 72, 80]. As shown by SAXS analysis (cf. Chapter II. 2.9.), these open structures must be crystallographic artefacts. Therefore, in the present analysis these open structures were considered as binary complexes. At the time of the analyses there were only two structures that exhibited really closed conformations, namely, the abortive ternary complex of TbPGK with PG and MgADP [74] and the ternary complex of TmPGK with PG and MgAMP-PNP (an unreactive analogue of MgATP) [73]. The atomic contacts in the main hinge β-strand L, exclusively characteristic of the closed conformations of Tb and TmPGK-s (2.5 and 2.0 Å resolutions, respectively) are selected on the basis of comparison to the open structures of the two available binary complexes with PG (2.0 Å resolution, pig muscle PGK [15]) and with MgADP (1.6 Å resolution, Bs PGK [6]). The structure-based sequence alignment was performed to compare of PGK-s from various sources. The contacts characteristic of each binary complex and of the closed conformations are listed in Table 28 and depicted in Figure 83. For comparison, the contacts characteristics of the recently published HsPGK TSA complex are also shown in Table 29.

It is shown that, in addition to the permanent invariant H-bonds (dashed lines), further contacts are formed in and around β-strand L upon binding of either PG (Figure 83A) or MgADP (Figure 83B) in the open conformations of their binary complexes. As noted earlier [15, 108] (cf. Figure 7), upon binding of PG a new contact is formed between one of the two NH atoms of Arg 38/Bs36/Tm36/Tb39 (the residue involved in electrostatic interaction with the carboxylate of PG) and the peptide O-atom of Thr 393/Bs371/ Tm374/Tb396. This important new interaction essentially connects the N- and C-terminal domains. As an extension of this H-bond network, a new strong H-bond is formed between the OH-group of Ser 392/Bs370/Tm373/Tb395 and the peptide N-atom of Gly 394/Bs372/Tm375/Tb397. These H-bonds are characteristic of all the known PGK crystal structures that bind PG, independent of their conformational (open or closed) states. There is another noticeable H-bond between the peptide O-atom of Gly 371/Bs349/Tm352/Tb374 and the OH–group of Ser 398/Bs376/Tm379/Tb401, but it can only be observed in the PG binary complex [15] among the PG binding structures. Thus, this H-bond on itself may not be relevant for the domain closure, but it still indicates a close distance between the N-termini of helices 13 and 14, which is not the case for the substrate-free enzyme. Movement of helix 14 upon PG binding, indeed, was reported [15, 74].

Table 29. Comparison of the atomic contacts of HsPGK (open and closed structures) in the main hinge region at β-stand L to those operating in the open and closed conformations of pig muscle and Tb PGKs, respectively.

Structural elements involved	Pig muscle PGK (open)				HsPGK		TbPGK (closed)		
	Atom1	Atom2	Substrate free	PG-binary	PG-binary Open	TS-ternary Closed	Atom1	Atom2	ternary
H1-βL	R38:NE	T393:O	5.81	4.04	**2.96**	**3.01**	R39:NE	T396:O	**2.88**
	R38:NH1	T393:O	6.12	**3.53**	4.60	4.69	R39:NH1	T396:O	4.64
	R38:NH2	T393:O	5.16	**2.68**	**2.77**	**2.87**	R39:NH2	T396:O	**2.93**
	R38:NE	T393:OG1	8.21	5.70	5.27	5.38	R39:NE	T396:OG2	5.33
3PG-H1	3-PG:O1	R38:NH1		**3.52**	**2.74**	**2.86**	3-PG:O1	R39:NH1	**2.74**
	3-PG:O2	R38:NH1		4.06	3.17	**3.37**	3-PG:O1	R39:NH2	**3.21**
Nucleotide –H13 and βJ	ADP:βO1	T375:N				2.92	ADP:βO1	S378:N	2.95
	ADP:βO1	T375:OG1				2.63	ADP:βO1	S378:OG	2.72
	ADP:βO3	N336:OD1				3.13	ADP:βO3	N338:ND2	3.04
	ADP:βO2	G373:N				3.41	ADP:βO2	G376:N	3.42
	ADP:βO1	G374:N				2.80			
βK-H14	G371:O	S398:OG	5.45	2.76	3.21	6.28	G374:O	S401:OG	6.41
βK-βJ	G371:O	N336:ND2	6.55	4.21	**3.03**	**2.91**	G374:O	N338:ND2	**2.72**
βK-βL	I370:O	S392:N	2.97	2.90	2.82	2.81	I373:O	S395:N	2.70
	G372:N	S392:O	4.62	4.16	3.76	**3.23**	G375:N	S395:O	**3.46**
	G372:N	G394:O	4.47	4.90	5.15	**3.18**	G375:N	G397:O	**3.24**
within βL	T393:OG1	G394:N	3.30	3.37	3.40	3.48	T396:OG1	G397:N	3.42
	S392:OG	T393:N	3.51	3.10	3.01	3.01	S395:OG	T396:N	3.01
	S392:OG	G394:N	4.03	**2.97**	**3.01**	**3.01**	S395:OG	G397:N	**2.84**
H13-βL	G373:N	T393:O	6.40	6.54	5.88	4.59	G376:N	T396:O	5.10
βL-H14	S392:OG	S398:OG	4.38	4.58	4.61	**3.29**	S395:OG	S401:OG	**3.00**
	G394:O	S398:OG	3.04	2.63	2.91	2.85	G397:O	S401:OG	2.59

Atomic distances are given in Å-s.
The H-bonding and ionic interactions that are formed upon substrate binding or upon domain closure are indicated in bold.

In the MgADP-bound structure (Figure 83B) the β-phosphate O-atoms of the nucleotide are linked through well-formed H-bonds to the OH-group of Thr/Ser 375/Bs353/Tm356/Tb378, the residue located in helix 13 (sequentially preceding β-strand L), as well as to the peptide N-atoms of the same residue and of Gly 373/Bs351/Tm354/Tb376. There are also further contacts of MgADP with helix 13, noted elsewhere [6]. Furthermore, due to the simultaneous interaction of the β-phosphate of MgADP with the conserved Asn 336/Bs316/Tm317/Tb338 from the β-strand J, this strand is fixed together with the connected β-strand K and with the sequentially following helix α13 within the C-domain upon the nucleotide binding. As a consequence, Asn 336/Bs316/Tm317/Tb338 directly interacts with the peptide O-atom of Gly 371/Bs349/Tm352/Tb374 at the N-terminus of helix 13. These new interactions are also characteristic of other PGK complexes, in which the bound nucleotide interacts with helix 13 and thereby inducing a typical conformational change at the N-terminus of this helix. Due to this change, the N-atom of the sequentially next Gly 372/Bs350/Tm353/Tb375 is now directed towards the β-strand L, but no detectable interaction is yet formed with it. Thus, in contrast to the PG binary complex, no definitive new contact is formed with β-strand L in the binary complex with the nucleotide. In spite of this, a new strong H-bond is observed at the end of β-strand L between the OH-groups of the Ser 392/Bs370/Tm373/Tb395 and Ser 398/Bs376/Tm379/Tb401. This bond can stabilize a certain conformational state of β-strand L. The H-bond between the OH-group of Ser 392/Bs370/Tm373/Tb395 and the peptide N-atom of Gly 394/Bs372/Tm375/Tb397, characteristic of PG binding, may be also present here, but in a weaker form. The exceptional behaviour of certain nucleotide complexes (such as of MgATP or MgAMP-PCP) may be related to the different binding modes of these nucleotides to PGK in some crystals, showing no interactions with helix 13 [7, 72, 107].

Figure 83C illustrates the atomic interactions of the same region as determined in the closed crystal structure of TbPGK ternary complex. Most of the contacts are formed separately in PG and MgADP binary complexes; only few additional contacts (Table 28) are formed. Among them there is a new contact between the NE atom of Arg 38/Bs36/Tm36/Tb39 and the peptide O-atom of Thr 393/Bs371/Tm374/Tb396, which further strengthens the interaction between the two domains. This contact, however, is not really characteristics of the closed conformation, since this bond also exists in the open conformations of all three different pig muscle PGK ternary complexes. There are, however, two other, more interesting peptide H-bonds (denoted by the arrows in Figure 83C) from the same atom of Gly 372/Bs350/Tm353/Tb375(N) to Ser 392/Bs370/Tm373/Tb395(O) and to Gly 394/Bs372/Tm375/Tb397(O). The same contacts have been also indicated above in Figures 81B (TbPGK) and 81C (HsPGK). Neither of these bonds exists in any open PGK conformations (see Figures 81A and 81A showing pig muscle PGK ternary complex with PG and MgADP and binary complex with PG, respectively, as well as Figure 83B showing BsPGK binary complex with MgADP). The first bond creates a new contact of βL with the closest other strand, βK, i.e., between the β-strands before and after helix 13, in sequence. The second one connects the N-terminus of helix 13 to the end of βL, i.e., to the N-terminus of the following helix 14. In this way the helices before and after βL, in sequence, are connected. These new H-bonds, characteristic of the ternary complex, constitute an extension of the H-bond network (part of which already exists in the respective binary complexes), as a consequence of the simultaneously formed H-bonds on both sides of the hinge, under the concerted action of PG and MgADP. In spite of the small number of the new contacts, the shape of β-strand L is substantially changed as shown above (cf. Figure 16A). Thus, the bonds operating in the binary complexes now act together to bring βL into a conformation strikingly different from either of the binary complexes. Taken together, these changes actually can be considered as operation of a double-sided molecular switch, controlled by simultaneous action of the two substrates, which can fix β-strand L in a conformation optimal for the closed state.

In the case of the other known, not fully closed structure of the ternary complex of TmPGK with PG and the MgATP analogue, MgAMP-PNP [73] one can find a slightly different mechanism. Although the structure of the binary complex with MgAMP-PNP is lacking, from analysis of the contacts operating in the main molecular hinge of the closed TmPGK structure (Figure 83D) a variation of the mechanism may be deduced. In addition to the contact between Arg 38/Bs36/Tm36/Tb39(NH) and the peptide O-atom of Thr 393/Bs371/Tm374/Tb396, observed already in the open structure of PG binary complex, here again a new contact is formed between the NE atom of Arg 38/Bs36/Tm36/Tb39 and the OH-group of Thr 393/Bs371/Tm374/Tb396, i.e., in the way different from the case of the closed structure of TbPGK. This observation raises the possibility that besides the peptide O-atom of this Thr, its conserved side-chain may also have a direct role in domain closure. The new H-bonding interaction, found above for TbPGK between the atoms of Gly 372/Bs350/Tm353/Tb375(N) and Ser 392/Bs370/Tm373/Tb395(O), also operates in TmPGK (marked by an arrow) and completes the H-bonding interactions between β-strands L and K, similar to the closed structure of TbPGK. There is, however, a striking difference between the two closed structures: the H-bond between Gly 372/Bs350/Tm353/Tb375(N) and Gly 394/Bs372/Tm375/Tb397(O), characteristic of TbPGK, is replaced by the bond between Gly 373/Bs351/Tm354/Tb376(N) and Thr 393/Bs371/Tm374/Tb396(O) atoms in TmPGK (marked by an arrow). Thus, these two bonds are formed alternatively in the two closed structures. At the same time the contact between the OH-groups of the two serines (392/Bs370/Tm373/Tb395 and 398/Bs376/Tm379/Tb401), characteristic of the MgADP-bound structures of both binary and the ternary complexes, is suspended in case of MgAMPPNP binding to Tm PGK.

It is reasonable to assume that the formation of the new H-bond between Gly 373/Bs351/Tm354/Tb376(N) and Thr 393/Bs371/Tm374/Tb396(O) atoms is greatly helped by simultaneous interactions of both the β- and γ-phosphate O-atoms of MgAMP-PNP with the same Gly 373/Tm354(N) atom. In this way, from one side, the nucleotide effect is transmitted towards the peptide O-atom of Thr 393/Bs371/Tm374/Tb396 in βL, i.e., exactly where the effect of PG is transmitted from the other side through the above shown interactions with Arg 38/Bs36/Tm36/Tb39. These two-sided interactions of Thr 393/Bs371/Tm374/Tb396 in TmPGK, under the simultaneous control of both PG and the nucleotide, may be considered here as a variant of the above suggested double switch mechanism. The characteristic differences, that can be observed both in the molecular contacts (Figures 83C and 83D) and in the conformations of βL (cf. also Figure 16) in the two differently closed PGK structures direct our attention to the fact that this hinge region is a very mobile part of the molecule, which possibly adopts different conformations between the completely open and the closed states.

Molecular dynamics calculations by Chandra et al. [106] could also generate domain closure and describe the process at the level of relative movements of the key helices (numbered 7, 13 and 14) of the interdomain region. Their model does not deal with either any atomic details or the conformational change of βL, and only qualitatively support the existence of the other hinges at the C- and N-termini of the interdomain helix 7, as deduced earlier from the X-ray data obtained for TbPGK [74]. From the presently shown description of the operation of the main molecular hinge it seems also plausible that the main hinge itself directs the operation of the other hinges at the termini of helix 7 through the permanent molecular contacts existing in the well-structured interdomain region. A specific way of propagation of the conformational effect through the atomic contacts of completely conserved side-chains has been described elsewhere [80, 108] (cf. Chapters II. 2.7. and II. 2.8.) and makes the above picture of the molecular events more complete. In accordance with the primary role of βL, no characteristic changes of molecular contacts of the conserved side-chains occur at the termini of helix 7, there are notable changes only in the backbone H-bond pattern, especially at its C-hinge [116]. Site-directed mutagenesis with yeast PGK at the termini of helix 7 either did not support [229] or did not provide clearly conclusive results [232] in favour of the importance of the hinges at helix 7, in agreement with their possible secondary role. Other mutations at the interdomain region of yeast PGK (according to the mammalian PGK numbering Glu 192 in helix 7 [230] and His 390 in βL [233, 234, 236] did not produce dramatic effects on the kinetic properties (cf. Table 12). It was concluded therefore, that cumulative effect of multiple interactions of the side-chains in the interdomain region (rather than individual residues) is likely to be responsible for the efficient mechanism of ligand induced domain movement [188].

Based on the results of the above molecular graphical analysis, the essential molecular event of domain closure may be summarized as follows: the two bound substrates simultaneously direct the operation of a double molecular switch at the β-strand L, leading to substantial conformational changes of this main molecular hinge. Thus, under the synergistic action of both bound substrates their separate minor effects are enlarged resulting in a major change in the shape of β-strand L, possibly at the expense of their binding energy. This primary conformational change at βL is transmitted through the interaction of the conserved residues towards the other important hinges (e.g., termini of helix 7) and drives their operation. Operation of this mechanism is entirely supported by the available experimental data of site-directed mutation (Chapter II. 2.7.), DSC calorimetric (Chapter II. 2.8) and SAXS analysis (Chapter II. 2.9.).

A central question is the transmission of the effects of substrates from the substrate binding sites towards the main hinge. PGK seems to be crucial in this respect since its two essential catalytic residues, Arg 38 and Lys 215, move about 2-3 Å and more than 10 Å, respectively, upon binding of the substrates, raising the possibility of their direct involvement in domain closure. The contacts of PG with Arg 38, indeed, plausibly mediate the effect of PG towards the main hinge at βL, as illustrated by the arrows in Figure 36 and Figure 83. Mutagenesis studies, indeed confirmed the direct role of Arg 38 in domain closure, but at the same time, excluded the participation of Lys 215 in it (Chapters II. 2.7. and 2.8.).

Therefore, further mutagenesis studies have been carried out by exchanging Lys 219, Asn 336, Glu 343 and Thr 375 at the nucleotide site into Ala [228]. Except T375 that seems to be irrelevant in this respect, all the other three residues have been proved to be involved in domain closure (Chapters II. 2.7.-2.9.).

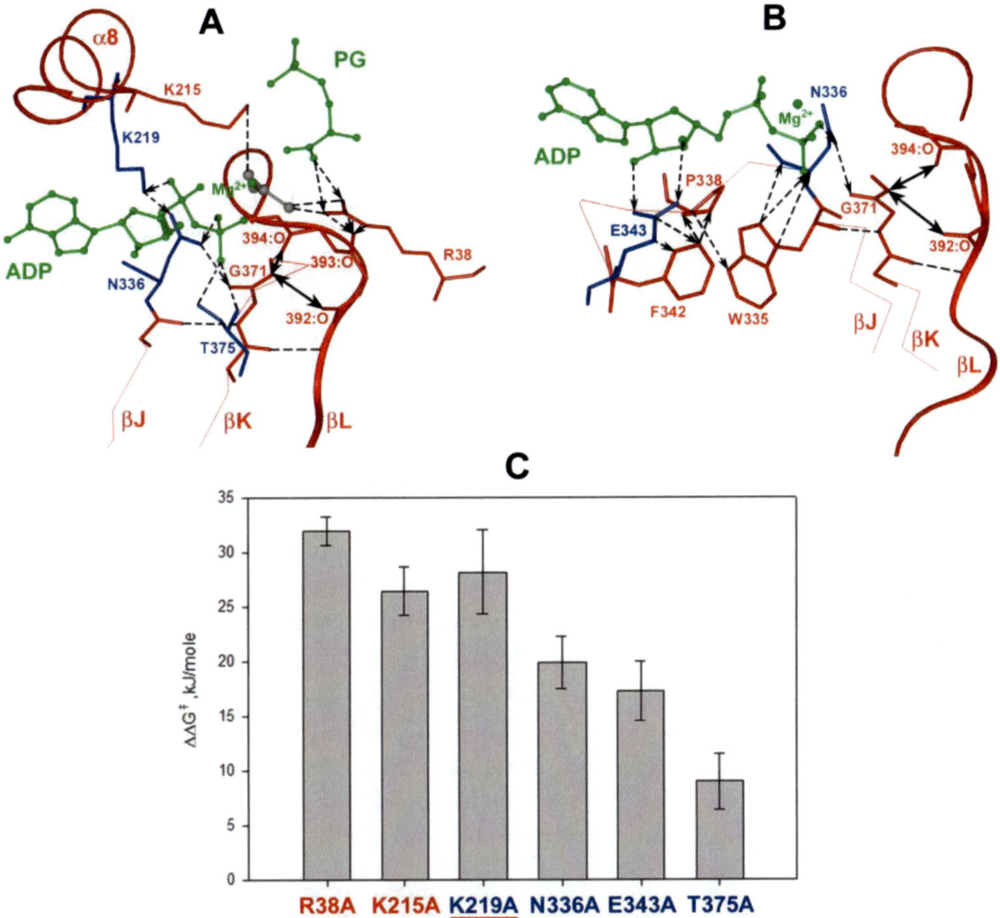

Figure 84. Role of the active site side-chains in the operation of the main hinge at βL (adapted with permission from [227]; Copyright 2008 American Chemical Society): In panels (A) and (B) the structural details (Cα traces, ribbons and stick models of certain side-chains) in the surroundings of the main hinge at βL, including the substrate-binding sites, are illustrated in case of the closed conformation of TbPGK. The transferring phospho-group (gray) is also modelled into the active site. The numbering of the side-chains corresponds to the mammalian PGK sequence numbering. The side-chains that have been replaced via mutagenesis with Ala are highlighted as blue stick models. Arrows indicate the transmission of the effects of PG and the nucleotides toward βL. The H-bonds formed only in the closed conformation of the ternary complex are shown by double sided black arrows. - The difference energy diagram as the effect of mutations is illustrated in panel (C). The activation free energy changes during the reaction catalysed by the various PGK mutants relative to the wild-type enzyme ($\Delta\Delta G^{*}$) are plotted for both the active site (labelled as red) and the nucleotide-binding (labelled as blue) mutants.

These residues interact with the nucleotide substrates. Panels A and B on Figure 84 depict (by arrows) the possible route of the conformational transmission between the nucleotide site and the main hinge region at βL in the case of the closed TbPGK crystal structure. In all nucleotide-binding structures, Lys 219/Tb223 (located in α-helix 8) and Asn 336/Tb338 (located in βJ) interact with the α- and β-phosphates of the nucleotide, respectively (Figure 84A). Furthermore, Lys 219/Tb223 and Asn 336/Tb338 are also H-bonded to each other. The side-chain of Asn 336/Tb338 is H-bonded further to the peptide

O atom of Gly 371/Tb374 close to the N-terminus of the α-helix 13. These interactions also strengthen the otherwise existing connections among β-strands J, K, and L; meanwhile, two new H-bonding connections (labelled with double-sided black arrows) are formed between β-strands K and L when both substrates are bound. The formation of the latter two H-bonds leads to an essential change in the conformation of βL, and as a consequence, domain closure is completed. This structural background clearly supports the role of both nucleotide binding Lys 219 and Asn 336 residues in domain closure. The arrows in Figure 84A designate the pathway of transmission of the nucleotide effect (through the contribution of Lys 219 and Asn 336) towards the hinge at βL. The much smaller activity of K219A compared to that of N336A, however, raised the question of whether Lys 219 may also have an additional role that was indeed approved later by ^{19}F-NMR and crystallographic studies, as shown above (Chapter IX.). As for Glu 343/Tb345, the carboxylate of this residue interacts with the ribose OH groups in all nucleotide-binding structures, including TbPGK (Figure 84B). The side-chain of Glu 343/Tb345 is also involved in the permanent hydrophobic interactions with the sequentially neighbouring Phe 342/Tb344, which interacts further with the carbon atoms of the ring of Pro 338/Tb340. This transmission of the nucleotide effect is further continued through the interaction between the rings of Pro 338/Tb340 and Trp 335/Tb337. The latter side-chain is contacted further with the main chain carbon and the Cα atoms of Asn 336/Tb338. This route is depicted by arrows in Figure 84B. In this way, the effect of the nucleotide is transmitted from both Glu 343 and Lys 219 to the main hinge β-strand L.

Figure 84C illustrates the activation free energy changes for the chemical step ($\Delta\Delta G^*$) of the PGK catalysed reaction. As expected, mutants of the three essential catalytic side-chains (Arg 38, Lys 215 and Lys 219) exhibit the highest differences. While the mutants of Asn 336 and Glu 343, which have cumulative roles in transmission of the nucleotide effect towards the main hinge, shows somewhat decreased differences. The smallest difference is due to mutation of the less important Thr 375.

It has to be noted that in addition to the above-described induced effects, the substrates probably also affect more directly the relative positions of helices α8 and α13 (through temporal interactions with the mobile terminal phosphate of the bound MgATP, cf. Chapters I. 1.4. and II. 2.5.) as well as of helices α1 and α14 (through simultaneous interactions with PG or bPG). Namely, the 1-phosphate of the bound bPG (that has a certain extent of mobility even within the closed active site cavity), in addition to its interaction with Arg 38 at the N-terminus of helix 1, also interacts with Gly 396 at the N-terminus of helix 14, thus, the 1-phosphate of the bound bPG may also contribute to the movement of helix 14 closer to helix 1 during domain closure. Similar relative movement of these helices (in about 3 Å) indeed has been observed in the PG binary complex of pig muscle PGK (compared to the substrate-free enzyme), leading to approximately the first 8° of domain rotation [15] (cf. Chapter I. 1.7.).

In summary, these studies with PGK have helped to uncover in detail the molecular events that are initiated at the substrate sites, are transmitted to the hinge at β-strand L, and finally lead to domain closure. Thus, the results with HsPGK are consistent with the involvement of both substrates (each bound to separate domains) in the realization of domain movement during the catalytic cycle. A similar contribution of the substrates, especially of the nucleotides, to the induced fit closure of the active site has been concluded for other kinases, e.g., arginine kinase [420] or adenylate kinase [421, 422]. The work with PGK is entirely consistent with the ligand-driven domain closure model (e.g., [422]), which is in apparent contradiction with other literature data (e.g., [423] and 424). There are no experimental data that indicate the occurrence of the conformational transition of PGK in the absence of substrates. However, molecular dynamic simulations indicate the occurrence of domain motion on the time scale of nanoseconds even for substrate-free PGK [425]. It is indeed possible that the equilibrium of these fast conformational fluctuations is perturbed by the bound substrates, which may lead to formation of the domain-closed conformations on a much longer time scale.

In: Phosphoglycerate Kinase: A Hinge-Bending Enzyme ISBN: 978-1-62808-836-6
Editor: Mária Vas

Chapter XI

Structural-Temporal Description of the Reaction Cycle

Matthew W. Bowler[1,2] and Mária Vas[3]
[1]EMBL Grenoble, Grenoble, France
[2]Unit for Virus Host Cell Interactions, UMI 3265 (CNRS-EMBL-UJF), Grenoble, France
[3]Institute of Enzymology, Research Centre for Natural Sciences, Hungarian Academy of Sciences, Budapest, Hungary

The previous chapters fully reconcile the long story of discovering the active closed conformation of PGK and its formation under the concerted action of the two bound substrates. However, based on this knowledge we are still not able to report on the occurrence of a complete catalytic cycle of PGK-catalysed phospho-transfer. Although previous studies have pointed to an induction of hinge-bending upon ligand-binding, this model does not adequately describe how the products are subsequently released. From transient kinetic studies (cf. Chapter V.) there are indications of slow, rate-limiting release of the products during the catalytic cycle of PGK, but its possible structural background and energetic aspects are not yet considered.

Recent solution X-ray scattering (SAXS) studies, in combination with the various crystal structures of HsPGK, however, have revealed a new conformation for the resting state of the enzyme and have allowed the complete reaction cycle of PGK to be resolved both structurally and temporally [426]. SAXS was employed, as X-ray structures can only define snapshots during the catalytic cycle and loose details on the extent and timing of the domain movements that occur in solution. Scattering curves of the apoenzyme, the PG- and MgADP-binary complexes, the MgATP·PG ternary complex the (equivalent to the enzyme in catalytic turnover), the MgADP·PG ternary complex (a dead-end complex), and the PG·AlF_4^-·MgADP complex (the fully closed conformation) were measured (Figure 85). Although similar experiments have been performed previously [109, 227] (cf. Chapter II. 2.9.), the experiments described by Zerrad et al. benefit from the ability to trap the fully closed conformation in solution by inhibition with aluminium fluoride (the PG·AlF_4^-·MgADP complex) and compare the curves to the crystal structure of this complex and to the various open forms from the same species. In this respect, the study provides a full explanation of domain movement during catalysis.

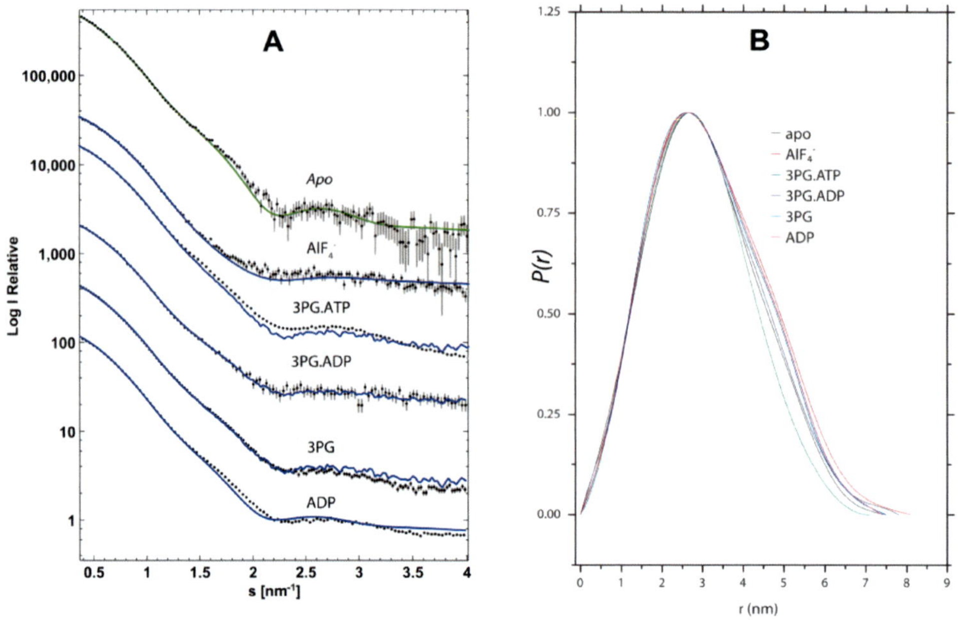

Figure 85. Scattering curves (A) of the apo- form and the various substrate complexes of HsPGK and the pair distance distribution functions (B) (adapted from [426]): (A) The experimental SAXS curves for the samples are shown with the two component fits from OLIGOMER (*blue lines*). For the apo data, the fit of the DEN refined structure as calculated using CRYSOL is also shown (*green line*). The curves are displaced for visualization and therefore log (I) values are relevant for the apo sample only. (B) Pair distances distribution function, $P(r)$, plots calculated by an indirect Fourier transformation of the SAXS data using GNOM.

Table 30. SAXS experimental data obtained with various complexes of HsPGK, their oligomer fits and comparison with crystal structures (adapted from [426])

Protein complex	*Rg nm*	% Open	% Closed	Chi-square of fit
Apo-enzyme	2.42	N.D.	N.D.	1.08 (DEN model)
PG·AlF_4^-·ADP	2.24	39.3	60.6	1.36
PG·ATP	2.34	92.5	7.5	7.55
PG·ADP	2.37	96.5	3.5	0.9
PG	2.40	96.6	3.4	2.75
ADP	2.42	96.7	3.3	5.15
PG·AlF_4^-·ADP crystal structure	2.17			
Binary complex crystal structure	2.33			

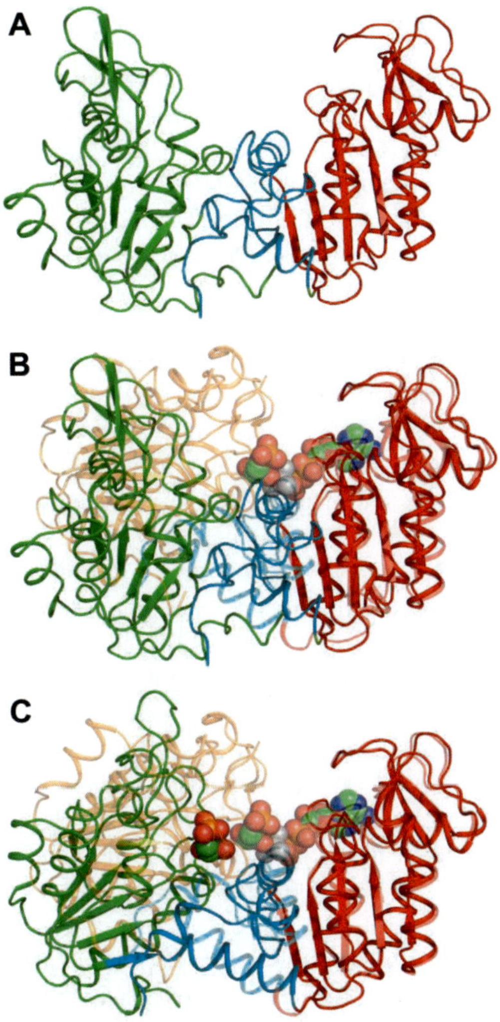

Figure 86. Comparison of the conformations of the fully open apo HsPGK (A) with the fully closed HsPGK·PG·MgF_3^-·ADP complex (B) and with the half-open form of the HsPGK·PG binary complex (**C**) (adapted from [426])**:** The molecules are represented by ribbon diagrams: C-terminal domain (red), the interdomain region (cyan) and the N-terminal domain (green for the open conformations and orange for the closed conformation). The substrates are shown as spheres, coloured by atoms. The structures are superimposed according to the C-terminal domains.

The radius of gyration (*R*g) of the apoprotein was found to be larger than that calculated from crystal structures of the open conformation (Table 30). Ab initio modelling for the apoenzyme produced a molecular envelope close to that obtained from crystal structures of the open conformation but with a greater angle between the two domains. To define the domain orientation in this new conformation, a homology model of the HsPGK apoenzyme was constructed from the crystal structure of the mouse apoenzyme [427], and the N- and C-terminal domains refined as rigid bodies against the scattering curves to define their relative orientation in solution. DEN (deformable elastic network) refinement [428, 429] was then used to further characterize the resting state of the enzyme (Figure 86A). The model obtained gives an excellent fit to the data (Table 30) and shows a conformation considerably more open than observed in crystal structures, with a 56° rotation required to form the catalytically active conformation (Figure 86B), as opposed to a maximum of 33° from the most open crystal structure [7]. Without the constraints of a crystal lattice, the protein thus adopts its fully open resting state. Scattering curves of the binary complexes, with either 3PG or ADP alone, show an *Rg* close to that of the apoprotein implying a requirement for both substrates to bind before domain closure, in agreement with previous suggestions [74, 109]. Scattering curves obtained from the PG·AlF_4^-·MgADP TSA complex have a similar *Rg* to the crystal structure of the same complex (Table 29), and *ab initio* modelling confirms a more closed conformation of this complex. SAXS data therefore define both the fully open and fully closed states of PGK in solution.

The curves obtained from of the MgATP·PG ternary complex, however, do not match any of the crystal structures, even though the *Rg* is similar to the half-open crystal structure (Table 30). The *ab initio* model calculated from these data presents a shape reminiscent of both open and closed conformations as SAXS scattering curves originate from an average of all the particles in solution, and the enzyme was in full catalytic turnover throughout the measurements. It was possible, however, to calculate the contribution of conformational states from each stage of the catalytic cycle to the overall scattering curves observed. In this manner, the relative time that the enzyme spends in each conformation during catalysis could be determined. The degree of inhibition in the TSA complex was also determined (Table 30) by calculating the percentage of particles inhibited in the closed conformation.

The results of the fit suggest that when the enzyme is in full catalytic turnover it spends 92% of its time in the fully open conformation and only 8% of the time in the catalytically competent fully closed conformation. This figure is even lower for the binary and the MgADP·PG dead end complexes with 97% of time spent in the open conformation, presumably because these complexes cannot form all of the interactions needed to stabilize the closed conformation. The TSA complex was found to be only 60% closed. This would account for the larger *Rg* compared with the crystal structure of the same complex and is probably due to incomplete inhibition by aluminium fluoride (Table 30).

These results imply that although the enzyme moves between the open and closed state, it has an energetic preference for the open conformation. A model of cycling and catalysis has been proposed where unliganded PGK remains open (ready to bind substrates or release products) most of the time. This model is fully consistent with the results of isothermal calorimetric titration experiments, where a very small energy difference could be calculated between the open and the closed conformations of the enzyme [430]. The binding of substrates induces structural changes that allow the stabilization of the closed state for sufficient time to enable catalysis.

In order to investigate the effects of ligand binding on the open conformation of PGK of the same origin, Zerrad and coworkers have first crystallized the binary complex of HsPGK with PG. As found previously in case of the PG binary complex of pig muscle PGK [15] (cf. Chapters I. 1.6. and 1.7.) the new complex also exhibited a half-open conformation with an opening angle between the two domains of 28° (Figure 86C) relative to the closed conformation observed in the crystal structure of the PG·MgF_3^-·ADP TSA complex. Reorientation of Arg 65 and Arg 170 upon binding of PG was also observed with HsPGK, indicating the alignment of these residues to be able to form the essential salt bridge with Asp 218 when the closed conformation is formed upon binding of the second substrate. – Second, they introduced either ADP or AMP-PCP (the non-hydrolysable analogue of ATP) to this crystal in order to investigate their effects on this open conformation stabilized by the present crystal lattice. This conformation is very similar to the previously observed crystal structures of open forms of PGK from various species [6, 80] (cf. also the black structure in Figure 79A). It was observed that on binding ADP, the loop region at the N terminus of the helix α8 (C-terminal domain) composed of residues 214–219 adopts two conformations, one moving about 2 Å towards the nucleotide. This motion may lead to the positioning of the catalytic residue Lys 215 to interact with the 3-phosphate of PG (or both the 1- and 3-phosphate groups of bPG), bound in the N-terminal domain, upon closure and stabilize this conformation (an arrangement not observed previously in other open conformation structures). A reorientation of Asp 218 to a position ready to form a salt bridge with Arg 65, another interaction important in stabilizing the closed conformation, was also observed (not shown). All of these changes prepare the C-terminal domain for stabilizing the closed conformation. The structure with bound AMP-PCP was found to be very similar to the equivalent structure previously determined with pig muscle PGK [72] with the γ-phosphate showing high extent of mobility (cf. Chapter I. 1.4.) and consequently supports the previous proposal of its temporal interactions with Lys 215 [107], i.e., the suggestion for preparation of this side-chain to interact with PG if the ternary complex is formed in solution.- Thus, both PG and the nucleotide, bound to the open conformation, act to prepare the interactions that will exist only in the catalytically competent closed form.

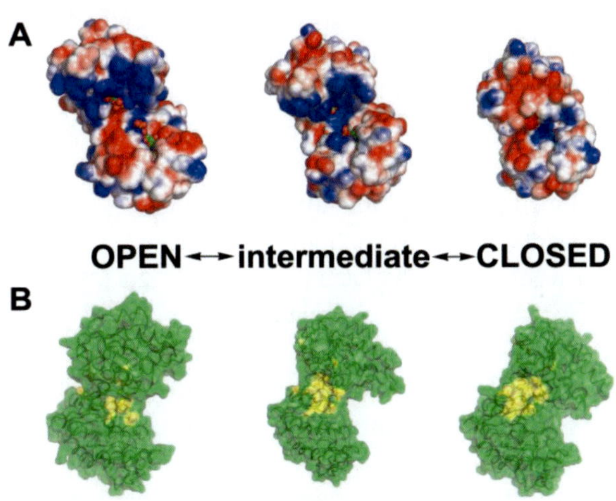

Figure 87. Changes in the solvent exposed surface upon domain closure (adapted from [426]): (A) Electrostatic potential representation (*blue*, positive; *red*, negative; *white*, neutral; values from +2 kcal/mol to -2 kcal/mol) of the solvent accessible surface (probe radius, 1.4 Å) of the fully open apoenzyme conformation, the half-open binary complex, and the fully closed conformation. In the fully open conformation, the binding sites for ligands are well separated and exposed to the bulk solvent. The complementary charged areas that stabilize the closed conformation can be seen easily. Ligands are shown as *spheres* (*green* carbon atoms). (B) Surface representation of the transition from fully open to fully closed state. Residues involved in hydrophobic interactions are coloured *yellow*. As the enzyme moves to the fully closed state, a hydrophobic patch is exposed.

Taken together, the SAXS and X-ray data have led to a new model of catalysis in PGK. The protein remains mostly in an open conformation with (even in the apo-form) short burst of closure, whereas ligand binding prepares the interactions needed to stabilize the closed conformation for the 8% of time spent on catalysis, rather than actually inducing closure. Analysis of the solvent accessible area and the electrostatic potential map of the different conformations (Figure 87A) show that the fully open SAXS-observed conformation is needed for efficient substrate exchange. Only in this conformation are the substrate binding sites fully accessible to solvent with large positive and negative paths available for binding. The transition to the crystallographically observed half-open conformation occludes the binding sites implying that this conformation would not be able to efficiently exchange ligands. No significant time is spent in this conformation in solution, and it is therefore part of the transition between open and closed and not functionally significant itself. Movement to the catalytically active conformation completely excludes the ligands from the solvent. The fully open conformation is clearly optimal for ligand exchange, but how is this conformation maintained/readopted in the presence of the strong electrostatic interactions present when ligands are bound? A detailed comparison of the fully open and fully closed structures points to a possible "spring" that drives the preference for the open conformation. The N- and C-terminal domains are linked by a single helix (helix 7, residues 187–198) that is highly conserved across all kingdoms of life. Many of these residues are hydrophobic. Analysis of the fully open conformation shows extensive interaction between these residues and other conserved hydrophobic residues in the core of the protein (for the full list of residues, cf. Table 31). When the enzyme moves to the half-open and fully closed conformations, some of these interactions are disrupted, and many residues form a solvent exposed hydrophobic patch (Figure 87B). This hydrophobic patch has a solvent exposed area of 502 $Å^2$ in the fully closed conformation, which reduces to 150 $Å^2$ in the half-open conformation and has a minimum of 92 $Å^2$ in the open state (masking 82% of the area from solvent). These residues only become buried when the opening angle between the domains is about 50°. This masking of the hydrophobic region from solvent could account for the preference of PGK to adopt an open conformation, as this will be thermodynamically more stable.

The exposure to solvent of this hydrophobic patch in the catalytically competent closed conformation, stabilized when both substrates are bound, suggests that this conformation will exist only long enough for catalysis to occur and that the overall thermodynamic preference for the open conformation will eventually destabilize the closed conformation, causing it to spring open, leading to the release of products and rebinding of substrates. The hydrophobic patch could act as a spring by constantly applying pressure to the closed conformation, leading to the release of products after catalysis.

However, the question arises, if the enzyme has a stable state that is noncatalytic and spends most of its time in this state, how can it act as an efficient catalyst? As the actual phospho-transfer reaction occurs extremely fast [431], the closed state does not need to be stabilized for too long. Therefore, for catalysis to be efficient, the open conformation needs to be dominant. The destabilizing effect of the hydrophobic patch on the closed conformation may act like a powerful spring, allowing the protein to efficiently bind and, importantly, release substrates and products. Competition between the higher affinities for ADP over ATP and bPG over PG [107] may also play a part in the release of products. As the catalytic trap will only be primed when all ligands are bound, this prevents any time lost in futile stabilization of the closed conformation.

Table 31. Hydrophobic residues and their solvent exposed area in different conformations of HsPGK (adapted from [426)]

Residue	Solvent exposed area closed (Å^2)	Solvent exposed area half open (Å^2)	Solvent exposed area open (Å^2)
Leu8	0	4.0	1.7
Phe48	54	31.5	24.7
Phe165	0.2	0.1	2.4
Phe187	53.1	15.6	7.8
Leu188	15.8	4.3	3.1
Met189	0	1.2	0
Leu193	23.4	5.2	4.1
Tyr195	58.8	22.4	17.0
Phe196	3.6	0.7	0.2
Ala197	61.5	10.2	3.4
Leu200	41.3	7.8	1.9
Ile369	1.3	0.5	0.2
His390	20.3	0.4	2.7
Leu401	26.2	4.6	1.7
Leu402	5.9	1.1	0
Val406	108.1	30.4	17.7
Leu407	3.2	1.2	0.7
Val410	18.3	0.2	1.8
Leu413	7.1	8.8	1.3
Total	**502.1 Å^2**	**150.2 Å^2**	**92.4 Å^2**

Based on the above considerations, the temperature-driven internal motion of protein structure seems to be essential for functioning of PGK, as also suggested for more complex allosteric proteins [432, 433]. Although it has been assumed that the open, or unliganded, conformations of proteins that act through domain movements will be favoured [52], a mechanism for this equilibrium has not been proposed. Here, it was shown that the equilibrium between open and closed is greatly biased toward the open conformation of PGK, and a hydrophobic spring may be responsible. This mechanism may prove general for some of the large number of enzymes that act by domain movements.

Part IV

Pathological Anatomy (Misfolding) and Physiology (Misfunction)

In: Phosphoglycerate Kinase: A Hinge-Bending Enzyme ISBN: 978-1-62808-836-6
Editor: Mária Vas

Chapter XII

Pathological Anatomy of PGK: Misfolding and Aggregation

Jeannine Yon-Kahn
Institute De Biochimie Et Biophysique Moleculaire Et Cellulaire,
Université de Paris-Sud, Orsay, France

Protein aggregation is a widespread phenomenon. Its industrial and medical importance has been recognised for many years [434-436]. Both ordered and amorphous aggregate structures have been described. A substantial body of information supports the idea that protein aggregation often arises from partially folded intermediate states as a result of a kinetic competition between proper folding and misfolding [434]. The aggregates are formed from the non-native protein due to intermolecular interactions which overcome the intramolecular contacts. Aggregation can occur under certain conditions *in vivo* as well as *in vitro*, or due to destabilising mutations [437]. The aggregation process depends on such conditions as protein concentration, temperature, pH, ionic strength, and redox environment [434, 436]. Aggregation has been reported to occur during the *in vitro* refolding of monomeric as well as oligomeric proteins. Several types of aggregates have been observed. The formation of transient aggregates has been reported during the *in vitro* refolding of proteins. Such a situation has been described for phosphoglycerate kinase [325], and is detailed in Chapter III. This process was found to be reversible. Most aggregation processes, however, are irreversible and they are the subject of the present chapter.

12.1. Mechanism of Protein Aggregation

Destabilisation or partial unfolding of proteins is frequently accompanied by the irreversible formation of aggregates [434, 435]. The destabilisation of the protein can be achieved by changing almost any intensive thermodynamic parameter, e.g., temperature, pressure, pH, concentration of salts or alcohols [438, 439]. Conditions of optimal stability depend on the balance between hydrogen bonds and hydrophobic interactions. According to this balance, cold denaturation could also occur. Aggregation is always concentration dependent. Most examples of irreversible denaturation for monomeric proteins are related to multidomain proteins [276]. Another degree of complexity appears in the folding of oligomeric proteins [276]. In this case, the kinetics of aggregation is complex and multiphasic. An attempt to characterise these aggregates has shown that non-covalent interactions occur between monomeric species with partially restored secondary structures. In many cases, they can be resolubilised in high concentration of GuHCl into monomeric species and then refolded under optimal conditions to yield a functional protein [434, 435].

Irreversible aggregation can generate different types of aggregates. The morphology of the aggregates could be either amorphous, resulting in inclusion bodies, or ordered, generating amyloid fibrils.

Irreversible Aggregation

The formation of aggregates has been often considered as a trivial phenomenon consisting of a non-specific association of partially folded polypeptide chains to form a disordered precipitate [275]. However, several analyses indicate that protein aggregation occurs by specific intermolecular associations involving the recognition of a sequence partner in another molecule rather than in the same molecule, as demonstrated by Mitraki and King studying the folding process of P22 tail spike protein [439]. Analyses of the folding mechanism of various proteins have permitted identification of specific sites that are critical for association [441].

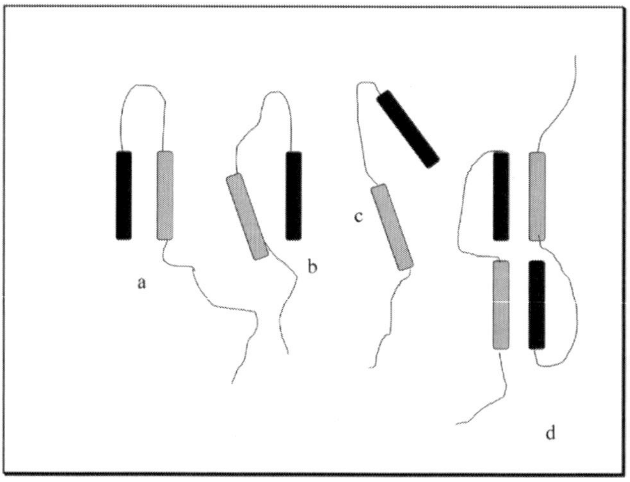

Figure 88. Schematic representation of domain swapping (a) monomeric native protein; (b and c) partially unfolded monomeric protein; (d) domain swapped dimer.

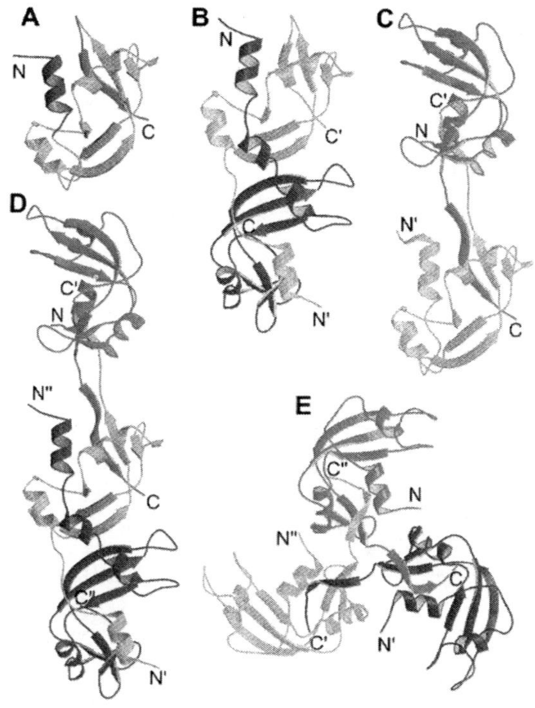

Figure 89. Domain swapping in ribonuclease (based on [446]): (A) ribbon diagram of the ribonuclease A monomer; (B) the N-terminal domain-swapped dimer; (C) the C-terminal domain swapped dimer; (D) the N- and C-terminal domain-swapped trimer; (E) the cyclic C-terminal domain swapped trimer.

One mechanism that accounts for the formation of aggregates during the refolding of multidomain proteins is domain swapping. This was first suggested by Monod and later proposed by Goldberg and coworkers to account for the refolding of tryptophanase [442]. The concept was foreshadowed by the results of Crestfield and colleagues with ribonuclease [443]. From experiments based on chemical modification of this enzyme, the authors proposed that the active dimer they obtained is formed by exchanging the N-terminal fragments. The term "three-dimensional domain swapping" was introduced by Bennett et al. to describe the structure of a diphteric toxin dimer [444]. The mechanism as illustrated in Figure 88 involves the replacement of one domain of a monomeric protein by the same domain of an identical neighbouring molecule, thus resulting in an intertwined dimer or oligomer. When the exchange is reciprocated domain-swapped dimers are formed. When the exchange is not reciprocated, but propagated along multiple polypeptide chains, higher order assemblies or aggregates may form. Eisenberg and co-workers have obtained two types of dimers of ribonuclease A upon domain swapping in mild acid conditions and high protein concentration showing that the same protein may be able to swap several different parts of the sequence [444, 445]. The formed aggregates can differ in both structure and oligomerisation state: the major type results from the swapping of its terminal β-strand, whereas the minor type is formed by swapping of the N-terminal helix with that of an identical molecule. The authors have also obtained trimers as shown in Figure 89 [446]. Such a phenomenon has been also observed with other proteins. It was proposed that three-dimensional domain-swapping may also play role in amyloid formation.

As detailed above, several mechanisms may be involved in the formation of aggregates. The energy landscape description of protein folding allows to rationalise the different mechanisms in a single view [284]. The presence of local energy minima separated by an activation barrier allows the accumulation of non-native intermediate states. If the barrier is high enough, these intermediates reach the native state extremely slowly, and kinetic competition favours the aggregation process.

Morphology of the Aggregates

As mentioned previously, aggregates can exhibit several types of morphology, mainly inclusion bodies which are amorphous, and amyloids which show ordered fibrillar structure [435].

Table 32. Amyloidogenic proteins and the corresponding diseases
(according to Kelly [451])

Clinical syndrome	*Precursor protein*	Fibril component
Alzeimer's disease	*APP*	β-peptide 1-40 to 1-43
Primary systemic amyloidosis	*Immunoglobulin light chain*	intact light chain or fragments
Secondary systemic amyloidosis	*Serum amyloid*	Amyloid A (76 residue fragment)
Senile systemic amyloidosis	*Transthyretin*	transthyretin or fragments
Familial amyloid polyneuropathy I	*Transthyretin*	over 45 transthyretin variants
Hereditary cerebral amyloid angiopathy	*Cystatin C*	Cystatin C minus 10 residues
Hemodialysis-related amyloidosis	*β_2-microglobulin*	β_2-microglobulin
Familial amyloid polyneuropathy III	*Apolipoprotein A1*	fragments of Apolipoprotein A1
Finnish hereditory systemic amyloidosis	*Gelsosin*	71 amino acid fragment of gelsosin
Type II diabetes	*Islet amyloid polypeptide (IAPP)*	fragment of IAPP
Medullary carcinoma of the thyroid	*calcitonin*	fragments of calcitonin
Spongiform encephalopathies	*Prion*	Prion or fragments there of
Atrial amyloidosis	*Atrial natriuretic factor (ANF)*	ANF
Hereditary non-neuropathic systemic amyloidosis	*Lysozyme*	Lysozyme or fragments thereof
Injection-localized amyloidosis	*Insulin*	Insulin
Hereditary renal amyloidosis	*Fibrinogen*	Fibrinogen fragments
Parkinson disease	*α-synucleine**	

* from J.C.Rochet & P.T.Lansbury [452]

Historically, the term amyloid was introduced by Virchow in 1854 to describe fibrillar protein deposits associated with diseases known as amyloidoses that involve the extracellular deposit of plaques with the aspect of starch [447]. Protein aggregation resulting in the formation of amyloid fibrils plays a key role in the origin of several pathologies such as spongiform encephalopathy. Deposition of amyloid plaques was observed in several diseases (e.g., Creutzfeld-Jacob disease, Alzheimer disease, Parkinson disease, type II diabetes or Kuru) and it can lead to tissue degradation and loss of organ function [448, 449]. At least twenty human amyloidogenic proteins are known to form amyloid plaques *in vivo*. These are responsible for several pathologies illustrated in Table 32. Although these proteins have no sequence homology all of them form similar amyloid structures, suggesting a common sequence independent self-assembly pathway. Some of these proteins formed amyloid *in vitro* as well [450].

In all proteins known to form amyloid fibrils, there is a conversion of α-helices to β-structures. Amyloid fibrils are insoluble and generally protease resistant structures. The most commonly used method to detect amyloid deposits is by staining with Congo red which exhibits a green birefringence upon amyloid binding [447]. Amyloid fibrils are generally 60 to 100Å in diameter and of variable length. X-ray diffraction studies on fibrils, NMR studies, cryoelectron microscopy and infrared Fourier transform experiments have shown that amyloid fibrils are made of two or more β-sheet filaments wound around one another. The fibrils have a characteristic cross-β structure, the individual β-sheets being oriented perpendicular to the long axis of the fibril.

The mechanisms involved in the formation of amyloid fibrils have been extensively studied [451, 452]. Oligomeric protofibril intermediates have been characterised with respect to their structure and temporal evolution. A well-documented example is provided by the studies on transthyretine by Kelly and his group [451, 453].

A number of observations indicate that the mutations leading to amyloid fibril formation result in a decreased stability of the native state [451, 452]. In all cases, the formation of fibrils occurs from a partially structured molecule via nucleation dependent oligomerisation [451] (Figure 90). It was shown that fibril formation takes place only after a lag phase which can be abolished upon seeding [452]. Nucleation is followed by the formation of protofibrils, whose characteristics have been determined. Atomic force microscopy and fluorescence correlation spectroscopy have been used to monitor transitions among the different types of assemblies [452].

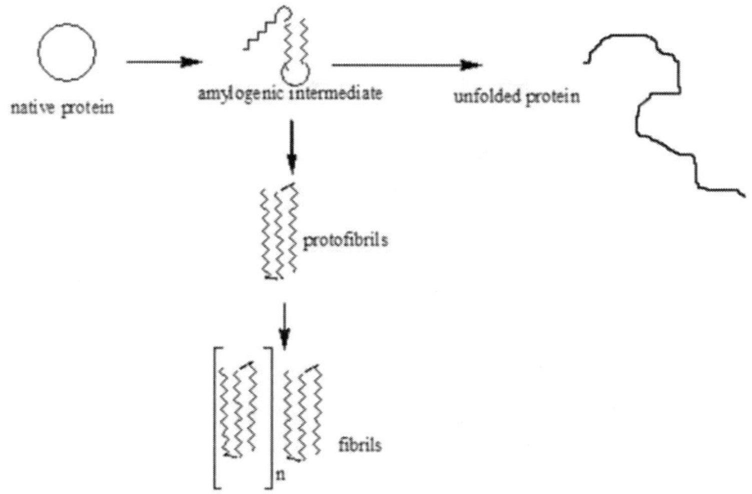

Figure 90. Schematic representation of amyloid fibril formation from a partially unfolded protein.

Several observations have shown that different proteins unrelated to amyloid diseases are able to aggregate *in vitro* into amyloid fibrils when exposed to mild denaturing conditions. These fibrils are indistinguishable from those found under pathological conditions. Such amyloid formation has been demonstrated for several proteins, such as normal lysozyme, an acyl phosphatase, and SH3 domain of a phosphatidylinositol kinase (Figure 91) [454-456]. It has been suggested that the ability to form amyloids under appropriate conditions is a property common to many if not all proteins [455]. Interactions between the amino acid side-chains stabilise the native state. Amyloids are stabilised by hydrogen bonding between the atoms of the polypeptide backbone. This indicates that amyloid formation reflects a general polymerisation tendency of polypeptide chains rather than a property restricted to definite chameleon sequences capable of adopting either a β or an α helicoidal structure depending on the conditions [457].

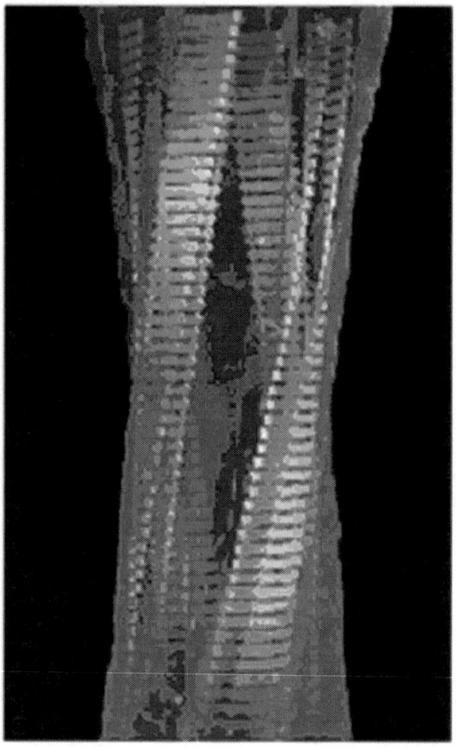

Figure 91. Molecular model of amyloid fibril derived from cryoelectron microscopy analysis (based on [458]): Fibrils grown from an SH3 domain upon incubation of a solution containing the protein at low pH.

12.2. Aggregation of Phosphoglycerate Kinase

Similar to the non-pathological proteins just cited, phosphoglycerate kinase can generate amyloid structures. Its native structure is stabilised by numerous long-range interactions that can be disrupted under destabilising conditions. Damaschun and coworkers [459] have tried to refold PGK unfolded at pH 2. At low ionic strength, the acid-unfolded protein consists of an ensemble of molecules with different continually exchanging conformations. These conformations are random-walk structures with roughly 3 nm persistence length, but more than 90% of the amino acids in the ensemble have Φ and Ψ angles corresponding to those of a strait β chain. The addition of sodium chloride or the sodium salt of trichloracetic acid induces a monomeric collapsed conformation. The anion induced partially folded state has a significant amount of α helical and some intra-molecular β structure. Although this non-native state has a high content of secondary structure, it is less compact than the native state [459]. Soon after the initiation of the collapse, as assessed by dynamic light scattering, the molecules begin to associate, first into dimers, then in octamers, which grow over a period of many hours to form fibrils or rods with amyloid-like structure. The presence of the amyloid structures was confirmed by Congo-red staining. The growth of the amyloid structures was assessed by several methods: dynamic light scattering, solution X-ray scattering, electron microscopy. Changes in the secondary structure of PGK were followed by spectroscopic means: far and near UV, circular dichroism as well as infra-red spectroscopy. These methods indicated an increasing content of β structures over a period of many hours [460].

The kinetic study of Damaschun and coworkers has shown that the formation of cross-β structures is practically finished after a few hours, whereas the elongation of the amyloid fibrils monitored by small angle scattering was not yet completed after 1300 hours [460]. This indicates that cross-β structures are formed only during the association of monomers to small aggregates. These aggregates assemble to amyloid fibrils without transformation of the polypeptide chain regions into cross-β structures [460]. Electron microscopy and X-ray scattering data showed that the obtained amyloid fibrils had a persistence length of 73 nm. The cross section of the amyloids was elliptical with dimensions of 10.2 nm x 5.1 nm. About half of the amino acids were found to be organised in cross-β structures, which gave rise to typical infrared spectra.

The group of Damaschun has proposed a model of amyloid formation of PGK, applying Smoluchowski's coagulation theory using size-independent association rate [461]. Their model consists of a two-step process, with an on-path intermediate. The presence of the intermediate is of extreme importance, since the on-path intermediates of the fibrils growth are though to be toxic, causing cell death and disease. The two steps are denoted as CtCo and CoCo in the Figure 92. During the first step, there is a formation of critical oligomers. These oligomers –the on-path intermediate in the fibril growth- formed in the first stage have an average mass of ten monomeric units. The fibrillar morphology typical to amyloids arises in the second step of the coagulation process, in which protofibrils are formed by clustering of the critical oligomers (CoCO stage). Electron microscopy visualises protofibrils as 50-150 nm long curved structures. Protofibrils can transform later into mature fibrils, which are straighter, and show different morphology. Both steps of amyloid formation were found to be irreversible. Increasing grey shading in Figure 92 indicates the increase in β-sheet content. The formed β-strands are stabilised by intermolecular interactions. Since

Smoluchowski's coagulation theory accounts for their data, the authors concluded that protofibril formation of PGK is under kinetic control far away from thermodynamic equilibrium. Interestingly, the transient formation of reversible oligomers during the refolding of PGK, under conditions that favour the correct folding, has been shown to be also under kinetic control [325].

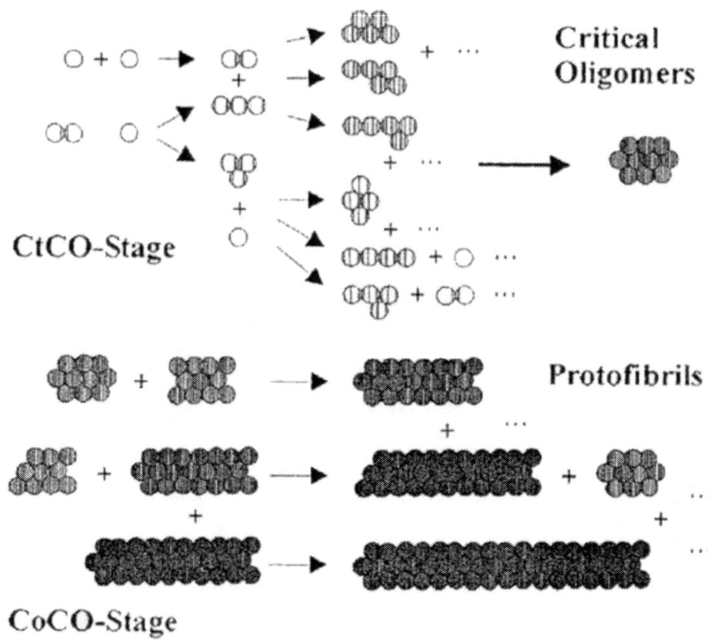

Figure 92. Model of amyloid assembly of PGK: The increase of β-strand content is represented by the increasing number of secants drawn within the circles (see text) (adapted from [461]).

The role of structural domains in the amyloid formation of PGK has been investigated by Osváth et al. [327]. There are proteins that are built of two structural domains (e.g., the prion protein) and are deposited full-length in amyloid plaques formed *in vivo*. Differences in the mechanisms of folding of single- and multi-domain proteins are well known. Using PGK as a model system, Osváth and coworkers have shown that domain interactions also direct the misfolding pathway and the tertiary structure of the amyloid formed. The experiments were made with wild type PGK and single tryptophan mutants of the individual N- and C-domains, and of the complete protein. The four mutants are logically organised in two pairs. Both pairs contain one individual domain (either the N- or the C-domain) and a corresponding mutant of the intact protein. With the pair, the single tryprophan residues are placed at identical positions. Fluorescence measurements allowed the comparison of the structural changes in the vicinity of the tryptophan residues of the individual domain and that of the complete protein during misfolding and amyloid formation. Misfolding and amyloid formation were initiated with the method described by Damaschun and coworkers [459]. Electron microscopy images recorded 5 days after initialising amyloid growth show that similar fibrils grew from the wild type PGK and the single-tryptophan variants (Figure 93). Tryptophan fluorescence was used to detect the conformational changes accompanying the formation of amyloidogenic intermediates and the growth of amyloid fibrils. As in the previous study, the Smoluchowski coagulation theory was applied to interpret the kinetic data. The kinetics of misfolding was found to be different for the individual domains, whereas misfolding of the domains within the complete protein was synchronised. These finding indicate that the interactions between the polypeptide chains in the two domains direct the misfolding process from the early steps of amyloid formation.

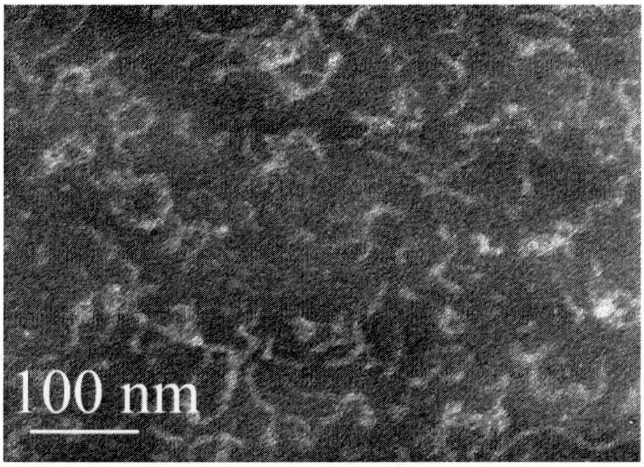

Figure 93. Electron microscopy image of PGK amyloid fibrils (by courtesy of Dr. Szabolcs Osváth).

In: Phosphoglycerate Kinase: A Hinge-Bending Enzyme
Editor: Mária Vas
ISBN: 978-1-62808-836-6

Chapter XIII

Pathological Physiology of PGK

Jeannine Yon-Kahn
Institute De Biochimie Et Biophysique Moleculaire Et Cellulaire,
Université de Paris-Sud, Orsay, France

Human phosphoglycerate kinase deficiencies have been described to be associated with serious diseases, such as mild to severe haemolytic anaemia, and neurological disorders the latter being variable, often including mental retardation, behavioural aberrations, and neurological symptoms; a few cases are associated with myopathy [462]. Human PGK has two isoenzymes, referred to as PGK1 and PGK2. PGK1 is a ubiquitous enzyme expressed in all somatic cells. PGK2 is a testicular tissue-specific enzyme. To date, severe PGK deficiency has been reported in 31 families and the responsible mutation characterised in 20 among them as shown in Table 33 [463-495]. Most cases (68%) had haemolytic anaemia, 48% central nervous system disorders and 48% myopathy.

In 1968, a first case of PGK deficiency associated with non-spherocytic anaemia in a 63 years old Caucasian female was reported by Kraus et al. [461]. The diagnosis was well established as well as the evidence that the red cells were deficient in PGK. Since the substrate affinity was normal, it was suggested that reduced enzyme levels rather than enzymatically altered enzyme are responsible of the disease. The hereditary nature of this defect was not possible to prove due to the absence of living members of this family. A second case reported in 1969 by Valentine et al. [464] was a large Chinese family and a pedigree and biochemical studies indicated that PGK deficiency was linked to the X-chromosome. Indeed, the PGK structural gene is located on the X-chromosome q13; therefore the mode of transmission of the variant enzymes indicated a X-linkage [465]. Thus, male hemizygotes with little or no PGK activity have chronic haemolysis that can be severe, often requiring erythrocytes transfusion. In contrast, females who are mosaic, have normal and PGK deficient erythrocytes and they may display variable degrees of haemolysis. As the PGK gene is linked to the X-chromosome, deficiency is expressed at the clinical level only in males whereas females have a normal phenotype or minor disorder associated to low PGK activity. In this case, the haemolytic anaemia was accompanied with central nervous system dysfunction. In the 1970s several other cases of haemolytic anaemia associated with PGK deficiency were reported [466-468] but the mutations responsible for the pathologies remain unknown.

When the PGK sequence was known, the mutations responsible for the diseases were determined for several cases. Thus, the correlation between the functional and structural abnormalities of these variants was discussed. Normal human PGK1, which consists of 416 amino acids, has been completely sequenced, but its three dimensional structure remains unknown. Since there are only 14 amino acid differences between horse and human PGK, a close structural similarity between the two enzymes was suggested. Based on this similarity, Figure 94 illustrates the positions of the various mutations. They appear in both domains, with a major frequency in the C-domain. Single amino acid substitutions were identified in most of them.

A variant referred to as PGK Uppsala found in a Swedish patient was reported to be associated with severe erythrocyte deficiency and chronic non spherocytic haemolytic anaemia, accompanied with neurological disorders [469]. It results from the amino acid substitution R205P in the C-domain, located in a loop (203-207) linking an α-helix and a β-strand. It is likely that the presence of a Pro residue induces a dislocation of the β-strand (208-213). According to the authors, the mutant enzyme had lower affinity for all substrates and lower specific activity for both forward and backward reactions. In order to analyse the structural and functional consequences of this mutation, the corresponding mutant was constructed in yeast PGK [470]. The three-dimensional structure of this mutant was resolved at 2.9 Å. The overall structure was not found to be modified; only small local changes were observed. The most significant difference compared with the wild type enzyme was a decrease in stability by about 3kcal/mol while the kinetic parameters were not found greatly affected, the catalytic constant being lowered by only 10-20%.

In contrast to these variants, PGK München is not associated with haemolytic anaemia, or with neurological disorders or myopathy. The enzyme deficiency found in blood from a PGK München subject results from amino acid substitution Asp267Asn. It was characterised by a decrease in red cell enzyme activity which is about 20% of normal, a decrease in heat stability and a light acidic shift of pH optimum, although its affinity for substrates was similar to that of the normal enzyme [471, 472].

A variant, PGK Tokyo, was described by Fujii et al.., in a patient with chronic non spherocytic anaemia and neurological disorders including mental retardation. The abnormality was found to be a single amino acid substitution from Val to Met at position 265 (V265M) which is proximal to the nucleotide binding site. Compared to the normal enzyme, PGK Tokyo had a lower specific activity, 31% of normal in the backward reaction and 15% of normal in the forward reaction. This variant had also an increased thermal instability probably caused by the insertion of a large hydrophobic residue at this position.

In 1983, DiMauro and coworkers [473] described the case of a 14 year-old boy who had myoglobinuria and renal failure after intense exercise. It was the first case of myopathy associated with PGK deficiency but without haemolytic anaemia. PGK activity was found to be 5% of the normal mean and all other glycolytic enzymes were found to be normal. The mutation of this variant referred to as PGK New-Jersey is not known.

Another variant, PGK San Francisco, was discovered in a boy with severe haemolytic anaemia, but without neuromuscular dysfunction or developmental retardation [474]. The mutation responsible for the pathology was not determined.

Table 33. Pathologies associated with PGK deficiency

Reference	Variant name	PGK activity in red cells*	Pathology			Mutation
			HA	CNS	M	
Kraus et al. [463]		75	+	-	-	NA
Valentine et al. [464]		5	+	+	-	NA
Cartier et al. [466]		20	+	-	-	NA
Arese et al. [467]		42	+	-	-	NA
Konrad et al. [468]		10	+	+	-	NA
Fujii & Yoshida [469]	Uppsala	10	+	+	-	K205P
Fujii et al. [471]	Munchen	21	-	-	-	D267N
Fujii et al. [472]	Tokyo	15	+	+	-	V265M
DiMauro et al..[473]	New-Jersey	5	-	-	+	NA
Guis et al. [474]	SanFrancisco	13	+	-	-	NA
Maeda & Yoshida [475]	Matsue	5	+	+	-	L88P
Maeda et al. [476]	Michigan	10	+	+	-	C315R
Fujii et al. [477]	Shizuoka	1	+	-	+	G157V
Tonin et al. [478]	Alberta	NA	-	-	+	NA
Cohen-Solal et al. [479]	Creteil	3	-	-	+	D314N
Cohen-Solal et al. [479]	Amiens	10	+	+	-	D163V
Sugie et al. [481]		8	-	+	+	NA
Sugie et al. [483]		1	+	+	+	NA
Tsujino et al. [484]	North Carolina	3	-	+	+	10 aa insertions between 138 and 139
Turner et al. [481]	Amiens/ New-York	2	+	+	-	D163V
Yoshida et al..[485]	Alabama	4	+	-	-	deletion of K190
Schroder et al..[486]		12	-	-	+	NA
Ookawara et al.; [487]	Antwerp	6	-	-	+	E251A or frameshift at 251 and termination at 261 aa
Valentin et al..[488]	Herlev	50	+	-	-	D284V
Sugie et al. [489]	Hamamatsu	8	-	+	+	I252T
Hamano et al. [490]	Fukui	4	-	-	+	frameshift at 213 and termination at 231
Aasly et al.[491]		4	-	-	+	NA
Morimoto et al. [492]	Kyoto	6	+	+	+	A353P
Noel et al. [493]	Murcia	36	+	+	-	S319D
Noel et al. [493]	Barcelona	6	+	+	-	I46N
Flanagan et al. [482]	Amiens/ New-York		+	+	-	D163V
Spiegel et al. [495]	Afula	0.02	-	-	+	T377P

*percentage of normal activity

HA: Haemolytic anaemia
CNS Central nervous system dysfunction
M Myopathy
NA: not available
The numbering of amino acid residues refers to the numbering of human PGK sequence.

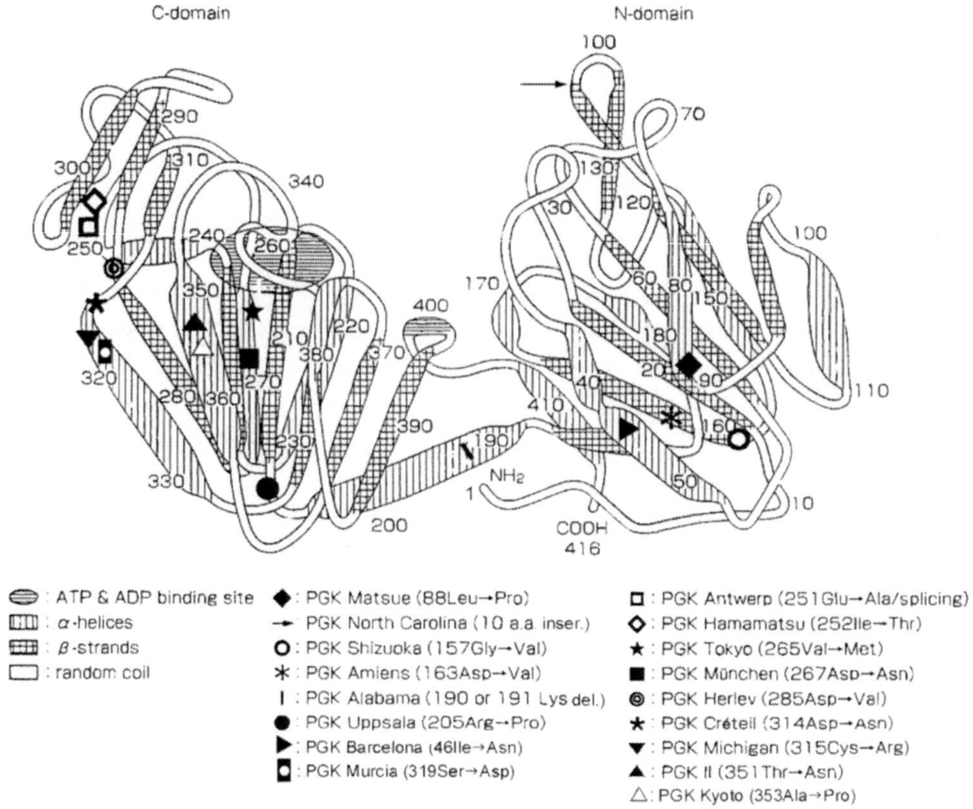

Figure 94. Positions of the different mutations in the variants human PGK (adapted from Fujii and Miwa [462]): Three newly discovered mutants are also added: ◀PGK Kyoto (Ala353Pro); ▶PGK Barcelona (I46N); ▫ PGK Murcia: (S319N). The numbering of amino acid residues refers to the numbering of human PGK sequence. It is interesting to remark that most mutations are localised in the C-domain.

A Japanese patient with PGK Matsue had severe chronic haemolytic anaemia accompanied with mental disorders; he died at age 9 from complications of pneumonia [475]. The responsible mutation was identified to be a L88P substitution in the N-domain caused by T/A →C/G transition in exon 3 of the variant gene. The activity of the enzyme in red blood cells, muscles and fibroblasts of the patient was found to be 5% of those of normal tissues. The specific activity of PGK was reported to be 35 to 40% of that of normal PGK with higher K_m values for all substrates. The substitution position laid in a random coil between helix 2 and β-strand C in the N-domain of the molecule. The presence of a Pro residue may cause a substantial change in the torsion angle of the polypeptide chain, resulting in a structural instability. As suggested by the authors [475], the major cause of a diminished concentration of PGK Matsue in the red cells and other tissues is most likely a more rapid denaturation and degradation of the variant enzyme.

A PGK deficiency case of X-chromosome-linked abnormality was described in a 14-year-old boy with mental retardation, a behaviour disorder and episodic haemolytic anaemia [476]. The enzyme activity in red blood cells of this patient was about 10% of normal. The variant PGK referred to as PGK Michigan was characterised by a lower affinity for substrates and a decrease in thermostability. A single T/A→C/G base transition in PGK exon 9 was found in the patient's gene. This mutation should induce an amino acid substitution Cys→Arg at position 315 in helix 11 (315-325) in the C-domain of the molecule. The substitution caused an instability of the molecule *in vitro* and presumably also *in vivo*. Because the variant gene was not detected in the proband's mother and siblings, it must have been generated by spontaneous mutation during oogenesis.

Another case of PGK deficiency was described by Fujii et al. [477]. The patient was a 27-year-old Japanese male who had chronic haemolytic anaemia and myoglobinuria, manifested by nauseas, anorexia and muscle weakness after exercise beginning at the age of 10. There was no family history of anaemia or neuromuscular diseases; both parents and brother were normal. It was the first reported case of PGK deficiency associated with both haemolysis and muscle symptoms. The pathology was the consequence of a single amino acid substitution G157V in a loop of the N-terminal domain of the PGK variant called PGK Shizuoka. This variant enzyme displayed normal substrate affinity but a low catalytic activity. According to the authors, it is likely that the replacement of the smallest Gly residue by a large Val residue disturbs the conformation of the enzyme and affects enzyme catalysis. It might be suggested that the introduction of a large residue restricts the flexibility of the molecule limiting the internal motion required for the full expression of enzyme activity.

A new myopathic variant, PGK Alberta was described by Tonin et al. [478] in a 37-year-old man with exercise intolerance, myagia and recurrent myoglobinuria. Kinetic properties of the mutant differed from those of the two previously reported cases [473, 477].

A case of PGK deficiency that was associated with rhabdomyolysis but not with haemolytic anaemia was reported by Rosa et al.., in 1982 [480], and the variant was called PGK Creteil. In the patient, a 31-year-old man several, symptoms occurred during physical exercises since his childhood, with painful cramps, abdominal pains, vomiting and dizzincss. A severe PGK deficiency in muscle was discovered. It was the first case reported in which a severe PGK deficiency did not result in haemolysis, but was associated to muscular abnormalities. The reinvestigation at the gene level has shown that the PGK Creteil variant arises from G→A nucleotide interchange at position 1022 in cDNA (exon 9) resulting in amino acid substitution D314N in the C-terminal domain [479]. This position is at the adjacent residue in PGK Michigan. But the two variants induced very different pathologies. Two other PGK variants associated with muscular dysfunction, named PGK New-Jersey [473] and PGK Alberta [478] were described by DiMauro's group on the basis of enzyme deficiency and phenotype, but the molecular defects of the mutants remain unknown. Rosa et al. [480] explained the molecular disorders induced by PGK Creteil by the fact that the enzymatic defect in muscle is responsible for the decrease in glycogen content what is needed for exercise. But, as underlined by these authors, it is surprising that PGK deficiency was not generally associated with muscular disorders, because the enzyme is ubiquitous and the level of PGK activity is decreased in muscle as well as in other tissues.

Another variant, PGK Amiens, identified in 1974 was first reported by the same group, in a patient of 3 years old [479]. At this time, the proband had chronic haemolytic anaemia with occasionally acute haemolysis crisis and retardation in language development. Twenty years later, the proband had marked mental retardation with frequent seizure associated with haemolytic anaemia. The biological analyses showed a decrease in PGK activity in red blood cells and leukocytes. The genetic analysis showed a nucleotide substitution A571→T in cDNA, leading to the replacement of Asp 163 by Val in the N-domain close to the catalytic site [479].

The same variant was identified in two other cases. As previously mentioned, in a previous report by Valentine et al. [464], PGK deficiency was observed in a large Chinese kindred including relatives in Australia in the New-York area. Later, molecular analysis revealed that the responsible mutation in PGK is D163V [481]. As this family and the French one came from a diverse population and had no obvious familial ancestry, the D163V was thought to be a recurrent mutation. Recently, Flanagan et al. [482] reported a case of PGK deficiency in a white American family with chronic haemolytic anaemia and neurological dysfunction as speech delay; furthermore the patient had also a retinal dystrophy previously unreported in PGK deficiency. The proband of mainly European ancestry was hospitalised at 9 years of age for hemiplegic migraine and presented seizure and fever and chronic haemolytic anaemia. His brother has also haemolytic anaemia and speech delay. The two brothers were found to be PGK deficient. The parents and sister were healthy. Molecular analyses had revealed that the responsible mutation in this family is D163V. Asp 163 has been found to be a highly conserved residue in all PGK of known sequence. It is located in the βE-strand of the N-domain and establishes contact with helix 7 of the same domain. As shown in the pig PGK molecule, Asp 163 is important for side-chain hydrogen bonding and electrostatic interactions between βE-strand and helix 7 [108]. As pig PGK is highly homologous to human PGK, the reported D163V mutation would be expected to introduce a new hydrophobic side-chain in a solvent exposed position and remove the favourable contacts between strand βE and helix 7, leading to a decreased stability of the molecule.

In 1994 Sugie et al. [483] described the cases of two patients with PGK deficiency, the one having both neuromuscular and central nervous system dysfunctions without haemolytic anaemia; the other presented the same dysfunction but also haemolytic anaemia. The mutation responsible for the disease is not known.

Another mutant, referred to as North Carolina, was identified on a 12-year-old boy with the myopathic form of PGK deficiency [484]. A G to A substitution at the 5' end of intron 4 was identified in the PGK gene of this patient. This mutation destroys the consensus sequence GT at the 5'splice junction of the intron, causing the insertion into the transcript of a 30 bp fragment at the 5' end of the intron. This insertion results in ten additional amino acids between residue 138 and residue 139 within the "nose" of the N-domain of PGK. It is likely that such an insertion would increase conformational fluctuations in this region resulting in a destabilisation of the molecule.

In a 37-year-old white male school teacher who had had infrequent episodes of haemolysis with jaundice prompting to a diagnosis of hepatitis, Yoshida et al. [485] demonstrated a molecular abnormality of a PGK variant associated with severe red cell enzyme deficiency. The variant was called PGK Alabama based on the regional origin. The genetic analysis revealed the deletion of Lys 190 located in helix 7 between the N- and C-domains. PGK activity of the patient's red blood cells was only about 4% of normal, but the Michaelis constants for substrates were found to be normal. The region around Lys 190 is important for maintaining the interdomain conformation of the enzyme, and it is likely that the deletion of Lys 190 causes molecular instability of this variant PGK, as revealed by the rapid *in vitro* inactivation of the enzyme.

The case of a 23-year-old man who presented episodes of muscular weakness, cramps, rhabdomyolysis, and renal failure was reported by Schroder et al. [486]. Biochemical analysis has shown a reduced PGK activity in muscle, and accumulations of glycogen in muscle mitochondria. The authors concluded that mitochondria are involved in that case of PGK deficiency.

Ookawara et al. [487] reported a novel PGK variant in a patient from unrelated Belgian parents, called PGK Antwert based in its regional origin. The patient presented occasional muscle contractures and myoglobinuria after vigorous exercise. Although the patient was associated with severe tissue PGK deficiency, no neurological abnormalities and no significant haematological problems were observed. The variant gene has been transmitted from the heterozygous mother to the patient and his brother. Genetic analysis has revealed that the variant results from a single amino acid substitution E251A of the enzyme. The substitution position is located between helix 9 and helix 10 in a region that is not involved in substrate binding. Glu 251 is not evolutionary conserved from bacteria to humans. Therefore, it is unlikely that this mutation could induce a modification of the catalytic activity of the enzyme. The authors attributed the low PGK activity to retarded and aberrant pre-mRNA splicings caused by the change of the consensus 5'splice sequence AGgt to a non consensus sequence CGgt at the junction between exon 7 and intron 7 in the variant gene

In a 69-year old Danish patient with isolated chronic haemolysis, but who had no neurological or muscular disorders, Valentin et al. [488] described a new mutation D284V in the PGK gene. The mutated gene was expressed only partially; both normal and substituted nucleotides were found at the same position in a ratio of approximately 1:9. The authors presume that the most likely explanation for this observation is based on the occurrence of a somatic mutation of the PGK gene.

Another mutant called Hamamatsu was found by Sugie et al. [489] in a 11-year-old retarded boy who had recurrent episodes of convulsions followed by generalised myalgia, muscle weakness associated with PGK deficiency. Genetic analysis revealed that the mutation was an 853 T-C inversion leading to a single amino acid substitution I252T.

Hamano et al. [490] reported a 36-year-old man with exertional myoglobinuria and muscle cramps without haemolytic anaemia or central nervous system disorders. They found a deficiency of PGK activity in muscles and erythrocytes. The genetic analysis has shown a base pair deletion in exon 6 of the PGK gene, leading to a frame shift at residue 213 with a termination at 231, producing a truncated PGK. This variant was called PGK Fukui.

Aasly et al. described the case of two young brothers who both had muscle pain, cramps and stiffness shortly after heavy exercise, resulting from very low PGK level in muscle [491]. These episodes were accompanied by rhabdomyolysis and myoglobinuria. The mutation responsible for the disease is not known.

The case of a 3-year-old Japanese boy with PGK deficiency was reported by Morimoto et al. [492]. The patient had anaemia and jaundice at birth necessitating red cells and haptoglobin transfusions. The patient developed the full clinical symptoms of PGK deficiency including haemolytic anaemia, myopathy, central nervous system disorders and growth retardation which was unusual. PGK deficiency was confirmed by enzyme assays as well as by the sequence of the patient's PGKI gene. A novel missense (1060G→C) PGKI gene mutation resulting in A353P amino acid substitution was identified and the variant was referred to as PGK Kyoto. This mutation is located in a helix on the surface of the C-domain.

Two previously undescribed mutations of the PGK gene leading to enzyme deficiency were reported by Noel et al. [493] in patients from Spanish origin. These cases were the first description of this enzymopathy in Spain. The patient exhibited a severe chronic haemolytic anaemia associated with progressive neurologic impairment. In the first patient the diagnosis of PGK deficiency was established at the age of three years during a haemolytic crisis. At 7 years of age, the haemolytic crises were associated with a progressive neurologic impairment leading to mental retardation. No muscular dystrophy was observed. Sequence analysis of the first patient's gene revealed a novel missense mutation (140T→A) resulting in an amino acid change I46N in the N-domain. This variant was called PGK Barcelona based on the origin of the patient. Ile 46 in the N-domain is located in helix 1 and establishes close contact with Leu 88 located in helix 2 facing helix 1. Noel et al. [493] have evaluated to 3.32 Kcal/mol the energy of destabilisation induced by the mutation.

The second case was a boy, the second of a three children of a healthy unrelated couple from Murcia. The patient required frequent transfusions from birth. The diagnosis of PGK deficiency was made when he was 6 years old. The anaemia was associated with a severe and progressive encephalopathy with cortical and subcortical atrophy and epileptic crisis. The child died of severe encephalopathy at 7 years of age. In this second patient a G to A transversion was discovered at nucleotide 958 in the PGK gene leading to a S319N amino acid substitution in the C-domain. This variant was named PGK Murcia. Ser 319 is located in helix 11. This amino acid maintains close contacts with neighbouring residues and the –OH group establishes a hydrogen bond with the N main backbone atom of Asp 284 located in the facing helix. This mutation decreases the enzyme stability by 1.24 Kcal/mol. These two mutations did not significantly modify the substrate binding sites, but these residues are important for the conformational stability of PGK.

Most of the PGK variants result from single amino acid substitutions distributed in both domains. Noel et al. [493] have remarked that these substitution occur in three specific regions of the enzyme molecule. The first region located in the N-domain includes PGK Barcelona, PGK Matsue, and PGK Amiens. In the second region located in the C-domain, are PGK Murcia, PGK Herlev, PGK Michigan and PGK Creteil. The third region in the same C-domain includes PGK München, PGK Hamamata, PGK Antwerp and PGK Tokyo. Again, a novel mutation within the C-terminal domain (T377P) has been recently identified by Spiegel et al. [495] (called variant PGK Afula) and described in a case of an 18-year-old man with muscle cramps and exertional myoglobinuria, but without haemolytic anaemia or brain dysfunction.

In general, no correlation could be inferred between these different locations of the mutations and the clinical expression of PGK deficiency. Furthermore, in most cases these various pathologies result from a lowering in enzyme activity in tissues. In haemolytic anaemia, the abnormal enzymes with decreased activity slow down the glycolytic flux, preventing cells from producing energy necessary to maintain their viability. In normal situation the erythrocyte volume is kept almost constant owing to operate of transport Na,K ATPase which utilises energy provided by glycolysis. If transport is abnormal, the volume and rheological properties of erythrocytes will be abnormal resulting in osmotic lysis. Martinov and coworkers [494] have developed a mathematical model to calculate the critical minimum values of Na,K ATPase and glycolytic enzyme activities at which the erythrocyte viability is lost. Their model includes all reactions of glycolysis, adenylate metabolism, ionic balance, and osmotic regulation of erythrocyte volume. The main conclusion of these authors is that in phosphofructokinase, glucose phosphate isomerase and PGK deficiencies, haemolytic anaemia can arise because of the instability of the enzyme. It is in agreement with what was reported for most cases of PGK deficiencies but not all; when evaluated, specific activity and substrate binding were not significantly modified, but the conformational instability of the enzyme was observed. A more recent systematic study on the molecular insight on the pathogenic effects of mutations causing PGK deficiency has been carried out by Chiarelli et al. [496] at the time of preparing this book. In order to provide a molecular framework to the different pathological manifestations these authors produced 16 different known mutants in recombinant forms and characterized them both functionally and structurally. Most mutations heavily affect thermal stability and to a different extent catalytic efficiency, in line with the remarkably low PGK activity clinically observed in the patients. Mutations grossly impairing protein stability, but moderately affecting kinetic properties (I46N, L88P, C315R, S319N and A353P) show rather good correlation with the clinical phenotypes of haemolytic anaemia and neurogical disorders. Variants highly perturbed in both catalytic efficiency (G157V, D163V, K190del, D284V, D315N and T377P) and heat stability (all, but T377P) result to be mainly associated with myopathy alone. Finally, mutations faintly affecting molecular properties (R205P, E251A, I252T, V265M and D267N) correlate with a wide spectrum of clinical symptoms. These findings may indicate that the different clinical manifestations associated with PGK deficiency mainly depend on distinctive type of perturbations caused by mutations in the PGK gene.

When PGK deficiency is accompanied by central nervous system disorders, it might be possible that the instable enzyme undergoes an aggregation process in brain. However, until now nothing indicates the presence of amyloid fibrils in the patients with mental dysfunctions induced by the various human PGK mutants. Nevertheless, taking into account the high structural similarities in PGKs from different organisms, such a possibility that has not been investigated cannot be excluded.

References

[1] Negelein, E. and Brömel, H. (1939) *Biochem. Z. 301*, 135-.

[2] Bücher, T. (1947) Über ein phosphatübertragendes garungsferment *Biochem Biophys Acta 1*, 292-314.

[3] Larsson-Raznikiewicz, M. (1970) The phosphoglycerate kinase reaction and its metal ion specificity *Eur J Biochem 17*, 183-92.

[4] Banks, R. D., Blake, C. C. F., Evans, P. R., Haser, R., Rice, D. W., Hardy, G. W., Merrett, M. and Phillips, A. W. (1979) Sequence, structure and activity of phosphoglycerate kinase: a possible hinge-bending enzyme *Nature 279*, 773-777.

[5] Watson, H. C., Walker, N. P. C., Shaw, P. J., Bryant, T. N., Wendell, P. L., Fothergill, L., Perkin, R. E., Conroy, S. C., Dobson, M. J., Tuite, M. F., Kingsman, A. J. and Kingsman, S. M. (1982) Sequence and structure of yeast phosphoglycerate kinase *EMBO J. 1*, 1635-1640.

[6] Davies, G. J., Gamblin, S. J., Littlechild, J. A., Dauter, Z., Wilson, K. S. and Watson, H. C. (1994) Structure of the ADP complex of the 3-phosphoglycerate kinase from *Bacillus stearothermophilus* at 1.65 A *Acta Crystallogr. D50*, 202-209.

[7] Flachner, B., Kovári, Z., Varga, A., Gugolya, Z., Vonderviszt, F., Náray-Szabó, G. and Vas, M. (2004) Role of phosphate chain mobility of MgATP in completing the 3-phosphoglycerate kinase catalytic site: binding, kinetic, and crystallographic studies with ATP and MgATP *Biochemistry 43*, 3436-49.

[8] Axelrod, B. and Bandurski, R. S. (1953) Phosphoglyceryl kinase in higher plants *J Biol Chem 204*, 939-48.

[9] Ali, M. and Brownstone, Y. S. (1976) A study of phosphoglycerate kinase in human erythrocytes. I. Enzyme isolation, purification and assay *Biochim Biophys Acta 445*, 74-88.

[10] Gosselin-Rey, C. (1963) Purification and properties of phosphoglycerate kinase from chicken breast muscle *Biochim Biophys Acta 67*, 140-2.

[11] Rao, D. R. and Oesper, P. (1961) Purification and properties of muscle phosphoglycerate kinase *Biochem J 81*, 405-11.

[12] Krietsch, W. K. and Bücher, T. (1970) 3-phosphoglycerate kinase from rabbit sceletal muscle and yeast *Eur. J. Biochem. 17*, 568-575.

[13] Johnson, P. E., Maister, S. G. and Knowles, J. R. (1976) Crystalline 3-phospho-d-glycerate kinase from horse muscle *Biochemistry 15*, 2899-901.

[14] Huskins, K. R., Bernhard, S. A. and Dahlquist, F. W. (1982) Halibut muscle 3-phosphoglycerate kinase. Chemical and physical properties of the enzyme and its substrate complexes *Biochemistry 21*, 4180-8.

[15] Harlos, K., Vas, M. and Blake, C. C. F. (1992) Crystal structure of the binary complex of pig muscle phosphoglycerate kinase and its substrate 3-phospho-D-glycerate *Proteins 12*, 133-144.

[16] Lay, A. J., Jiang, X. M., Kisker, O., Flynn, E., Underwood, A., Condron, R. and Hogg, P. J. (2000) Phosphoglycerate kinase acts in tumour angiogenesis as a disulphide reductase *Nature 408*, 869-73.

[17] Lay, A. J., Jiang, X. M., Daly, E., Sun, L. and Hogg, P. J. (2002) Plasmin reduction by phosphoglycerate kinase is a thiol-independent process *J Biol Chem 277*, 9062-8.

[18] Davies, G. J., Gamblin, S. J., Littlechild, J. A. and Watson, H. C. (1992) Purification, crystallization and preliminary X-ray analysis of the 3-phosphoglycerate kinase from Bacillus stearothermophilus *J Mol Biol 227*, 1263-4.

[19] Fleming, T. M., Jones, C. E., Piper, P. W., Cowan, D. A., Adams, M. W. W. and Littlechild, J. A. (1996) The purification and crystallization of phosphoglycerate kinase from the hyperthermophilic eubacterium Thermotoga maritima *Prot. Pep. Lett. 3*, 213-218.

[20] Lee, J. H., Im, Y. J., Bae, J., Kim, D., Kim, M. K., Kang, G. B., Lee, D. S. and Eom, S. H. (2006) Crystal structure of Thermus caldophilus phosphoglycerate kinase in the open conformation *Biochem Biophys Res Commun 350*, 1044-9.

[21] Littlechild, J. A., Davies, G. J., Gamblin, S. J. and Watson, H. C. (1987) Phosphoglycerate kinase from the extreme thermophile *Thermus thermophilus FEBS Lett. 225*, 123-126.

[22] Crowhurst, G., McHarg, J. and Littlechild, J. A. (2001) Phosphoglycerate kinases from bacteria and archaea *Methods Enzymol 331*, 90-104.

[23] Zomer, A. W., Allert, S., Chevalier, N., Callens, M., Opperdoes, F. R. and Michels, P. A. (1998) Purification and characterisation of the phosphoglycerate kinase isoenzymes of Trypanosoma brucei expressed in Escherichia coli *Biochim Biophys Acta 1386*, 179-88.

[24] Mori, N., Singer-Sam, J. and Riggs, A. D. (1986) Evolutionary conservation of the substrate binding cleft of 3-phosphoglycerate kinases *FEBS Lett. 204*, 313-317.

[25] Watson, H. C. and Littlechild, J. A. (1990) Isoenzymes of phosphoglycerate kinase: evolutionary conservation of the structure of this glycolytic enzyme *Biochem Soc Trans 18*, 187-90.

[26] Fleming, T. and Littlechild, J. (1997) Sequence and structural comparison of thermophilic phosphoglycerate kinases with a mesophilic equivalent *Comp Biochem. Physiol A Physiol 118*, 439-451.

[27] Hogg, P. J. (2003) Disulfide bonds as switches for protein function *Trends Biochem Sci 28*, 210-4.

[28] Krishnan, P., Fu, Q., Lam, W., Liou, J. Y., Dutschman, G. and Cheng, Y. C. (2002) Phosphorylation of pyrimidine deoxynucleoside analog diphosphates: selective phosphorylation of L-nucleoside analog diphosphates by 3-phosphoglycerate kinase *J Biol Chem 277*, 5453-9.

[29] Krishnan, P., Liou, J. Y. and Cheng, Y. C. (2002) Phosphorylation of pyrimidine L-deoxynucleoside analog diphosphates. Kinetics of phosphorylation and dephosphorylation of nucleoside analog diphosphates and triphosphates by 3-phosphoglycerate kinase *J Biol Chem 277*, 31593-600.

[30] Krishnan, P., Gullen, E. A., Lam, W., Dutschman, G. E., Grill, S. P. and Cheng, Y. C. (2003) Novel role of 3-phosphoglycerate kinase, a glycolytic enzyme, in the activation of L-nucleoside analogs, a new class of anticancer and antiviral agents *J Biol Chem 278*, 36726-32.

[31] Gallois-Montbrun, S., Faraj, A., Seclaman, E., Sommadossi, J. P., Deville-Bonne, D. and Veron, M. (2004) Broad specificity of human phosphoglycerate kinase for antiviral nucleoside analogs *Biochem Pharmacol 68*, 1749-56.

[32] Deville-Bonne, D., El Amri, C., Meyer, P., Chen, Y., Agrofoglio, L. A. and Janin, J. (2010) Human and viral nucleoside/nucleotide kinases involved in antiviral drug activation: structural and catalytic properties *Antiviral Res 86*, 101-20.

[33] Varga, A., Chaloin, L., Sági, G., Sendula, R., Gráczer, É., Liliom, K., Závodszky, P., Lionne, C. and Vas, M. (2011) Nucleotide promiscuity of 3-phosphoglycerate kinase is in focus: implications for the design of better anti-HIV analogues *Mol Biosyst 7*, 1863-73.

[34] Myre, M. A. and O'Day, D. H. (2004) Calmodulin binds to and inhibits the activity of phosphoglycerate kinase *Biochim Biophys Acta 1693*, 177-83.

[35] Scopes, R. K. (1973)3-Phosphoglycerate kinase in *The Enzymes* (Boyer, P., Ed.) pp 335-351, Acad. Press, New York, London.

[36] Joao, H. C. and Williams, R. J. P. (1993) The anatomy of a kinase and the control of phosphate transfer *Eur. J. Biochem. 216*, 1-18.

[37] Lowe, G., Cullis, P. M., Jarvest, R. L., Potter, B. V. and Sproat, B. S. (1981) Stereochemistry of phosphoryl transfer *Philos Trans R Soc Lond B Biol Sci 293*, 75-92.

[38] Knowles, J. R. (1980) Enzyme-catalyzed phosphoryl transfer reactions *Annu Rev Biochem 49*, 877-919.

[39] Herschlag, D. and Jencks, W. P. (1990) Catalysis of the hydrolysis of phosphorylated pyridines by Mg(OH)+: a possible model for enzymatic phosphoryl transfer *Biochemistry 29*, 5172-9.

[40] Admiraal, S. J. and Herschlag, D. (1995) Mapping the transition state for ATP hydrolysis: implications for enzymatic catalysis *Chem Biol 2*, 729-39.

[41] Admiraal, S. J., Meyer, P., Schneider, B., Deville-Bonne, D., Janin, J. and Herschlag, D. (2001) Chemical rescue of phosphoryl transfer in a cavity mutant: a cautionary tale for site-directed mutagenesis *Biochemistry 40*, 403-13.

[42] Aimes, R. T., Hemmer, W. and Taylor, S. S. (2000) Serine-53 at the tip of the glycine-rich loop of cAMP-dependent protein kinase: role in catalysis, P-site specificity, and interaction with inhibitors *Biochemistry 39*, 8325-32.

[43] Yang, J., Ten Eyck, L. F., Xuong, N. H. and Taylor, S. S. (2004) Crystal structure of a cAMP-dependent protein kinase mutant at 1.26A: new insights into the catalytic mechanism *J Mol Biol 336*, 473-87.

[44] Delbaere, L. T., Sudom, A. M., Prasad, L., Leduc, Y. and Goldie, H. (2004) Structure/function studies of phosphoryl transfer by phosphoenolpyruvate carboxykinase *Biochim Biophys Acta 1697*, 271-8.

[45] Williams, N. H. (2004) Models for biological phosphoryl transfer *Biochim Biophys Acta 1697*, 279-87.

[46] Mildvan, A. S. (1981) Conformations and arrangement of substrates at active sites of ATP-utilizing enzymes *Philos Trans R Soc Lond B Biol Sci 293*, 65-74.

[47] Tari, L. W., Matte, A., Pugazhenthi, U., Goldie, H. and Delbaere, L. T. (1996) Snapshot of an enzyme reaction intermediate in the structure of the ATP- Mg2+-oxalate ternary complex of Escherichia coli PEP carboxykinase *Nat Struct Biol 3*, 355-63.

[48] Kack, H., Gibson, K. J., Lindqvist, Y. and Schneider, G. (1998) Snapshot of a phosphorylated substrate intermediate by kinetic crystallography *Proc Natl Acad Sci U S A 95*, 5495-500.

[49] Janin, J. and Wodak, S. J. (1983) Structural domains in proteins and their role in the dynamics of protein function *Prog Biophys Mol Biol 42*, 21-78.

[50] Bennett, W. S. and Huber, R. (1984) Structural and functional aspects of domain motions in proteins *CRC Crit Rev Biochem 15*, 291-384.

[51] Lesk, A. M. and Chothia, C. (1984) Mechanisms of domain closure in proteins *J Mol Biol 174*, 175-91.

[52] Gerstein, M., Lesk, A. M. and Chothia, C. (1994) Structural mechanisms for domain movements in proteins *Biochemistry 32*, 6739-6749.

[53] Watson, H. C., Duee, E. and Mercer, W. D. (1972) Low resolution structure of glyceraldehyde 3-phosphate dehydrogenase *Nat New Biol 240*, 130-3.

[54] Kendrew, J. C., Bodo, G., Dintzis, H. M., Parrish, R. G., Wyckoff, H. and Phillips, D. C. (1958) A three-dimensional model of the myoglobin molecule obtained by x-ray analysis *Nature 181*, 662-6.

[55] Pickles, B., Jeffrey, B. A. and Rossman, M. G. (1964) Some Preliminary Results for the Crystal Structure of Lactic Dehydrogenase *J Mol Biol 93*, 598-600.

[56] Fifis, T. and Scopes, R. K. (1978) Purification of 3-phosphoglycerate kinase from diverse sources by affinity elution chromatography *Biochem J 175*, 311-9.

[57] Blake, C. C., Evans, P. R. and Scopes, R. K. (1972) Structure of horse-muscle phosphoglycerate kinase at 6 angstrom resolution *Nat New Biol 235*, 195-8.

[58] Blake, C. C. and Evans, P. R. (1974) Structure of horse muscle phosphoglycerate kinase. Some results on the chain conformation, substrate binding and evolution of the molecule from a 3 angstrom Fourier map *J Mol Biol 84*, 585-601.

[59] Wendell, P. L., Bryant, T. N. and Watson, H. C. (1972) Low resolution structure of yeast phosphoglycerate kinase *Nat New Biol 240*, 134-6.

[60] Bryant, T. N., Watson, H. C. and Wendell, P. L. (1974) Structure of yeast phosphoglycerate kinase *Nature 247*, 14-7.

[61] Vas, M. and Batke, J. (1981) Evidence for absence of an interaction between purified 3-phosphoglycerate kinase and glyceraldehyde-3-phosphate dehydrogenase *Biochim Biophys Acta 660*, 193-8.

[62] Vas, M. and Batke, J. (1990) Kinetic misinterpretation of a coupled enzyme reaction can lead to the assumption of an enzyme-enzyme interaction. The example of 3-phospho-D-glycerate kinase and glyceraldehyde-3-phosphate dehydrogenase couple *Eur J Biochem 191*, 679-83.

[63] Fokina, K. V., Dainyak, M. B., Nagradova, N. K. and Muronetz, V. I. (1997) A study on the complexes between human erythrocyte enzymes participating in the conversions of 1,3-diphosphoglycerate *Arch Biochem Biophys 345*, 185-92.

[64] Prabhakar, P., Malhotra, O. P. and Kayastha, A. M. (1999) Phosphoglycerate kinase--glyceraldehyde-3-phosphate dehydrogenase interaction: reaction rate studies *Indian J Biochem Biophys 36*, 88-100.

[65] Levitt, M. and Chothia, C. (1976) Structural patterns in globular proteins *Nature 261*, 552-558.

[66] Orengo, C. A., Bray, J. E., Buchan, D. W., Harrison, A., Lee, D., Pearl, F. M., Sillitoe, I., Todd, A. E. and Thornton, J. M. (2002) The CATH protein family database: a resource for structural and functional annotation of genomes *Proteomics 2*, 11-21.

[67] Vinayagam, A., Shi, J., Pugalenthi, G., Meenakshi, B., Blundell, T. L. and Sowdhamini, R. (2003) DDBASE2.0: updated domain database with improved identification of structural domains *Bioinformatics 19*, 1760-4.

[68] Schleif, R. (1999) Arm-domain interactions in proteins: a review [editorial] *Proteins 34*, 1-3.

[69] May, A., Vas, M., Harlos, K. and Blake, C. C. F. (1996) 2.0 A resolution structure of a ternary complex of pig muscle phosphoglycerate kinase containing 3-phospho-D-glycerate and the nucleotide Mn adenylylimidodiphosphate *Proteins 24*, 292-303.

[70] McPhillips, T. M., Hsu, B. T., Sherman, M. A., Mas, M. T. and Rees, D. C. (1996) Structure of the R65Q mutant of yeast 3-phosphoglycerate kinase complexed with Mg-AMP-PNP and 3-phospho-D-glycerate *Biochemistry 35*, 4118-4127.

[71] Davies, G. J., Gamblin, S. J., Littlechild, J. A. and Watson, H. C. (1993) The structure of a thermally stable 3-phosphoglycerate kinase and a comparison with its mesophilic equivalent. *Proteins 15*, 283-289.

[72] Kovári, Z., Flachner, B., Náray-Szabó, G. and Vas, M. (2002) Crystallographic and thiol-reactivity studies on the complex of pig muscle phosphoglycerate kinase with ATP analogues: correlation between nucleotide binding mode and helix flexibility *Biochemistry 41*, 8796-806.

[73] Auerbach, G., Huber, R., Grättinger, M., Zaiss, K., Schurig, H., Jaenicke, R. and Jacob, U. (1997) Closed structure of phosphoglycerate kinase from *Thermotoga maritima* reveals the catalytic mechanism and determinants of thermal stability *Structure 5*, 1475-1483.

[74] Bernstein, B. E., Michels, P. A. M. and Hol, W. G. J. (1997) Synergistic effects of substrate-induced conformational changes in phosphoglycerate kinase activation *Nature 385*, 275-278.

[75] Gondeau, C., Chaloin, L., Lallemand, P., Roy, B., Périgaud, C., Barman, T., Varga, A., Vas, M., Lionne, C. and Arold, S. T. (2008) Molecular basis for the lack of enantioselectivity of human 3-phosphoglycerate kinase *Nucleic Acids Res 36*, 3620-9.

[76] Cliff, M. J., Bowler, M. W., Varga, A., Marston, J. P., Szabó, J., Hounslow, A. M., Baxter, N. J., Blackburn, G. M., Vas, M. and Waltho, J. P. (2010) Transition state analogue structures of human phosphoglycerate kinase establish the importance of charge balance in catalysis *J Am Chem Soc 132*, 6507-16.

[77] Merrett, M. (1981) Primary structure of 3-phosphoglycerate kinase from horse muscle. II. Amino acid sequence of cyanogen bromide peptides CB1-CB4 and CB6-CB14, sequence of methionine-containing regions, and complete sequence of the enzyme *J Biol Chem 256*, 10293-305.

[78] Hardy, G. W., Darbre, A. and Merrett, M. (1981) Primary structure of 3-phosphoglycerate kinase from horse muscle. I. Purification of cyanogen bromide peptides and amino acid sequence of peptide CB5 (104 residues) *J Biol Chem 256*, 10284-92.

[79] Perkins, R. E., Conroy, S. C., Dunbar, B., Fothergill, L. A., Tuite, M. F., Dobson, M. J., Kingsman, S. M. and Kingsman, A. J. (1983) The complete amino acid sequence of yeast phosphoglycerate kinase *Biochem J 211*, 199-218.

[80] Szilágyi, A. N., Ghosh, M., Garman, E. and Vas, M. (2001) A 1.8 A resolution structure of pig muscle 3-phosphoglycerate kinase with bound MgADP and 3-phosphoglycerate in open conformation: new insight into the role of the nucleotide in domain closure *J Mol Biol 306*, 499-511.

[81] Orioli, P., Cini, R., Donati, D. and Mangani, S. (1980) X-ray structure of the ternary complex Zn(II)-ATP-2,2'-bipyridyl and possible model for ATP transport *Nature 283*, 691-3.

[82] Yount, R. G., D. Babcock, Ballantyne, W. and Ojala, D. (1971) Adenylyl Imidodiphosphate, an Adenosine Triphosphate Analog Containing a P-N-P Linkage *Biochemistry 10*, 2484-2489.

[83] Hol, W. G., van Duijnen, P. T. and Berendsen, H. J. (1978) The alpha-helix dipole and the properties of proteins *Nature 273*, 443-6.

[84] Bernstein, B. E. and Hol, W. G. (1998) Crystal structures of substrates and products bound to the phosphoglycerate kinase active site reveal the catalytic mechanism *Biochemistry 37*, 4429-4436.

[85] Murray-Rust, P. and Motherwell, S. (1978) Computer retrieval and analysis of molecular geometry. III. Geometry of the beta-1' aminofuranoside fragment. *Acta Crystallogr B34*, 2534-2546.

[86] Olson, W. K. and Sussman, J. L. (1982) How flexible is the furanose ring? I. A comparison of experimental and theoretical studies. *J. Am. Chem. Soc. 104*, 270-278.

[87] Harding, M. M. (1999) The geometry of metal-ligand interactions relevant to proteins *Acta Crystallogr D Biol Crystallogr 55*, 1432-43.

[88] Hol, W. G. (1985) Effects of the alpha-helix dipole upon the functioning and structure of proteins and peptides *Adv Biophys 19*, 133-65.

[89] Cini, R., Sabat, M., Sundaralingam, M., Burla, M. C., Nunzi, A., Polidori, G. and Zanazzi, P. F. (1983) Interaction of adenosine 5'-triphosphate with metal ions X-ray structure of ternary complexes containing Mg(II), Ca(II), Mn(II), Co(II), ATP and 2,2'-dipyridylamine *J Biomol Struct Dyn 1*, 633-7.

[90] Sabat, M., Cini, R., Haromy, T. and Sundaralingam, M. (1985) Crystal structure of the alpha, beta, gamma-tridentate manganese complex of adenosine 5'-triphosphate cocrystallized with 2,2'-dipyridylamine *Biochemistry 24*, 7827-33.

[91] Vas, M. and Csanády, G. (1987) The two fast-reacting thiols of 3-phosphoglycerate kinase are structurally juxtaposed. Chemical modification with bifunctional reagents *Eur J Biochem 163*, 365-8.

[92] Cserpán, I. and Vas, M. (1983) Effects of substrates on the heat stability and on the reactivities of thiol groups of 3-phosphoglycerate kinase *Eur. J. Biochem. 131*, 157-162.

[93] Tompa, P., Hong, P. T. and Vas, M. (1986) The phosphate group of 3-phosphoglycerate accounts for conformational changes occurring on binding to 3-phosphoglycerate kinase. Enzyme inhibition and thiol reactivity studies. *Eur. J. Biochem. 154*, 643-649.

[94] Hopfner, K. P., Karcher, A., Shin, D. S., Craig, L., Arthur, L. M., Carney, J. P. and Tainer, J. A. (2000) Structural biology of Rad50 ATPase: ATP-driven conformational control in DNA double-strand break repair and the ABC-ATPase superfamily *Cell 101*, 789-800.

[95] Xu, R. M., Carmel, G., Sweet, R. M., Kuret, J. and Cheng, X. (1995) Crystal structure of casein kinase-1, a phosphate-directed protein kinase *Embo J 14*, 1015-23.

[96] Fu, Z., Wang, M., Potter, D., Miziorko, H. M. and Kim, J. J. (2002) The structure of a binary complex between a mammalian mevalonate kinase and ATP: insights into the reaction mechanism and human inherited disease *J Biol Chem 277*, 18134-42.

[97] Schulz-Gahmen, U., DeBondt, H. L. and Kim, S. H. (1996) High-resolution crystal structures of human cyclin-dependent kinase 2 with and without ATP: bound waters and natural ligand as guides for inhibitor design *J Med Chem 39*, 4540-4546.

[98] Retailleau, P., Huang, X., Yin, Y., Hu, M., Weinreb, V., Vachette, P., Vonrhein, C., Bricogne, G., Roversi, P., Ilyin, V. and Carter, C. W., Jr. (2003) Interconversion of ATP binding and conformational free energies by tryptophanyl-tRNA synthetase: structures of ATP bound to open and closed, pre-transition-state conformations *J Mol Biol 325*, 39-63.

[99] Zarembinski, T. I., Hung, L. W., Mueller-Dieckmann, H. J., Kim, K. K., Yokota, H., Kim, R. and Kim, S. H. (1998) Structure-based assignment of the biochemical function of a hypothetical protein: a test case of structural genomics *Proc Natl Acad Sci U S A 95*, 15189-93.

[100] Izard, T. (2003) A novel adenylate binding site confers phosphopantetheine adenylyltransferase interactions with coenzyme A *J Bacteriol 185*, 4074-80.

[101] Desogus, G., Todone, F., Brick, P. and Onesti, S. (2000) Active site of lysyl-tRNA synthetase: structural studies of the adenylation reaction *Biochemistry 39*, 8418-25.

[102] Pickover, C. A., McKay, D. B., Engelman, D. M. and Steitz, T. A. (1979) Substrate binding closes the cleft between the domains of yeast phosphoglycerate kinase *J. Biol. Chem. 254*, 11323-11329.

[103] Sinev, M. A., Razgulyaev, O. I., Vas, M., Timchenko, A. A. and Ptitsyn, O. B. (1989) Correlation between enzyme activity and hinge-bending domain displacement in 3-phosphoglycerate kinase *Eur. J. Biochem. 180*, 61-66.

[104] Chakrabarti, P. (1993) Anion binding sites in protein structures *J Mol Biol 234*, 463-82.

[105] Copley, R. R. and Barton, G. J. (1994) A structural analysis of phosphate and sulphate binding sites in proteins. Estimation of propensities for binding and conservation of phosphate binding sites *J. Mol. Biol. 242*, 321-329.

[106] Chandra, N. R., Muirhead, H., Holbrook, J. J., Bernstein, B. E., Hol, W. G. and Sessions, R. B. (1998) A general method of domain closure is applied to phosphoglycerate kinase and the result compared with the crystal structure of a closed conformation of the enzyme *Proteins 30*, 372-380.

[107] Flachner, B., Varga, A., Szabó, J., Barna, L., Hajdú, I., Gyimesi, G., Závodszky, P. and Vas, M. (2005) Substrate-Assisted Movement of the Catalytic Lys 215 during Domain Closure: Site-Directed Mutagenesis Studies of Human 3-Phosphoglycerate Kinase *Biochemistry 44*, 16853-65.

[108] Varga, A., Flachner, B., Gráczer, É., Osváth, S., Szilágyi, A. N. and Vas, M. (2005) Correlation between conformational stability of the ternary enzyme-substrate complex and domain closure of 3-phosphoglycerate kinase *Febs J 272*, 1867-85.

[109] Varga, A., Flachner, B., Konarev, P., Gráczer, É., Szabó, J., Svergun, D., Závodszky, P. and Vas, M. (2006) Substrate-induced double sided H-bond network as a means of domain closure in 3-phosphoglycerate kinase *FEBS Lett 580*, 2698-706.

[110] Merli, A., Szilágyi, A. N., Flachner, B., Rossi, G. L. and Vas, M. (2002) Nucleotide Binding to Pig Muscle 3-Phosphoglycerate Kinase in the Crystal and in Solution: Relationship between Substrate Antagonism and Interdomain Communication *Biochemistry 41*, 111-119.

[111] Mozzarelli, A. and Rossi, G. L. (1996) Protein function in the crystal *Annu Rev Biophys Biomol Struct 25*, 343-65.

[112] Carugo, O. and Argos, P. (1997) Protein-protein crystal-packing contacts *Protein Sci 6*, 2261-2263.

[113] Cao, Y., Musah, R. A., Wilcox, S. K., Goodin, D. B. and McRee, D. E. (1998) Protein conformer selection by ligand binding observed with crystallography *Protein Sci 7*, 72-8.

[114] Ptitsyn, O. B., Pavlov, M. Y., Sinev, M. A. and Timchenko, A. A. (1986) Study of domain displacements in proteins by diffuse x-ray scattering in *Multidomain proteins* (Patthy, L., and Friedrich, P., Eds.) pp 9-25, Akadémiai Kiadó, Budapest.

[115] Henderson, S. J., Serpersu, E. H., Gerhardt, B. S. and Bunick, G. J. (1994) Conformational changes in yeast phosphoglycerate kinase upon substrate binding *Biophys Chem 53*, 95-104.

[116] Kovári, Z. and Vas, M. (2004) Protein conformer selection by sequence-dependent packing contacts in crystals of 3-phosphoglycerate kinase *Proteins 55*, 198-209.

[117] Janin, J. and Rodier, F. (1995) Protein-Protein Interaction at Crystal Contacts *Proteins 23*, 580-587.

[118] Janin, J. (1997) Specific versus Non-Specific Contacts in Protein Crystals *Nature Struct. Biol. 4*, 973-974.

[119] Xu, D., Tsai, C.-J. and Nussinov, R. (1997) Hydrogen Bonds and Salt Bridges Across Protein-Protein Interfaces *Protein Eng. 10*, 999-1012.

[120] Rossi, G. L. (1992) Biological function in the crystalline state *Curr Op Struct Biol 2*, 816-820.

[121] Littlefield, O. and Nelson, H. C. M. (2001) Crystal Packing Interaction That Blocks Crystallization of a Site-Specific DNA Binding Protein-DNA Complex *Proteins 45*, 219-228.

[122] Raaijmakers, H., Törö, I., Birkenbihl, R., Kemper, B. and Suck, D. (2001) Conformational Flexibility in T4 Endonuclease VII Revealed by Crystallography: Implications for Substrate Binding and Cleavage *J. Mol. Biol. 308*, 311-323.

[123] Blake, C. C. F., Rice, D. W. and Cohen, F. E. (1986) A "helix-scissors" mechanism for the hinge-bending conformational change in phosphoglycerate kinase *Int J Pept Protein Res 27*, 443-448.

[124] Hol, W. G. (1985) The role of the alpha-helix dipole in protein function and structure *Prog Biophys Mol Biol 45*, 149-95.

[125] Sherman, M. A., Szpikowska, B. K., Dean, S. A., Mathiowetz, A. M., McQueen, N. L. and Mas, M. T. (1990) Probing the role of arginines and histidines in the catalytic function and activation of yeast 3-phosphoglycerate kinase by site-directed mutagenesis *J. Biol. Chem. 265*, 10659-10665.

[126] Varga, A., Lionne, C., Lallemand, P., Szabó, J., Adamek, N., Valentin, C., Vas, M., Barman, T. and Chaloin, L. (2009) Direct kinetic evidence that lysine 215 is involved in the phospho-transfer step of human 3-phosphoglycerate kinase *Biochemistry 48*, 6998-7008.

[127] Mas, M. T., Chen, H. H., Aisaka, K., Lin, L. N. and Brandts, J. F. (1995) Effects of C-terminal deletions on the conformational state and denaturation of phosphoglycerate kinase *Biochemistry 34*, 7931-7940.

[128] Ritco-Vonsovici, M., Minard, P., Desmadril, M. and Yon, J. M. (1995) Is the continuity of the domains required for the correct folding of a two domain protein? *Biochemistry 34*, 16543-16551.

[129] Szilágyi, A. N. and Vas, M. (1998) Anion activation of 3-phosphoglycerate kinase requires domain closure *Biochemistry 37*, 8551-8563.

[130] Tanswell, P., Westhead, E. W. and Williams, R. J. P. (1976) Nuclear-magnetic-resonance study of the active-site structure of yeast phosphoglycerate kinase *Eur. J. Biochem. 63*, 249-262.

[131] Wilson, H. R., Williams, R. J. P., Littlechild, J. A. and Watson, H. C. (1988) NMR analysis of the interdomain region of yeast phosphoglycerate kinase *Eur. J. Biochem. 170*, 529-538.

[132] Fairbrother, W. J., Walker, P. A., Minard, P., Littlechild, J. A., Watson, H. and Williams, R. J. P. (1989) NMR analysis of site-specific mutants of yeast phosphoglycerate kinase. An investigation of the triose-binding site *Eur. J. Biochem. 183*, 57-67.

[133] Scheffler, J. E. and Cohn, M. (1986) Photochemically induced dynamic nuclear polarization NMR study of yeast and horse muscle phosphoglycerate kinase *Biochemistry 25*, 3788-96.

[134] Fairbrother, W. J., Bowen, D., Hall, L. and Williams, R. J. (1989) One- and two-dimensional NMR studies of yeast phosphoglycerate kinase. *Eur. J. Biochem. 184*, 617-625.

[135] Chapman, B. E., O'Sullivan, W. J., Scopes, R. K. and Reed, G. H. (1977) Magnetic resonance studies on manganese-nucleotide complexes of phosphoglycerate kinase *Biochemistry 16*, 1005-1010.

[136] Ray, B. D. and Nageswara-Rao, B. D. (1988) 31P NMR studies of enzyme-bound substrate complexes of yeast 3-phosphoglycerate kinase. 1. Effects of sulfate and pH. Mg(II) affinity at the two ATP sites *Biochemistry 27*, 5574-5578.

[137] Nageswara-Rao, B. D. (1984) *Phosphorus-31 NMR: Principles and Application*, Academic, New York.

[138] Fairbrother, W. J., Graham, H. C. and Williams, R. J. P. (1990) The roles of ATP^{4-} and Mg^{2+} in control steps of phosphoglycerate kinase *Eur. J. Biochem. 190*, 407-414.

[139] Graham, H. C. and Williams, R. J. P. (1991) The roles of ADP^{2-} and Mg^{2+} in control steps of phosphoglycerate kinase *Eur. J. Biochem. 197*, 81-91.

[140] Fairbrother, W. J., Graham, H. C. and Williams, R. J. P. (1990) An NMR study of anion binding to yeast phosphoglycerate kinase *Eur. J. Biochem. 190*, 161-169.

[141] Ray, B. D. and Nageswara-Rao, B. D. (1988) 31P NMR studies of enzyme-bound substrate complexes of yeast 3-phosphoglycerate kinase. 2. Structure measurements using paramagnetic relaxation effects of Mn(II) and Co(II) *Biochemistry 27*, 5579-85.

[142] Ray, B. D., Moore, J. M. and Nageswara-Rao, B. D. (1990) 31P NMR studies of enzyme-bound substrate complexes of yeast 3-phosphoglycerate kinase: III. Two ADP binding sites and their Mg(II) affinity; effects of vanadate and arsenate on enzymic complexes with ADP and 3-P-glycerate *J. Inorg. Biochem. 40*, 47-57.

[143] Pappu, K. M. and Serpersu, E. H. (1994) Proton NMR studies of a large protein. pH, substrate titrations, and NOESY experiments with perdeuterated yeast phosphoglycerate kinase containing [1H]histidine residues. *J Magn Reson B 105*, 157-166.

[144] Shibata, C. G., Gregory, J. D., Gerhardt, B. S. and Serpersu, E. H. (1995) Kinetic, binding, and NMR studies of perdeuterated yeast phosphoglycerate kinase and its interactions with substrates. *Arch. Biochem. Biophys. 319*, 204-210.

[145] Jardetzky, O. and Roberts, G. C. K. (1981) *NMR in Molecular Biology*, Academic Press, New York.

[146] Raghunathan, V., Chau, M. H., Ray, B. D. and Nageswara-Rao, B. D. (1999) Structural characterization of manganese(II)-nucleotide complexes bound to yeast 3-phosphoglycerate kinase: 13C relaxation measurements using [U-13C]ATP and [U-13C]ADP *Biochemistry 38*, 15597-605.

[147] Boyle, H. A., Fairbrother, W. J. and Williams, R. J. (1989) An NMR analysis of the binding of inhibitors to yeast phosphoglycerate kinase *Eur J Biochem 184*, 535-43.

[148] Joao, H. C., Williams, R. J. P., Littlechild, J. A., Nagasuma, R. and Watson, H. C. (1992) An investigation of large inhibitors binding to phosphoglycerate kinase and their effect on anion activation *Eur. J. Biochem. 205*, 1077-1088.

[149] Scopes, R. K. (1978) Binding of substrates and other anions to yeast phosphoglycerate kinase. *Eur. J. Biochem. 91*, 119-129.

[150] Jakeman, D. L., Ivory, A. J., Blackburn, G. M. and Williamson, M. P. (2003) Orientation of 1,3-bisphosphoglycerate analogs bound to phosphoglycerate kinase *J Biol Chem 278*, 10957-62.

[151] Blackburn, G. M., Jakeman, D. L., Ivory, A. J. and Williamson, M. P. (1994) *Bioorg Med Chem Lett 4*, 2573-2578.

[152] Jakeman, D. L., Ivory, A. J., Williamson, M. P. and Blackburn, G. M. (1998) Highly potent bisphosphonate ligands for phosphoglycerate kinase *J Med Chem 41*, 4439-52.

[153] Orr, G. A. and Knowles, J. R. (1974) The interaction of the phosphonate analogue of 3-phospho-D-glycerate with phosphoglycerate kinase *Biochem J 141*, 721-3.

[154] Vas, M. and Batke, J. (1984) Adenine nucleotides affect the binding of 3-phosphoglycerate to pig muscle 3-phosphoglycerate kinase. *Eur. J. Biochem. 139*, 115-123.

[155] Williams, D. M., Jakeman, D. L., Vyle, J. S., Williamson, M. P. and Blackburn, G. M. (1998) Synthesis and binding of stable bisubstrate ligands for phosphoglycerate kinase *Bioorg Med Chem Lett 8*, 2603-8.

[156] Sherman, M. A., Fairbrother, W. J. and Mas, M. T. (1992) Characterization of the structure and properties of the His62-->Ala and Arg 38-->Ala mutants of yeast phosphoglycerate kinase: an investigation of the catalytic and activatory sites by site-directed mutagenesis and NMR *Protein Sci. 1*, 752-760.

[157] Moore, J. M. and Reed, G. H. (1985) Coordination scheme and stereochemical configuration of manganese(II) adenosine 5'-diphosphate at the active site of 3-phosphoglycerate kinase *Biochemistry 24*, 5328-33.

[158] Gregory, J. D. and Serpersu, E. H. (1993) Arrangement of substrates at the active site of yeast phosphoglycerate kinase. Effect of sulfate ion. *J. Biol. Chem. 268*, 3880-3888.

[159] Jaffe, E. K., Nick, J. and Cohn, M. (1982) Reactivity and metal-dependent stereospecificity of the phosphorothioate analogs of ADP and ATP and reactivity of Cr(III)ATP in the 3-phosphoglycerate kinase reaction. Structure of the metal nucleotide substrates *J Biol Chem 257*, 7650-6.

[160] Dunaway-Mariano, D. and Cleland, W. W. (1980) Investigations of substrate specificity and reaction mechanism of several kinases using chromium(III) adenosine 5'-triphosphate and chromium(III) adenosine 5'-diphosphate *Biochemistry 19*, 1506-15.

[161] Serpersu, E. H., Summitt, L. L. and Gregory, J. D. (1992) Inactivation of yeast phosphoglycerate kinase by Cr-ATP complexes and its implications on the conformation of the enzyme active site. *J Inorg Biochem. 48*, 203-215.

[162] Eckstein, F. (1985) Nucleoside phosphorothioates *Annu Rev Biochem 54*, 367-402.

[163] Webb, M. R., Ash, D. E., Leyh, T. S., Trentham, D. R. and Reed, G. H. (1982) Electron paramagnetic resonance studies of MN(II) complexes with myosin subfragment 1 and oxygen 17-labeled ligands *J Biol Chem 257*, 3068-72.

[164] Reed, G. H. and Leyh, T. S. (1980) Identification of the six ligands to manganese(II) in transition-state-analogue complexes of creatine kinase: oxygen-17 superhyperfine coupling from selectively labeled ligands *Biochemistry 19*, 5472-80.

[165] Cornelius, R. D. and Cleland, W. W. (1978) Substrate activity of (adenosine triphosphato)tetraamminecobalt(III) with yeast hexokinase and separation of diastereomers using the enzyme *Biochemistry 17*, 3279-86.

[166] Vlasova, I. (1997) Study of yeast phosphoglycerate kinase by electron paramagnetic resonance. 2. Mn2+ ion binding *Biofizika (Russian) 42*, 815-824.

[167] Roustan, C., Fattoum, A., Jeanneau, R. and Pradel, L. A. (1980) Yeast 3-phosphoglycerate kinase: sulfate and substrate binding, their effect on the conformational state of the enzyme *Biochemistry 19*, 5168-75.

[168] Simons, T. J. (1979) Vanadate--a new tool for biologists *Nature 281*, 337-8.

[169] Climent, F., Bartrons, R., Pons, G. and Carreras, J. (1981) Effect of vanadate on phosphoryl transfer enzymes involved in glucose metabolism *Biochem Biophys Res Commun 101*, 570-6.

[170] Macara, I. G., Kustin, K. and Cantley, L. C., Jr. (1980) Glutathione reduces cytoplasmic vanadate. Mechanism and physiological implications *Biochim Biophys Acta 629*, 95-106.

[171] Nour-Eldeen, A. F., Craig, M. M. and Gresser, M. J. (1985) Interaction of inorganic vanadate with glucose-6-phosphate dehydrogenase. Nonenzymic formation of glucose 6-vanadate *J Biol Chem 260*, 6836-42.

[172] Jaffe, E. K. and Cohn, M. (1978) 31P nuclear magnetic resonance spectra of the thiophosphate analogues of adenine nucleotides; effects of pH and Mg2+ binding *Biochemistry 17*, 652-7.

[173] Pecoraro, V. L., Hermes, J. D. and Cleland, W. W. (1984) Stability constants of Mg2+ and Cd2+ complexes of adenine nucleotides and thionucleotides and rate constants for formation and dissociation of MgATP and MgADP *Biochemistry 23*, 5262-71.

[174] Jaffe, E. K. and Cohn, M. (1979) Diastereomers of the nucleoside phosphorothioates as probes of the structure of the metal nucleotide substrates and of the nucleotide binding site of yeast hexokinase *J Biol Chem 254*, 10839-45.

[175] Tomasselli, A. G., Dougherty, J. J., Jr. and Noda, L. H. (1983) Fluorescence and chemical modification of tryptophan as probes of structure of GTP:AMP phosphotransferase from beef heart mitochondria *Eur J Biochem 136*, 297-302.

[176] Dragland-Meserve, C. J., Rothermel, J. D., Houlihan, M. J. and Botelho, L. H. (1985) Inhibition of cyclic AMP-dependent protein kinase-induced changes in the kinetic properties of hepatic pyruvate kinase by the specific cyclic AMP antagonist, the (Rp)-diastereomer of adenosine cyclic 3',5'-phosphorothioate *J Cyclic Nucleotide Protein Phosphor Res 10*, 371-82.

[177] Cohn, M., Shih, N. and Nick, J. (1982) Reactivity and metal-dependent stereospecificity of the phosphorothioate analogs of ATP in the arginine kinase reaction. Structure of the metal-nucleoside triphosphate substrate *J Biol Chem 257*, 7646-9.

[178] Burgers, P. M. and Eckstein, F. (1980) Structure of metal x nucleotide complex in the creatine kinase reaction. A study with diastereomeric phosphorothioate analogs of adenosine di- and triphosphate *J Biol Chem 255*, 8229-33.

[179] Janson, C. A. and Cleland, W. W. (1974) The kinetic mechanism of glycerokinase *J Biol Chem 249*, 2562-6.

[180] Janson, C. A. and Cleland, W. W. (1974) The inhibition of acetate, pyruvate, and 3-phosphogylcerate kinases by chromium adenosine triphosphate *J Biol Chem 249*, 2567-71.

[181] Janson, C. A. and Cleland, W. W. (1974) The specificity of chromium nucleotides as inhibitors of selected kinases *J Biol Chem 249*, 2572-4.

[182] Danenberg, K. D. and Cleland, W. W. (1975) Use of chromium-adenosine triphosphate and lyxose to elucidate the kinetic mechanism and coordination state of the nucleotide substrate for yeast hexokinase *Biochemistry 14*, 28-39.

[183] Dunaway-Mariano, D. and Cleland, W. W. (1980) Preparation and properties of chromium(III) adenosine 5'-triphosphate, chromium(III) adenosine 5'-diphosphate, and related chromium(III) complexes *Biochemistry 19*, 1496-505.

[184] Lester, L. M., Rusch, L. A., Robinson, G. J. and Speckhard, D. C. (1998) Mapping the active sites of 3-phosphoglycerate kinase and glycerol kinase with monoammine chromium(III) ATP *Biochemistry 37*, 5349-55.

[185] Pappu, K. M., Gregory, J. D. and Serpersu, E. H. (1994) Substrate activity of Rh(III)ATP with phosphoglycerate kinase and the role of the metal ion in catalysis *Arch Biochem Biophys 311*, 503-8.

[186] Pappu, K. M., Kunnumal, B. and Serpersu, E. H. (1997) A new metal-binding site for yeast phosphoglycerate kinase as determined by the use of a metal-ATP analog *Biophys J 72*, 928-935.

[187] Ritco-Vonsovici, M., Mouratou, B., Minard, P., Desmadril, M., Yon, J. M., Andrieux, M., Leroy, E. and Guittet, E. (1995) Role of the C-terminal helix in the folding and stability of yeast phosphoglycerate kinase *Biochemistry 34*, 833-841.

[188] Mas, M. T. and Resplandor, Z. E. (1988) Structure-function relationships in 3-phosphoglycerate kinase: role of the carboxy-terminal peptide *Proteins 4*, 56-62.

[189] Szilágyi, A. N., Kotova, N. V., Semisotnov, G. V. and Vas, M. (2001) Incomplete refolding of a fragment of the N-terminal domain of pig muscle 3-phosphoglycerate

kinase that lacks a subdomain: Comparison with refolding of the complementary C-terminal fragment *Eur J Biochem 268*, 1851-60.

[190] Provencher, S. W. and Glöckner, J. (1981) Estimation of protein secondary structure from circular dichroism. *Biochemistry 20*, 33-37.

[191] Nojima, H., Ikai, A. and Noda, H. (1976) Anomalous fluorescence of yeast 3-phosphoglycerate kinase *Biochim Biophys Acta 427*, 20-7.

[192] Privat, J. P., Wahl, P., Auchet, J. C. and Pain, R. H. (1980) Time resolved spectroscopy of tryptophyl fluorescence of yeast 3-phosphoglycerate kinase *Biophys Chem 11*, 239-48.

[193] Wasylewski, Z. and Eftink, M. R. (1987) Frequency domain fluorescence studies of yeast phosphoglycerate kinase and its ternary complex *Eur J Biochem 167*, 513-8.

[194] Dryden, D. T. and Pain, R. H. (1989) Assignment of the heterogeneous static and time-resolved tryptophan fluorescence of 3-phosphoglycerate kinase *Biochim Biophys Acta 997*, 313-21.

[195] Varley, P. G., Dryden, D. T. and Pain, R. H. (1991) Resolution of the fluorescence of the buried tryptophan in yeast 3-phosphoglycerate kinase using succinimide *Biochim Biophys Acta 1077*, 19-24.

[196] Cioni, P., Puntoni, A. and Strambini, G. B. (1993) Tryptophan phosphorescence as a monitor of the solution structure of phosphoglycerate kinase from yeast *Biophys Chem 46*, 47-55.

[197] Mouawad, L., Desmadril, M., Perahia, D., Yon, J. M. and Brochon, J. C. (1990) The effects of ligands on the conformation of phosphoglycerate kinase: fluorescence anisotropy decay and theoretical interpretation *Biopolymers 30*, 1151-1160.

[198] Dryden, D. T., Varley, P. G. and Pain, R. H. (1992) A study of the hinge-bending mechanism of yeast 3-phosphoglycerate kinase. *Eur. J. Biochem. 208*, 115-123.

[199] Cheung, C. W. and Mas, M. T. (1996) Substrate-induced conformational changes in yeast 3-phosphoglycerate kinase monitored by fluorescence of single tryptophan probes *Protein Sci. 5*, 1144-1149.

[200] Desmadril, M., Minard, P., Ballery, N., Gaillard-Miran, S., Hall, L. and Yon, J. M. (1991) Conformational changes in yeast phosphoglycerate kinase upon ligand binding: fluorescence of a linked probe and chemical reactivity of genetically introduced cysteinyl residues. *Proteins 10*, 315-324.

[201] Haran, G., Haas, E., Szpikowska, B. K. and Mas, M. T. (1992) Domain motions in phosphoglycerate kinase: determination of interdomain distance distributions by site-specific labelling and time-resolved fluorescence energy transfer *Proc Natl Acad Sci U S A 89*, 11764-8.

[202] White, E. M., Holland, A. R. and MacDonald, G. (2008) Infrared studies reveal unique vibrations associated with the PGK-ATP-3-PG ternary complex *Biochemistry 47*, 84-91.

[203] Hjelmgren, T., Strid, L. and Arvidsson, L. (1976) An essential arginyl residue in phosphoglycerate kinase from yeast *FEBS Lett 68*, 137-40.

[204] Markland, F. S., Bacharach, A. D., Weber, B. H., O'Grady, T. C., Saunders, G. C. and Umemura, N. (1975) Chemical modification of yeast 3-phosphoglycerate kinase *J Biol Chem 250*, 1301-10.

[205] Roustan, C., Fattoum, A. and Pradel, L. A. (1976) Physicochemical and kinetic properties of iodinated yeast 3-phosphoglycerate kinase *Biochemistry 15*, 2172-7.

[206] Hjelmgren, T., Arvidsson, L. and Larsson-Raznikiewicz, M. (1976) Spectrophotometric pH titrations and nitration with tetranitromethane of the tyrosyl residues in yeast phosphoglycerate kinase *Biochim Biophys Acta 445*, 342-9.

[207] Bacharach, A. D., Markland, F. S., Pellino, A. and Weber, B. H. (1977) Modification of yeast 3-phosphoglycerate kinase: isolation and sequence determination of a nitrated active-site peptide and isolation of a carboxyl modified active-site peptide *Biochem Biophys Res Commun 74*, 165-71.

[208] Desvages, G., Roustan, C., Fattoum, A. and Pradel, L. A. (1980) Structural studies on yeast 3-phosphoglycerate kinase. Identification by immuno-affinity chromatography of one glutamyl residue essential for yeast 3-phosphoglycerate kinase activity. Its location in the primary structure *Eur J Biochem 105*, 259-66.

[209] Arvidsson, L. and Larsson-Raznikiewicz, M. (1973) The SH group of yeast phosphoglycerate kinase *Biochim Biophys Acta 310*, 430-6.

[210] Stinson, R. A. (1974) Chromophoric labelling of yeast 3-phosphoglycerate kinase with an organomercurial *Biochemistry 13*, 4523-9.

[211] Sharma, H. K. (1982) The fast-reacting sulfhydryl group of rat muscle phosphoglycerate kinase is necessary for activity and maintenance of tertiary structure *Biochemistry 21*, 6661-8.

[212] Minard, P., Desmadril, M., Ballery, N., Perahia, D., Mouawad, L., Hall, L. and Yon, J. M. (1989) Study of the fast-reacting cysteines in phosphoglycerate kinase using chemical modification and site-directed mutagenesis. *Eur. J. Biochem. 185*, 419-423.

[213] Bunni, M. A., Baindur, S. R., Douglas, K. T. and Drake, A. F. (1987) Organothallium(III) reagents for modification of biomacromolecules: irreversible labelling of phosphoglycerate kinase from rabbit muscle *Biochim Biophys Acta 923*, 421-30.

[214] Dékány, K. and Vas, M. (1984) Inactivation of pig muscle 3-phosphoglycerate kinase by thiol modification depends on reagent size *Eur. J. Biochem. 139*, 125-130.

[215] Wiksell, E. and Larsson-Raznikiewicz, M. (1987) Affinity labelling of phosphoglycerate kinase by 5'-[p-(fluorosulfonyl)benzoyl]-1,N6-ethenoadenosine *J. Biol. Chem. 262*, 14472-14478.

[216] LaDine, J. R. and Cross, R. L. (1991) The adenine nucleotide-binding site on yeast 3-phosphoglycerate kinase. Affinity labelling of Lys-131 by pyridoxal 5'-diphospho-5'-adenosine *J Biol Chem 266*, 7194-8.

[217] Pineda, T., Kwon, O. S., Serpersu, E. H. and Churchich, J. E. (1993) Interaction of ATP analogs with yeast 3-phosphoglycerate kinase. Affinity labelling of the hinge region *Eur J Biochem 212*, 719-26.

[218] Wrobel, J. A. and Stinson, R. A. (1981) The effects of anions, substrates, metal ions and sulfhydryl reagents on the proteolytic susceptibility of yeast phosphoglycerate kinase *Biochim Biophys Acta 662*, 236-45.

[219] Young, T. A., Skordalakes, E. and Marqusee, S. (2007) Comparison of proteolytic susceptibility in phosphoglycerate kinases from yeast and E. coli: modulation of conformational ensembles without altering structure or stability *J Mol Biol 368*, 1438-47.

[220] Jiang, S. X. and Vas, M. (1988) Limited proteolysis of 3-phosphoglycerate kinase without loss of enzymic activity *FEBS Lett. 231*, 151-154.

[221] Fairbrother, W. J., Hall, L., Littlechild, J. A., Walker, P. A., Watson, H. C. and Williams, R. J. (1989) Site-directed mutagenesis of histidine 62 in the 'basic patch' region of yeast phosphoglycerate kinase *FEBS Lett. 258*, 247-250.

[222] Walker, P. A., Joao, H. C., Littlechild, J. A., Williams, R. J. and Watson, H. C. (1992) Characterisation of yeast phosphoglycerate kinase modified by mutagenesis at residue 21 *Eur J Biochem 207*, 29-37.

[223] Sherman, M. A., Dean, S. A., Mathiowetz, A. M. and Mas, M. T. (1991) Site-directed mutations of arginine 65 at the periphery of the active site cleft of yeast 3-

phosphoglycerate kinase enhance the catalytic activity and eliminate anion-dependent activation *Protein Eng. 4*, 935-940.

[224] Barber, M. D., Gamblin, S. J., Watson, H. and Littlechild, J. A. (1993) Site-directed mutagenesis of yeast phosphoglycerate kinase. Arginines 65, 121 and 168 *FEBS Lett. 320*, 193-197.

[225] Walker, P. A., Littlechild, J. A., Hall, L. and Watson, H. C. (1989) Site-directed mutagenesis of yeast phosphoglycerate kinase. The 'basic-patch' residue arginine 168 *Eur. J. Biochem. 183*, 49-55.

[226] Minard, P., Bowen, D. J., Hall, L., Littlechild, J. A. and Watson, H. C. (1990) Site-directed mutagenesis of aspartic acid 372 at the ATP binding site of yeast phosphoglycerate kinase: over-expression and characterization of the mutant enzyme *Protein Eng. 3*, 515-521.

[227] Szabó, J., Varga, A., Flachner, B., Konarev, P. V., Svergun, D. I., Závodszky, P. and Vas, M. (2008) Communication between the Nucleotide site and the Main Molecular Hinge of 3-Phosphoglycerate Kinase *Biochemistry 47*, 6735-6744.

[228] Szabó, J., Varga, A., Flachner, B., Konarev, P. V., Svergun, D. I., Závodszky, P. and Vas, M. (2008) Role of side-chains in the operation of the main molecular hinge of 3-phosphoglycerate kinase *FEBS Lett 582*, 1335-40.

[229] Bailey, J. M., Lin, L. N., Brandts, J. F. and Mas, M. T. (1990) Substitution of a proline for alanine 183 in the hinge region of phosphoglycerate kinase: effects on catalysis, activation by sulfate, and thermal stability *Journal Protein Chemistry 9*, 59-67.

[230] Mas, M. T., Resplandor, Z. E. and Riggs, A. D. (1987) Site-directed mutagenesis of glutamate-190 in the hinge region of yeast 3-phosphoglycerate kinase: implications for the mechanism of domain movement *Biochemistry 26*, 5369-5377.

[231] Joao, H. C., Taddei, N. and Williams, R. J. (1992) Investigating interdomain region mutants Phe194----Leu and Phe194----Trp of yeast phosphoglycerate kinase by 1H-NMR spectroscopy *Eur J Biochem 205*, 93-104.

[232] McHarg, J., Kelly, S. M., Price, N. C., Cooper, A. and Littlechild, J. A. (1999) Site-directed mutagenesis of proline 204 in the 'hinge' region of yeast phosphoglycerate kinase *Eur J Biochem 259*, 939-45.

[233] Wilson, C. A., Hardman, N., Fothergill-Gilmore, L. A., Gamblin, S. J. and Watson, H. C. (1987) Yeast phosphoglycerate kinase: investigation of catalytic function by site-directed mutagenesis *Biochem J 241*, 609-14.

[234] Graham, H. C., Williams, R. J., Littlechild, J. A. and Watson, H. C. (1991) A proton-NMR study of a site-directed mutation (His388----Glu) in the interdomain region of yeast phosphoglycerate kinase. Implications for domain movement *Eur J Biochem 196*, 261-9.

[235] Johnson, C. M., Cooper, A. and Brown, A. J. (1991) A comparison of the reactivity and stability of wild type and His388Gln mutant phosphoglycerate kinase from yeast *Eur J Biochem 202*, 1157-64.

[236] Mas, M. T., Bailey, J. M. and Resplandor, Z. E. (1988) Site-directed mutagenesis of histidine-388 in the hinge region of yeast 3-phosphoglycerate kinase: effects on catalytic activity and activation by sulfate *Biochemistry 27*, 1168-72.

[237] Hu, C. Q., and Sturtevant, J. M. (1987) Thermodynamic study of yeast phosphoglycerate kinase *Biochemistry 26*, 178-82.

[238] Brandts, J. F., Hu, C. Q., Lin, L. N. and Mas, M. T. (1989) A simple model for proteins with interacting domains. Applications to scanning calorimetry data *Biochemistry 28*, 8588-96.

[239] Griko, Y. V., Venyaminov, S. Y. and Privalov, P. L. (1989) Heat and cold denaturation of phosphoglycerate kinase (interaction of domains). *FEBS Lett. 244*, 276-278.

[240] Galisteo, M. L., Mateo, P. L. and Sanchez-Ruiz, J. M. (1991) Kinetic study on the irreversible thermal denaturation of yeast phosphoglycerate kinase *Biochemistry 30*, 2061-6.

[241] Freire, E., Murphy, K. P., Sanchez-Ruiz, J. M., Galisteo, M. L. and Privalov, P. L. (1992) The molecular basis of cooperativity in protein folding. Thermodynamic dissection of interdomain interactions in phosphoglycerate kinase *Biochemistry 31*, 250-256.

[242] Damaschun, G., Damaschun, H., Gast, K., Misselwitz, R., Muller, J. J., Pfeil, W. and Zirwer, D. (1993) Cold denaturation-induced conformational changes in phosphoglycerate kinase from yeast *Biochemistry 32*, 7739-46.

[243] Gast, K., Damaschun, G., Desmadril, M., Minard, P., Müller-Frohne, M., Pfeil, W. and Zirwer, D. (1995) Cold denaturation of yeast phosphoglycerate kinase: which domain is more stable? *FEBS Lett. 358*, 247-250.

[244] Zaiss, K. and Jaenicke, R. (1999) Thermodynamic study of phosphoglycerate kinase from *Thermotoga maritima* and its isolated domains: reversible thermal unfolding monitored by differential scanning calorimetry and circular dichroism spectroscopy *Biochemistry 38*, 4633-9.

[245] Bentahir, M., Feller, G., Aittaleb, M., Lamotte-Brasseur, J., Himri, T., Chessa, J. P. and Gerday, C. (2000) Structural, kinetic, and calorimetric characterization of the cold-active phosphoglycerate kinase from the antarctic *Pseudomonas sp.* TACII18 *J Biol Chem 275*, 11147-53.

[246] Osváth, S., Sabelko, J. J. and Gruebele, M. (2003) Tuning the heterogeneous early folding dynamics of phosphoglycerate kinase *J Mol Biol 333*, 187-99.

[247] Koch, M. H., Vachette, P. and Svergun, D. I. (2003) Small-angle scattering: a view on the properties, structures and structural changes of biological macromolecules in solution *Q. Rev. Biophys. 36*, 147-227.

[248] Bennett, W. S., Jr. and Steitz, T. A. (1978) Glucose-induced conformational change in yeast hexokinase *Proc Natl Acad Sci U S A 75*, 4848-52.

[249] Anderson, C. M., Zucker, F. H. and Steitz, T. A. (1979) Space-filling models of kinase clefts and conformation changes *Science 204*, 375-80.

[250] Varga, A., Szabó, J., Flachner, B., Roy, B., Konarev, P., Svergun, D., Závodszky, P., Perigaud, C., Barman, T., Lionne, C. and Vas, M. (2008) Interaction of human 3-phosphoglycerate kinase with L-ADP, the mirror image of D-ADP *Biochem Biophys Res Commun 366*, 994-1000.

[251] Svergun, D. I., Barberato, C. and Koch, M. H. J. (1995) CRYSOL - a program to evaluate X-ray solution scattering of biological macromolecules from atomic coordinates *J. Appl. Crystallogr. 28*, 768-773.

[252] Wu, H. (1931) Studies on denaturation of proteins. A theory of denaturation (Reproduced in Adv. Prot. Chem. 41, 6-26) *Chinese J Physiol 5*, 321-331.

[253] Kauzman, I. (1959) Some factors in the interpretation of protein denaturation *Adv Protein Chem 14*, 1-67.

[254] Anfinsen, C. B., Haber, E., Sela, M. and White, F. H. (1961) The kinetics of formation of native ribonuclease during oxidation of the reduced polypeptide chain *Proc Natl Acad Sci USA 47*, 1309-1314.

[255] Anfinsen, C. B. (1973) Principles that govern the folding of protein chains. *Science 181*, 223-230.

[256] Levinthal, C. (1968) Are there pathways for protein folding? *J Chim Phys 65*, 44-45.

[257] Tsong, T. Y. and Baldwin, R. L. (1972) A sequential model of nucleation-dependent protein folding: kinetic studies of ribonuclease A *J Mol Biol 63*, 453-69.

[258] Baldwin, R. L. (1975) Intermediates in protein folding reactions and the mechanism of protein folding *Annu Rev Biochem 44*, 453-75.

[259] Fersht, A. R. (1997) Nucleation mechanisms in protein folding *Curr Opin Struct Biol 7*, 3-9.

[260] Kim, P. S. and Baldwin, R. L. (1990) Intermediates in the folding reactions of small proteins *Annu Rev Biochem 59*, 631-660.

[261] Jaenicke, R. (1987) Folding and association of proteins *Prog Biophys Mol Biol 49*, 117-237.

[262] Ptitsyn, O. B. (1973) Stages in the mechanism of self-organization of protein molecules *Dokl Akad Nauk SSSR 210*, 1213-5.

[263] Wetlaufer, D. B. (1973) Nucleation, rapid folding, and globular intrachain regions in proteins *Proc Natl Acad Sci U S A 70*, 697-701.

[264] Karplus, M. and Weaver, D. L. (1976) Protein-folding dynamics *Nature 260*, 404-6.

[265] Karplus, M. and Weaver, D. L. (1994) Protein folding dynamics: the diffusion-collision model and experimental data *Protein Sci 3*, 650-668.

[266] Dill, K. A. (1985) Theory for the folding and stability of globular proteins *Biochemistry 24*, 1501-1509.

[267] Chan, H. S. and Dill, K. A. (1990) Origins of structure in globular proteins *Proc Natl Acad Sci U S A 87*, 6388-6392.

[268] Harrison, S. C. and Durbin, R. (1985) Is there a single pathway for the folding of a polypeptide chain? *Proc Natl Acad Sci U S A 82*, 4028-30.

[269] Yon, J. M. (2002) Protein folding in the post-genomic era *J Cell Mol Med 6*, 307-327.

[270] Ohgushi, M. and Wada, A. (1983) 'Molten-globule state': a compact form of globular proteins with mobile side-chains *FEBS Lett 164*, 21-4.

[271] Ptitsyn, O. B., Pain, R. H., Semisotnov, G. V., Zerovnik, E. and Razgulyaev, O. I. (1990) Evidence for a molten globule state as a general intermediate in protein folding *FEBS Lett. 262*, 20-24.

[272] Chaffotte, A. F., Cadieux, C., Guillou, Y. and Goldberg, M. E. (1992) A possible initial folding intermediate: the C-terminal proteolytic domain of tryptophan synthase beta chains folds in less than 4 milliseconds into a condensed state with non-native-like secondary structure *Biochemistry 31*, 4303-8.

[273] Hagen, S. J., Hofrichter, J., Szabo, A. and Eaton, W. A. (1996) Diffusion-limited contact formation in unfolded cytochrome c: estimating the maximum rate of protein folding *Proc Natl Acad Sci U S A 93*, 11615-7.

[274] McCammon, J. A. (1996) A speed limit for protein folding *Proc Natl Acad Sci U S A 93*, 11426-7.

[275] Ghélis, C. and Yon, J. M. (1982) *Protein Folding*, Acad. Press, New York.

[276] Jaenicke, R. (1991) Protein folding: local structures, domains, subunits, and assemblies *Biochemistry 30*, 3147-61.

[277] Slaby, I. and Holmgren, A. (1979) Structure and enzymatic functions of thioredoxin refolded by complementation of two tryptic peptide fragments *Biochemistry 18*, 5584-91.

[278] Ghélis, C., Tempete-Gaillourdet, M. and Yon, J. M. (1978) The folding of pancreatic elastase: independent domain refolding and inter-domain interaction *Biochem Biophys Res Commun 84*, 31-6.

[279] Shiba, K. and Schimmel, P. (1992) Functional assembly of a randomly cleaved protein *Proc Natl Acad Sci U S A 89*, 1880-4.

[280] Wetlaufer, D. B. (1981) Folding of protein fragments *Adv. Prot. Chem. 34*, 61-92.

[281] Vita, C., Fontana, A. and Jaenicke, R. (1989) Folding of thermolysin fragments. Hydrodynamic properties of isolated domains and subdomains *Eur J Biochem 183*, 513-8.

[282] Oas, T. G. and Kim, P. S. (1988) A peptide model of a protein folding intermediate *Nature 336*, 42-8.

[283] Sancho, J. and Fersht, A. R. (1992) Dissection of an enzyme by protein engineering. The N and C-terminal fragments of barnase form a native-like complex with restored enzymic activity *J Mol Biol 224*, 741-7.

[284] Wolynes, P. G., Onuchic, J. N. and Thirumalai, D. (1995) Navigating the folding routes *Science 267*, 1619-20.

[285] Dobson, C. M. and Karplus, M. (1999) The fundamentals of protein folding: bringing together theory and experiment *Curr Opin Struct Biol 9*, 92-101.

[286] Onuchic, J. N., Socci, N. D., Luthey-Schulten, Z. and Wolynes, P. G. (1996) Protein folding funnels: the nature of the transition state ensemble. *Fold. Des. 1*, 441-450.

[287] Lazaridis, T. and Karplus, M. (1997) "New view" of protein folding reconciled with the old through multiple unfolding simulations *Science 278*, 1928-31.

[288] Auerbach, G., Jacob, U., Grattinger, M., Schurig, H. and Jaenicke, R. (1997) Crystallographic analysis of phosphoglycerate kinase from the hyperthermophilic bacterium Thermotoga maritima *Biol. Chem. 378*, 327-329.

[289] Mandelman, D., Bentahir, M., Feller, G., Gerday, C. and Haser, R. (2001) Crystallization and preliminary X-ray analysis of a bacterial psychrophilic enzyme, phosphoglycerate kinase *Acta Crystallogr D Biol Crystallogr 57*, 1666-8.

[290] Betton, J. M., Desmadril, M., Mitraki, A. and Yon, J. M. (1984) Unfolding-refolding transition of a hinge bending enzyme: horse muscle phosphoglycerate kinase induced by guanidine hydrochloride. *Biochemistry 23*, 6654-6661.

[291] Missiakas, D., Betton, J. M., Minard, P. and Yon, J. M. (1990) Unfolding-refolding of the domains in yeast phosphoglycerate kinase: comparison with the isolated engineered domains. *Biochemistry 29*, 8683-8689.

[292] Mitraki, A., Betton, J. M., Desmadril, M. and Yon, J. M. (1987) Quasi-irreversibility in the unfolding-refolding transition of phosphoglycerate kinase induced by guanidine hydrochloride *Eur J Biochem 163*, 29-34.

[293] Yon, J. M., Betton, J. M., Desmadril, M., Mitraki, A., Minard, P., Gaillard, S., Ballery, N. and Missiakas, D. (1988) Survey of the folding pathway of a two-domain protein phosphoglycerate kinase. *J Chromatogr 440*, 421-437.

[294] Adams, B., Burgess, R. J. and Pain, R. H. (1985) The folding and mutual interaction of the domains of yeast 3-phosphoglycerate kinase. *Eur. J. Biochem. 152*, 715-720.

[295] Betton, J. M., Desmadril, M. and Yon, J. M. (1989) Detection of intermediates in the unfolding transition of phosphoglycerate kinase using limited proteolysis. *Biochemistry 28*, 5421-5428.

[296] Betton, J. M., Desmadril, M., Mitraki, A. and Yon, J. M. (1985) Kinetic studies of the unfolding-refolding of horse muscle phosphoglycerate kinase induced by guanidine hydrochloride. *Biochemistry 24*, 4570-4577.

[297] Semisotnov, G. V., Vas, M., Chemeris, V. V., Kashparova, N. J., Kotova, N. V., Razgulyaev, O. I. and Sinev, M. A. (1991) Refolding of pig muscle and yeast 3-phosphoglycerate kinases and of their proteolytic fragments *Eur. J. Biochem. 202*, 1083-1089.

[298] Szilágyi, A. N. and Vas, M. (1998) Sequential domain refolding of pig muscle 3-phosphoglycerate kinase: kinetic analysis of reactivation *Fold Des 3*, 565-75.

[299] Ballery, N., Minard, P., Desmadril, M., Betton, J. M., Perahia, D., Mouawad, L., Hall, L. and Yon, J. M. (1990) Introduction of internal cysteines as conformational probes in yeast phosphoglycerate kinase. *Protein Eng. 3*, 199-204.

[300] Ballery, N., Desmadril, M., Minard, P. and Yon, J. M. (1993) Characterization of an intermediate in the folding pathway of phosphoglycerate kinase: Chemical reactivity of genetically introduced cysteinyl residues during the folding process *Biochemistry 32*, 708-714.

[301] Garcia, P., Desmadril, M., Minard, P. and Yon, J. M. (1995) Evidence for residual structures in an unfolded form of yeast phosphoglycerate kinase *Biochemistry 34*, 397-404.

[302] Szpikowska, B. K., Beechem, J. M., Sherman, M. A. and Mas, M. T. (1994) Equilibrium unfolding of yeast phosphoglycerate kinase and its mutants lacking one or both native tryptophans: a circular dichroism and steady-state and time-resolved fluorescence study *Biochemistry 33*, 2217-2225.

[303] Szpikowska, B. K. and Mas, M. T. (1996) Urea-induced equilibrium unfolding of single tryptophan mutants of yeast phosphoglycerate kinase: evidence for a stable intermediate *Arch. Biochem. Biophys. 335*, 173-182.

[304] Garcia, P., Mérola, F., Receveur, V., Blandin, P., Minard, P. and Desmadril, M. (1998) Steady State and Time-Resolved Fluorescence Study of Residual Structures in an Unfolded Form of Yeast Phosphoglycerate Kinase *Biochemistry 37*, 7444-7455.

[305] Calmettes, P., Roux, B., Durand, D., Desmadril, M., and Smith, J. C. (1993) Configurational distribution of denatured phosphoglycerate kinase. *J. Mol. Biol. 231*, 840-848.

[306] Flory, P. (1989) *Statistical mechanics of chain molecules*, Oxford Univ. Press.

[307] Petrescu, A. J., Receveur, V., Calmettes, P., Durand, D., Desmadril, M., Roux, B. and Smith, J. C. (1997) Small-angle neutron scattering by a strongly denatured protein: analysis using random polymer theory. *Biophys J 72*, 335-342.

[308] Receveur, V., Calmettes, P., Smith, J. C., Desmadril, M., Coddens, G. and Durand, D. (1997) Picosecond dynamical changes on denaturation of yeast phosphoglycerate kinase revealed by quasielastic neutron scattering. *Proteins 28*, 380-387.

[309] Mas, M. T., Chen, C. Y., Hitzeman, R. A. and Riggs, A. D. (1986) Active human-yeast chimeric phosphoglycerate kinases engineered by domain interchange *Science 233*, 788-90.

[310] Vas, M., Sinev, M. A., Kotova, N. and Semisotnov, G. V. (1990) Reactivation of 3-phosphoglycerate kinase from its proteolytic fragments *Eur. J. Biochem. 189*, 575-579.

[311] Minard, P., Hall, L., Betton, J. M., Missiakas, D. and Yon, J. M. (1989) Efficient expression and characterization of isolated structural domains of yeast phosphoglycerate kinase generated by site-directed mutagenesis. *Protein Eng. 3*, 55-60.

[312] Fairbrother, W. J., Minard, P., Hall, L., Betton, J. M., Missiakas, D., Yon, J. M. and Williams, R. J. (1989) Nuclear magnetic resonance studies of isolated structural domains of yeast phosphoglycerate kinase *Protein Eng. 3*, 5-11.

[313] Parker, M. J., Spencer, J., Jackson, G. S., Burston, S. G., Hosszu, L. L. P., Craven, C. J., Waltho, P. and Clarke, A. R. (1996) Domain behaviour during refolding of a thermostable phosphoglycerate kinase *Biochemistry 35*, 15740-15752.

[314] Hosszú, L. L. P., Craven, C. J., Spencer, J., Parker, M. J., Clarke, A. R., Kelly, M. and Waltho, J. P. (1997) Is the structure of the N-domain of phosphoglycerate kinase affected by isolation from the intact molecule? *Biochemistry 36*, 333-340.

[315] Andrieux, M., Leroy, E., Guittet, E., Ritco-Vonsovici, M., Mouratou, B., Minard, P., Desmadril, M. and Yon, J. M. (1995) Transferred nuclear Overhauser effect study of

the C-terminal helix of yeast phosphoglycerate kinase: NMR solution structure of the C-terminal bound peptide. *Biochemistry 34*, 842-846.

[316] Sherman, M. A., Chen, Y. and Mas, M. T. (1997) An engineered amino-terminal domain of yeast phosphoglycerate kinase with native-like structure *Protein Sci. 6*, 882-891.

[317] Gast, K., Damaschun, G., Damaschun, H., Misselwitz, R. and Zirwer, D. (1993) Cold denaturation of yeast phosphoglycerate kinase: kinetics of changes in secondary structure and compactness on folding and refolding *Biochemistry 32*, 7747-7752.

[318] Missiakas, D., Betton, J. M., Chaffotte, A., Minard, P. and Yon, J. M. (1992) Kinetic studies of the refolding of yeast phosphoglycerate kinase: comparison with the isolated engineered domains *Protein Sci. 1*, 1485.

[319] Osváth, S., Kohler, G., Závodszky, P. and Fidy, J. (2005) Asymmetric effect of domain interactions on the kinetics of folding in yeast phosphoglycerate kinase *Protein Sci 14*, 1609-16.

[320] Pecorari, F., Minard, P., Desmadril, M. and Yon, J. M. (1993) Structure and functional complementation of engineered fragments from yeast phosphoglycerate kinase *Protein Eng. 6*, 313-325.

[321] Pecorari, F., Guilbert, C., Minard, P., Desmadril, M. and Yon, J. M. (1996) Folding and functional complementation of engineered fragments from yeast phosphoglycerate kinase *Biochemistry 35*, 3465-3476.

[322] Tougard, P., Bizebard, T., Ritco-Vonsovici, M., Minard, P. and Desmadril, M. (2002) Structure of a circularly permuted phosphoglycerate kinase *Acta Crystallogr D Biol Crystallogr 58*, 2018-23.

[323] Collinet, B., Garcia, P., Minard, P. and Desmadril, M. (2001) Role of loops in the folding and stability of yeast phosphoglycerate kinase *Eur J Biochem 268*, 5107-18.

[324] Collinet, B., Herve, M., Pecorari, F., Minard, P., Eder, O. and Desmadril, M. (2000) Functionally accepted insertions of proteins within protein domains *J Biol Chem 275*, 17428-33.

[325] Pecorari, F., Minard, P., Desmadril, M. and Yon, J. M. (1996) Occurrence of transient multimeric species during the refolding of a monomeric protein. *J. Biol. Chem. 271*, 5270-5276.

[326] Sabelko, J., Ervin, J. and Gruebele, M. (1999) Observation of strange kinetics in protein folding [see comments] *Proc Natl Acad Sci U S A 96*, 6031-6.

[327] Osváth, S., Jackel, M., Agocs, G., Závodszky, P., Kohler, G. and Fidy, J. (2006) Domain interactions direct misfolding and amyloid formation of yeast phosphoglycerate kinase *Proteins 62*, 909-17.

[328] Beaucamp, N., Schurig, H. and Jaenicke, R. (1997) The PGK-TIM fusion protein from Thermotoga maritima and its constituent parts are intrinsically stable and fold independently *J. Biol. Chem. 378*, 679-685.

[329] Kress, S., Stein, A., Maurer, P., Weber, B., Reichert, J., Buchmann, A., Huppert, P. and Schwarz, M. (1998) Expression of hypoxia-inducible genes in tumor cells *J Cancer Res Clin Oncol 124*, 315-20.

[330] Warburg, O. (1956) On the origin of cancer cells *Science 123*, 309-14.

[331] Stathakis, P., Lay, A. J., Fitzgerald, M., Schlieker, C., Matthias, L. J. and Hogg, P. J. (1999) Angiostatin formation involves disulfide bond reduction and proteolysis in kringle 5 of plasmin *J Biol Chem 274*, 8910-6.

[332] Shetty, S., Muniyappa, H., Halady, P. K. and Idell, S. (2004) Regulation of urokinase receptor expression by phosphoglycerate kinase *Am J Respir Cell Mol Biol 31*, 100-6.

[333] Shetty, P., Velusamy, T., Bhandary, Y. P., Liu, M. C. and Shetty, S. (2010) Urokinase receptor expression involves tyrosine phosphorylation of phosphoglycerate kinase *Mol Cell Biochem 335*, 235-47.

[334] Daly, E. B., Wind, T., Jiang, X. M., Sun, L. and Hogg, P. J. (2004) Secretion of phosphoglycerate kinase from tumour cells is controlled by oxygen-sensing hydroxylases *Biochimica Et Biophysica Acta-Molecular Cell Research 1691*, 17-22.

[335] Wang, J., Dai, J., Jung, Y., Wei, C. L., Wang, Y., Havens, A. M., Hogg, P. J., Keller, E. T., Pienta, K. J., Nor, J. E., Wang, C. Y. and Taichman, R. S. (2007) A glycolytic mechanism regulating an angiogenic switch in prostate cancer *Cancer Res 67*, 149-59.

[336] Jung, Y., Shiozawa, Y., Wang, J., Wang, Z., Pedersen, E. A., Lee, C. H., Hall, C. L., Hogg, P. J., Krebsbach, P. H., Keller, E. T. and Taichman, R. S. (2009) Expression of PGK1 by prostate cancer cells induces bone formation *Mol Cancer Res 7*, 1595-604.

[337] Zieker, D., Konigsrainer, I., Tritschler, I., Loffler, M., Beckert, S., Traub, F., Nieselt, K., Buhler, S., Weller, M., Gaedcke, J., Taichman, R. S., Northoff, H., Brucher, B. L. and Konigsrainer, A. (2010) Phosphoglycerate kinase 1 a promoting enzyme for peritoneal dissemination in gastric cancer *Int J Cancer 126*, 1513-20.

[338] Maury, G. (2000) The enantioselectivity of enzymes involved in current antiviral therapy using nucleoside analogues: a new strategy? *Antivir Chem Chemother 11*, 165-89.

[339] Sabini, E., Hazra, S., Konrad, M., Burley, S. K. and Lavie, A. (2007) Structural basis for activation of the therapeutic L-nucleoside analogs 3TC and troxacitabine by human deoxycytidine kinase *Nucleic Acids Res 35*, 186-92.

[340] Sabini, E., Hazra, S., Konrad, M. and Lavie, A. (2007) Nonenantioselectivity Property of Human Deoxycytidine Kinase Explained by Structures of the Enzyme in Complex with l- and d-Nucleosides *J Med Chem 50*, 3004-14.

[341] Sandrini, M. P. and Piskur, J. (2005) Deoxyribonucleoside kinases: two enzyme families catalyze the same reaction *Trends Biochem Sci 30*, 225-8.

[342] Johansson, K., Ramaswamy, S., Ljungcrantz, C., Knecht, W., Piskur, J., Munch-Petersen, B., Eriksson, S. and Eklund, H. (2001) Structural basis for substrate specificities of cellular deoxyribonucleoside kinases *Nat Struct Biol 8*, 616-20.

[343] Liou, J. Y., Dutschman, G. E., Lam, W., Jiang, Z. and Cheng, Y. C. (2002) Characterization of human UMP/CMP kinase and its phosphorylation of D- and L-form deoxycytidine analogue monophosphates *Cancer Res 62*, 1624-31.

[344] Alexandre, J. A., Roy, B., Topalis, D., Pochet, S., Perigaud, C. and Deville-Bonne, D. (2007) Enantioselectivity of human AMP, dTMP and UMP-CMP kinases *Nucleic Acids Res 35*, 4895-904.

[345] Miller, W. H. and Miller, R. L. (1982) Phosphorylation of acyclovir diphosphate by cellular enzymes *Biochem Pharmacol 31*, 3879-84.

[346] Matthews, T. and Boehme, R. (1988) Antiviral activity and mechanism of action of ganciclovir *Rev Infect Dis 10 Suppl 3*, S490-4.

[347] Koch, K., Chen, Y., Feng, J. Y., Borroto-Esoda, K., Deville-Bonne, D., Gallois-Montbrun, S., Janin, J. and Morera, S. (2009)Nucleoside diphosphate kinase and the activation of antiviral phosphonate analogs of nucleotides: binding mode and phosphorylation of tenofovir derivatives in *Nucleosides Nucleotides Nucleic Acids* pp 776-92, England.

[348] Varga, A., Gráczer, E., Chaloin, L., Liliom, K., Závodszky, P., Lionne, C. and Vas, M. (2013) Selectivity of kinases on the activation of tenofovir, an anti-HIV agent *Eur J Pharm Sci 48*, 307-15.

[349] Vas, M., Varga, A. and Gráczer, É. (2010) Insight into the mechanism of domain movements and their role in enzyme function: example of 3-phosphoglycerate kinase *Curr Protein Pept Sci 11*, 118-47.
[350] Gondeau, C., Chaloin, L., Varga, A., Roy, B., Lallemand, P., Perigaud, C., Barman, T., Vas, M. and Lionne, C. (2008) Differences in the transient kinetics of the binding of D-ADP and its mirror image L-ADP to human 3-phosphoglycerate kinase revealed by the presence of 3-phosphoglycerate *Biochemistry 47*, 3462-73.
[351] Lallemand, P., Chaloin, L., Roy, B., Barman, T., Bowler, M. W. and Lionne, C. (2011) Interaction of human 3-phosphoglycerate kinase with its two substrates: is substrate antagonism a kinetic advantage? *J Mol Biol 409*, 742-57.
[352] Vale, R. D. and Milligan, R. A. (2000) The way things move: looking under the hood of molecular motor proteins *Science 288*, 88-95.
[353] Cornish-Bowden, A. (1986) Why is uncompetitive inhibition so rare? A possible explanation, with implications for the design of drugs and pesticides *FEBS Lett 203*, 3-6.
[354] Robertson, J. G. (2005) Mechanistic basis of enzyme-targeted drugs *Biochemistry 44*, 5561-71.
[355] Duclert-Savatier, N., Poggi, L., Miclet, E., Lopes, P., Ouazzani, J., Chevalier, N., Nilges, M., Delarue, M. and Stoven, V. (2009) Insights into the enzymatic mechanism of 6-phosphogluconolactonase from Trypanosoma brucei using structural data and molecular dynamics simulation *J Mol Biol 388*, 1009-21.
[356] Blattner, J., Helfert, S., Michels, P. and Clayton, C. (1998) Compartmentation of phosphoglycerate kinase in Trypanosoma brucei plays a critical role in parasite energy metabolism *Proc Natl Acad Sci U S A 95*, 11596-600.
[357] Ladame, S., Castilho, M. S., Silva, C. H., Denier, C., Hannaert, V., Perie, J., Oliva, G. and Willson, M. (2003) Crystal structure of Trypanosoma cruzi glyceraldehyde-3-phosphate dehydrogenase complexed with an analogue of 1,3-bisphospho-d-glyceric acid *Eur J Biochem 270*, 4574-86.
[358] Larsson-Raznikiewicz, M. (1964) Kinetic studies on the reaction catalysed by phosphoglycerate kinase: I. The effect of Mg^{2+} and adenosine-5'-triphosphate *Biochim. Biophys. Acta 85*, 60-68.
[359] Larsson-Raznikiewicz, M. (1967) Kinetic studies on the reaction catalyzed by phosphoglycerate kinase. II. The kinetic relationships between 3-phosphoglycerate, $MgATP^{2-}$ and activating metal ion *Biochim. Biophys. Acta 132*, 33-40.
[360] Larsson-Raznikiewicz, M. and Arvidsson, L. (1971) Inhibition of phosphoglycerate kinase by products and product homologues *Eur. J. Biochem. 22*, 506-512.
[361] Scopes, R. K. (1978) The steady-state kinetics of yeast phosphoglycerate kinase. Anomalous kinetic plots and the effects of salts on activity *Eur. J. Biochem. 85*, 503-516.
[362] Knowles, J. R. (1976) Whither enzyme mechanisms? *FEBS Lett 62*, E53-61.
[363] Johnson, P. E., Abbott, S. J. and Knowles, J. R. (1976) The "phosphoryl-enzyme" from phosphoglycerate kinase *Biochemistry 15*, 2893-8.
[364] Walsh, C. T., Jr. and Spector, L. B. (1971) A phosphoenzyme intermediary in phosphoglycerate kinase action *J Biol Chem 246*, 1255-61.
[365] Brevet, A., Roustan, C., Desvages, G., Pradel, L. A. and van Thoai, N. (1973) Yeast 3-phosphoglycerate kinase: evidence for a glutamyl residue in the phosphoryl transfer *Eur J Biochem 39*, 141-7.
[366] Johnson, P. E., Abbott, S. J., Orr, G. A., Semeriva, M. and Knowles, J. R. (1975) On the "phosphoryl-enzyme" of phosphoglycerate kinase *Biochem Biophys Res Commun 62*, 382-9.

[367] Webb, M. R. and Trentham, D. R. (1980) Analysis of chiral inorganic (160, 170, 180) thiophosphate and the stereochemistry of the 3-phosphoglycerate kinase reaction *J. Biol. Chem. 255*, 1775-1779.

[368] Geerlof, A., Travers, F., Barman, T. and Lionne, C. (2005) Perturbation of Yeast 3-Phosphoglycerate Kinase Reaction Mixtures with ADP: Transient Kinetics of Formation of ATP from Bound 1,3-Bisphosphoglycerate *Biochemistry 44*, 14948-55.

[369] Maegley, K. A., Admiraal, S. J. and Herschlag, D. (1996) Ras-catalyzed hydrolysis of GTP: a new perspective from model studies *Proc Natl Acad Sci U S A 93*, 8160-6.

[370] Schlichting, I. and Reinstein, J. (1997) Structures of active conformations of UMP kinase from *Dictyostelium discoideum* suggest phosphoryl transfer is associative *Biochemistry 36*, 9290-9296.

[371] Hummel, J. P. and Dreyer, W. J. (1962) Measurement of protein-binding phenomena by gel filtration *Biochim Biophys Acta 63*, 530-2.

[372] Vas, M., Merli, A. and Rossi, G. L. (1994) Antagonistic binding of substrates to 3-phosphoglycerate kinase monitored by the fluorescent analogue 2'(3')-O-(2,4,6-trinitrophenyl)adenosine 5'-triphosphate. *Biochem. J. 301*, 885-891.

[373] Geerlof, A., Schmidt, P. P., Travers, F. and Barman, T. (1997) Cryoenzymic studies on yeast 3-phosphoglycerate kinase. Attempt to obtain the kinetics of the hinge-bending motion *Biochemistry 36*, 5538-5545.

[374] Burbaum, J. J. and Knowles, J. R. (1989) Internal thermodynamics of enzymes determined by equilibrium quench: values of Kint for enolase and creatine kinase *Biochemistry 28*, 9306-17.

[375] Gutfreund, H. and Trentham, D. R. (1975) Energy changes during the formation and interconversion of enzyme-substrate complexes *Ciba Found Symp*, 69-86.

[376] Nageswara-Rao, B. D., Cohn, M. and Scopes, R. K. (1978) 31P NMR study of bound reactants and products of yeast 3-phosphoglycerate kinase at equilibrium and the effect of sulfate ion *J. Biol. Chem. 253*, 8056-8060.

[377] Jaffe, E. K. and Cohn, M. (1980) Shift of the equilibrium constant of the 3-P-glycerate kinase reaction towards 1,3-bis-P-glycerate with adenosine 5'-O-(2-thiotriphosphate) (ATP beta S) as substrate *J Biol Chem 255*, 3240-1.

[378] Schmidt, P. P., Travers, F. and Barman, T. (1995) Transient and equilibrium kinetic studies on yeast 3-phosphoglycerate kinase. Evidence that an intermediate containing 1,3-bisphosphoglycerate accumulates in the steady state *Biochemistry 34*, 824-832.

[379] Lionne, C., Iorga, B., Candau, R., Piroddi, N., Webb, M. R., Belus, A., Travers, F. and Barman, T. (2002) Evidence that phosphate release is the rate-limiting step on the overall ATPase of psoas myofibrils prevented from shortening by chemical cross-linking *Biochemistry 41*, 13297-308.

[380] Biosca, J. A., Travers, F. and Barman, T. E. (1983) A jump in an Arrhenius plot can be the consequence of a phase transition. The binding of ATP to myosin subfragment 1 *FEBS Lett 153*, 217-20.

[381] Thomas, T. M. and Scopes, R. K. (1998) The effects of temperature on the kinetics and stability of mesophilic and thermophilic 3-phosphoglycerate kinases *Biochem. J 330*, 1087-1095.

[382] Gutfreund, H. (1995)Transient kinetics of enzyme reactions in *Kinetics for the Life Sciences* (Gutfreund, H., Ed.) pp 138-180, Cambridge University Press, Cambridge.

[383] Barman, T. E., Bellamy, S. R., Gutfreund, H., Halford, S. E. and Lionne, C. (2006) The identification of chemical intermediates in enzyme catalysis by the rapid quench-flow technique *Cell Mol Life Sci 63*, 2571-83.

[384] Douzou, P. (1977) Enzymology at subzero temperatures *Adv Enzymol Relat Areas Mol Biol 45*, 157-272.

[385] Millar, N. C. and Geeves, M. A. (1983) The limiting rate of the ATP-mediated dissociation of actin from rabbit skeletal muscle myosin subfragment 1 *FEBS Lett 160*, 141-8.
[386] Biosca, J. A., Barman, T. E. and Travers, F. (1984) Transient kinetics of the binding of ATP to actomyosin subfragment 1: evidence that the dissociation of actomyosin subfragment 1 by ATP leads to a new conformation of subfragment 1 *Biochemistry 23*, 2428-36.
[387] Tesi, C., Kitagishi, K., Travers, F. and Barman, T. (1991) Cryoenzymic studies on actomyosin ATPase: kinetic evidence for communication between the actin and ATP sites on myosin *Biochemistry 30*, 4061-7.
[388] Barman, T. E., Brun, A. and Travers, F. (1980) A flow-quench apparatus for cryoenzymic studies. Application to the creatine kinase reaction *Eur J Biochem 110*, 397-403.
[389] Koshland, D. E., Jr. (1963) Properties of the active site of enzymes *Ann N Y Acad Sci 103*, 630-42.
[390] Negelein, E. (1957) Synthesis, Determination and Properties of 1,3-Diphosphoglyceric acid in *Methods in Enzymology* (Colowick, S. P. a. K., N.O., Ed.) pp 216-220, Acad. Press inc., New York.
[391] Kvassman, J. and Pettersson, G. (1989) Evidence that 1,3-bisphosphoglycerate dissociation from phosphoglycerate kinase is an intrinsically rapid reaction step *Eur J Biochem 186*, 261-4.
[392] Fisher, H. F. (2005) Transient-state kinetic approach to mechanisms of enzymatic catalysis *Acc Chem Res 38*, 157-66.
[393] Lionne, C., Stehle, R., Travers, F. and Barman, T. (1999) Cryoenzymic studies on an organized system: myofibrillar ATPases and shortening *Biochemistry 38*, 8512-20.
[394] Deville-Bonne, D., Laine, R. and Garel, J. R. (1991) Substrate antagonism in the kinetic mechanism of E. coli phosphofructokinase-1 *FEBS Lett 290*, 173-6.
[395] Khamis, M. M. and Larsson-Raznikiewicz, M. (1981) Activation and inhibition of phosphoglycerate kinase by sulphate ion *Biochim. Biophys. Acta 657*, 190-194.
[396] Larsson-Raznikiewicz, M. and Schierbeck, B. (1977) The possibility that the substrates of phosphoglycerate kinase may specifically regulate the direction of its reversible reaction [proceedings] *Biochem Soc Trans 5*, 770-1.
[397] Ali, M. and Brownstone, Y. S. (1976) A study of phosphoglycerate kinase in human erythrocytes. II. Kinetic properties *Biochim. Biophys. Acta 445*, 89-103.
[398] Larsson-Raznikiewicz, M. (1973) Graphical analyses on binding of ligands to a two-sited system. Theoretical treatments exemplified on yeast phosphoglycerate kinase *Arch. Biochem. Biophys. 158*, 754-762.
[399] Tanswell, P., Westhead, E. W. and Williams, R. J. P. (1974) Inhibition of yeast phosphoglycerate kinase by lanthanide-ATP complexes. *FEBS Lett. 48*, 60-63.
[400] Roustan, C., Brevet, A., Pradel, L. A. and Thoai, N. V. (1973) Yeast 3-phosphoglycerate kinase. Interaction of enzyme with substrates studied by partial isotopic exchange and difference spectrophotometry *Eur. J. Biochem. 37*, 248-255.
[401] Wiksell, E. and Larsson-Raznikiewicz, M. (1982) Substrate binding to phosphoglycerate kinase monitored by 1-anilino-8-naphthalenesulfonate *J. Biol. Chem. 257*, 12672-12677.
[402] Molnár, M. and Vas, M. (1993) Mg^{2+} affects the binding of ADP but not ATP to 3-phosphoglycerate kinase. Correlation between equilibrium dialysis binding and enzyme kinetic data *Biochem. J. 293*, 595-599.
[403] Gupta, R. K. and Moore, R. D. (1980) 31P NMR studies of intracellular free Mg2+ in intact frog skeletal muscle *J. Biol. Chem. 255*, 3987-3993.

[404] Strivastava, D. K. and Bernard, S. A. (1987) Biophysical Chemistry of Metabolic Reaction Sequences in Concentrated Enzyme Solution and in the Cell *Ann. Rev. Biophys. Chem. 16*, 175-204.

[405] Hiratsuka, T. and Uchida, K. (1973) Preparation and properties of 2'(or 3')-O-(2,4,6-trinitrophenyl) adenosine 5'-triphosphate, an analog of adenosine triphosphate *Biochim Biophys Acta 320*, 635-47.

[406] Rao, R., Al-Shawi, M. K. and Senior, A. E. (1988) Trinitrophenyl-ATP and -ADP bind to a single nucleotide site on isolated beta-subunit of Escherichia coli F1-ATPase. In vitro assembly of F1-subunits requires occupancy of the nucleotide-binding site on beta-subunit by nucleoside triphosphate *J Biol Chem 263*, 5569-73.

[407] Faller, L. D. (1990) Binding of the fluorescent substrate analogue 2',3'-O-(2,4,6-trinitrophenylcyclohexadienylidene)adenosine 5'-triphosphate to the gastric H+,K(+)-ATPase: evidence for cofactor-induced conformational changes in the enzyme *Biochemistry 29*, 3179-86.

[408] Larsson-Raznikiewicz, M. and Malström, B. G. (1961) The Metal-Ion Activation of 3-Phosphoglycerate Kinase in Correlation with Metal-Binding Studies *Arch Biochem Biophys 92*, 94-99.

[409] Herschlag, D. and Jencks, W. P. (1987) Effect of Divalent Metal Ions on the Rate and Transition-State Structure of Phosphoryl-Transfer Reactions *J Am Chem Soc 109*, 4665-4674.

[410] Morrison, J. F. and Uhr, M. L. (1966) The function of bivalent metal ions in the reaction catalysed by ATP:creatine phosphotransferase *Biochim Biophys Acta 122*, 57-74.

[411] Varga, A., Pálmai, Z., Gugolya, Z., Gráczer, É., Vonderviszt, F., Závodszky, P., Balog, E. and Vas, M. (2012) *Biochemistry 51, 10197-207*

[412] Wrobel, J. A. and Stinson, R. A. (1980) Metal ion binding to yeast phosphoglycerate kinase. Interaction between metal ion binding and anion binding *Eur J Biochem 104*, 249-54.

[413] Wittinghofer, A. (1997) Signaling mechanistics: Aluminum fluoride for molecule of the year *Current Biology 7*, R682-R685.

[414] Baxter, N. J., Olguin, L. F., Golicnik, M., Feng, G., Hounslow, A. M., Bermel, W., Blackburn, G. M., Hollfelder, F., Waltho, J. P. and Williams, N. H. (2006) A Trojan horse transition state analogue generated by MgF3- formation in an enzyme active site *Proc Natl Acad Sci U S A 103*, 14732-14737.

[415] Baxter, N. J., Blackburn, G. M., Marston, J. P., Hounslow, A. M., Cliff, M. J., Bermel, W., Williams, N. H., Hollfelder, F., Wemmer, D. E. and Waltho, J. P. (2008) Anionic charge is prioritized over geometry in aluminum and magnesium fluoride transition state analogs of phosphoryl transfer enzymes *J Am Chem Soc 130*, 3952-3958.

[416] Baxter, N. J., Hounslow, A. M., Bowler, M. W., Williams, N. H., Blackburn, G. M. and Waltho, J. P. (2009) MgF(3)(-) and alpha-Galactose 1-Phosphate in the Active Site of beta-Phosphoglucomutase Form a Transition State Analogue of Phosphoryl Transfer *J Am Chem Soc 131*, 16334-+.

[417] Baxter, N. J., Bowler, M. W., Alizadeh, T., Cliff, M. J., Hounslow, A. M., Wu, B., Berkowitz, D. B., Williams, N. H., Blackburn, G. M. and Waltho, J. P. (2010) Atomic details of near-transition state conformers for enzyme phosphoryl transfer revealed by MgF-3 rather than by phosphoranes *Proc Natl Acad Sci U S A 107*, 4555-60.

[418] Graham, D. L., Lowe, P. N., Grime, G. W., Marsh, M., Rittinger, K., Smerdon, S. J., Gamblin, S. J. and Eccleston, J. F. (2002) MgF3- as a transition state analog of phosphoryl transfer *Chem Biol 9*, 375-381.

[419] Lee, J. Y. and Yang, W. (2006) UvrD helicase unwinds DNA one base pair at a time by a two-part power stroke *Cell 127*, 1349-1360.

[420] Yousef, M. S., Clark, S. A., Pruett, P. K., Somasundaram, T., Ellington, W. R. and Chapman, M. S. (2003) Induced fit in guanidino kinases--comparison of substrate-free and transition state analog structures of arginine kinase *Protein Sci 12*, 103-11.

[421] Muller, C. W., Schlauderer, G. J., Reinstein, J. and Schulz, G. E. (1996) Adenylate kinase motions during catalysis: an energetic counterweight balancing substrate binding *Structure 4*, 147-56.

[422] Hayward, S. (2004) Identification of specific interactions that drive ligand-induced closure in five enzymes with classic domain movements *J Mol Biol 339*, 1001-21.

[423] Stillman, T. J., Migueis, A. M., Wang, X. G., Baker, P. J., Britton, K. L., Engel, P. C. and Rice, D. W. (1999) Insights into the mechanism of domain closure and substrate specificity of glutamate dehydrogenase from Clostridium symbiosum *J Mol Biol 285*, 875-85.

[424] Henzler-Wildman, K. and Kern, D. (2007) Dynamic personalities of proteins *Nature 450*, 964-72.

[425] Balog, E., Laberge, M. and Fidy, J. (2007) The influence of interdomain interactions on the intradomain motions in yeast phosphoglycerate kinase: a molecular dynamics study *Biophys J 92*, 1709-1716.

[426] Zerrad, L., Merli, A., Schroder, G. F., Varga, A., Gráczer, É., Pernot, P., Round, A., Vas, M. and Bowler, M. W. (2011) A spring-loaded release mechanism regulates domain movement and catalysis in phosphoglycerate kinase *J Biol Chem 286*, 14040-8.

[427] Sawyer, G. M., Monzingo, A. F., Poteet, E. C., O'Brien, D. A. and Robertus, J. D. (2008) X-ray analysis of phosphoglycerate kinase 2, a sperm-specific isoform from Mus musculus *Proteins 71*, 1134-44.

[428] Schröder, G. F., Brunger, A. T. and Levitt, M. (2007) Combining efficient conformational sampling with a deformable elastic network model facilitates structure refinement at low resolution *Structure 15*, 1630-41.

[429] Schröder, G. F., Levitt, M. and Brunger, A. T. (2010) Super-resolution biomolecular crystallography with low-resolution data *Nature 464*, 1218-22.

[430] Varga, A., Szabó, J., Flachner, B., Gugolya, Z., Vonderviszt, F., Závodszky, P. and Vas, M. (2009) Thermodynamic analysis of substrate induced domain closure of 3-phosphoglycerate kinase *FEBS Lett 583*, 3660-4.

[431] Mildvan, A. S. (1997) Mechanisms of signaling and related enzymes *Proteins 29*, 401-16.

[432] Frederick, K. K., Marlow, M. S., Valentine, K. G. and Wand, A. J. (2007) Conformational entropy in molecular recognition by proteins *Nature 448*, 325-9.

[433] Henzler-Wildman, K. A., Thai, V., Lei, M., Ott, M., Wolf-Watz, M., Fenn, T., Pozharski, E., Wilson, M. A., Petsko, G. A., Karplus, M., Hubner, C. G. and Kern, D. (2007) Intrinsic motions along an enzymatic reaction trajectory *Nature 450*, 838-44.

[434] Yon, J. M. (2004) Aggregation Protein in *Encyclopedia of Molecular Cell Biology and Molecular Medicine* (Mayers, R. A., Ed.) pp 23-52, Wiley-VCH.

[435] Fink, A. L. (1998) Protein aggregation: folding aggregates, inclusion bodies and amyloid *Fold Des 3*, R9-23.

[436] Georgiou, G. and Bernadez-Clark, E. (1990) in *199th National Meeting of the American Chemical Society*, American Chemical Society, Washington, Boston, Massachusetts.

[437] Dobson, C. M. (2003) Protein folding and misfolding *Nature 426*, 884-90.

[438] Smeller, L. (2002) Pressure-temperature phase diagrams of biomolecules *Biochim Biophys Acta 1595*, 11-29.

[439] Wiencek, J. M. (1999) New strategies for protein crystal growth *Annu Rev Biomed Eng 1*, 505-34.
[440] Mitraki, A. and King, J. (1992) Amino acid substitutions influencing intracellular protein folding pathways *FEBS Lett 307*, 20-5.
[441] Yon, J. M. (1996) The specificity of protein aggregation [new; comment] *Nat Biotechnol 14*, 1231.
[442] London, J., Skrzynia, C. and Goldberg, M. E. (1974) Renaturation of Escherichia coli tryptophanase after exposure to 8 M urea. Evidence for the existence of nucleation centers *Eur J Biochem 47*, 409-15.
[443] Crestfield, A. M., Stein, W. H. and Moore, S. (1962) On the aggregation of bovine pancreatic ribonuclease *Arch Biochem Biophys Suppl 1*, 217-22.
[444] Bennett, M. J., Choe, S. and Eisenberg, D. (1994) Domain swapping: entangling alliances between proteins *Proc Natl Acad Sci U S A 91*, 3127-31.
[445] Liu, Y., Gotte, G., Libonati, M. and Eisenberg, D. (2001) A domain-swapped RNase A dimer with implications for amyloid formation *Nat Struct Biol 8*, 211-4.
[446] Liu, Y., Gotte, G., Libonati, M. and Eisenberg, D. (2002) Structures of the two 3D domain-swapped RNase A trimers *Protein Sci 11*, 371-80.
[447] Cohen, A. S. (2000) in *Amylose* (Grateau, G., Benson, M. D., and Delpech, M., Eds.), Flammarion, Paris.
[448] Serpell, L. C., Sunde, M. and Blake, C. C. (1997) The molecular basis of amyloidosis *Cell Mol Life Sci 53*, 871-87.
[449] Ross, C. A. and Poirier, M. A. (2004) Protein aggregation and neurodegenerative disease *Nat Med 10 Suppl*, S10-7.
[450] Dobson, C. M. (2001) The structural basis of protein folding and its links with human disease *Philos Trans R Soc Lond B Biol Sci 356*, 133-45.
[451] Kelly, J. W. (1996) Alternative conformations of amyloidogenic proteins govern their behavior *Curr Opin Struct Biol 6*, 11-7.
[452] Rochet, J. C. and Lansbury, P. T., Jr. (2000) Amyloid fibrillogenesis: themes and variations *Curr Opin Struct Biol 10*, 60-8.
[453] Green, N. S., Palaninathan, S. K., Sacchettini, J. C. and Kelly, J. W. (2003) Synthesis and characterization of potent bivalent amyloidosis inhibitors that bind prior to transthyretin tetramerization *J Am Chem Soc 125*, 13404-14.
[454] Guijarro, J. I., Sunde, M., Jones, J. A., Campbell, I. D. and Dobson, C. M. (1998) Amyloid fibril formation by an SH3 domain *Proc Natl Acad Sci U S A 95*, 4224-8.
[455] Chiti, F., Webster, P., Taddei, N., Clark, A., Stefani, M., Ramponi, G. and Dobson, C. M. (1999) Designing conditions for in vitro formation of amyloid protofilaments and fibrils *Proc Natl Acad Sci U S A 96*, 3590-4.
[456] Bucciantini, M., Giannoni, E., Chiti, F., Baroni, F., Formigli, L., Zurdo, J., Taddei, N., Ramponi, G., Dobson, C. M. and Stefani, M. (2002) Inherent toxicity of aggregates implies a common mechanism for protein misfolding diseases *Nature 416*, 507-11.
[457] Dobson, C. M. (2002) Getting out of shape *Nature 418*, 729-30.
[458] Dobson, C. M. (1999) Protein misfolding, evolution and disease *Trends Biochem Sci 24*, 329-32.
[459] Damaschun, G., Damaschun, H., Gast, K. and Zirwer, D. (1999) Proteins can adopt totally different folded conformations *J Mol Biol 291*, 715-25.
[460] Damaschun, G., Damaschun, H., Fabian, H., Gast, K., Krober, R., Wieske, M. and Zirwer, D. (2000) Conversion of yeast phosphoglycerate kinase into amyloid-like structure *Proteins 39*, 204-11.

[461] Modler, A. J., Gast, K., Lutsch, G. and Damaschun, G. (2003) Assembly of amyloid protofibrils via critical oligomers--a novel pathway of amyloid formation *J Mol Biol 325*, 135-48.

[462] Fujii, H. and Miwa, S. (2000) Other erythrocyte enzyme deficiencies associated with non-haematological symptoms: phosphoglycerate kinase and phosphofructokinase deficiency *Baillieres Best Pract Res Clin Haematol 13*, 141-8.

[463] Kraus, A. P., Langston, M. F., Jr. and Lynch, B. L. (1968) Red cell phosphoglycerate kinase deficiency. A new cause of non-spherocytic hemolytic anemia *Biochem Biophys Res Commun 30*, 173-7.

[464] Valentine, W. N., Hsieh, H. S., Paglia, D. E., Anderson, H. M., Baughan, M. A., Jaffe, E. R. and Garson, O. M. (1969) Hereditary hemolytic anemia associated with phosphoglycerate kinase deficiency in erythrocytes and leukocytes. A probable X-chromosome-linked syndrome *N Engl J Med 280*, 528-34.

[465] Chen, S. H., Malcolm, L. A., Yoshida, A. and Giblett, E. R. (1971) Phosphoglycerate kinase: an X-linked polymorphism in man *Am J Hum Genet 23*, 87-91.

[466] Cartier, P., Habibi, B., Leroux, J. P. and Marchand, J. C. (1971) Congenital hemolytic anemia associated with phosphoglycerate kinase deficiency in erythrocytes, polynuclear cells and lymphocytes *Nouv Rev Fr Hematol 11*, 565-78.

[467] Arese, P., Bosia, A., Gallo, E., Mazza, U. and Pescarmona, G. P. (1973) Red cell glycolysis in a case of 3-phosphoglycerate kinase deficiency *Eur J Clin Invest 3*, 86-92.

[468] Konrad, P. N., McCarthy, D. J., Mauer, A. M., Valentine, W. N. and Paglia, D. E. (1973) Erythrocyte and leukocyte phosphoglycerate kinase deficiency with neurologic disease *J Pediatr 82*, 456-60.

[469] Fujii, H. and Yoshida, A. (1980) Molecular abnormality of phosphoglycerate kinase-Uppsala associated with chronic nonspherocytic hemolytic anemia *Proc Natl Acad Sci U S A 77*, 5461-5.

[470] Tougard, P., Le, T. H., Minard, P., Desmadril, M., Yon, J. M., Bizebard, T., Lebras, G. and Dumas, C. (1996) Structural and functional properties of mutant Arg203Pro from yeast phosphoglycerate kinase, as a model of phosphoglycerate kinase-Uppsala. *Protein Eng. 9*, 181-187.

[471] Fujii, H., Krietsch, W. K. and Yoshida, A. (1980) A single aminoacid substitution (Asp 268 leads to Asn) in 3-phosphoglycerate kinase variant (PGK Munchen) associated with enzyme deficiency *J. Biol. Chem. 255*, 6421-6423.

[472] Fujii, H., Chen, S. H., Akatsuka, J., Miwa, S. and Yoshida, A. (1981) Use of cultured lymphoplastoid cells for the study of abnormal enzymes: molecular abnormality of PGK variant (Val 266 to Met) associated with hemolytic anemia *Proc. Natl. Acad. Sci. USA 78*, 2587-2590.

[473] DiMauro, S., Dalakas, M. and Miranda, A. F. (1983) Phosphoglycerate kinase deficiency: another cause of recurrent myoglobinuria *Ann Neurol 13*, 11-9.

[474] Guis, M. S., Karadsheh, N. and Mentzer, W. C. (1987) Phosphoglycerate kinase San Francisco: a new variant associated with hemolytic anemia but not with neuromuscular manifestations *Am J Hematol 25*, 175-82.

[475] Maeda, M. and Yoshida, A. (1991) Molecular defect of a phosphoglycerate kinase variant (PGK-Matsue) associated with hemolytic anemia: Leu----Pro substitution caused by T/A----C/G transition in exon 3 *Blood 77*, 1348-52.

[476] Maeda, M., Bawle, E. V., Kulkarni, R., Beutler, E. and Yoshida, A. (1992) Molecular abnormalities of a phosphoglycerate kinase variant generated by spontaneous mutation *Blood 79*, 2759-62.

[477] Fujii, H., Kanno, H., Hirono, A., Shiomura, T. and Miwa, S. (1992) A single amino acid substitution (Gly 157 to Val) in phosphoglycerate kinase variant (PGK Shizuoka) associated with chronic hemolysis and myoglobinuria *Blood 79*, 1582-1585.

[478] Tonin, P., Shanske, S., Miranda, A. F., Brownell, A. K., Wyse, J. P., Tsujino, S. and DiMauro, S. (1993) Phosphoglycerate kinase deficiency: biochemical and molecular genetic studies in a new myopathic variant (PGK Alberta) *Neurology 43*, 387-91.

[479] Cohen-Solal, M., Valentin, C., Plassa, F., Guillemin, G., Danze, F., Jaisson, F. and Rosa, R. (1994) Identification of new mutations in two phosphoglycerate kinase (PGK) variants expressing different clinical syndromes: PGK Creteil and PGK Amiens *Blood 84*, 898-903.

[480] Rosa, R., George, C., Fardeau, M., Calvin, M. C., Rapin, M. and Rosa, J. (1982) A new case of phosphoglycerate kinase deficiency: PGK Creteil associated with rhabdomyolysis and lacking hemolytic anemia *Blood 60*, 84-91.

[481] Turner, G., Fletcher, J., Elber, J., Yanagawa, Y., Dave, V. and Yoshida, A. (1995) Molecular defect of a phosphoglycerate kinase variant associated with haemolytic anaemia and neurological disorders in a large kindred *Br J Haematol 91*, 60-5.

[482] Flanagan, J. M., Rhodes, M., Wilson, M. and Beutler, E. (2006) The identification of a recurrent phosphoglycerate kinase mutation associated with chronic haemolytic anaemia and neurological dysfunction in a family from USA *Br J Haematol 134*, 233-7.

[483] Sugie, H., Sugie, Y., Tsurui, S. and Ito, M. (1994) Phosphoglycerate kinase deficiency *Neurology 44*, 1364-5.

[484] Tsujino, S., Tonin, P., Shanske, S., Nohria, V., Boustany, R. M., Lewis, D., Chen, Y. T. and DiMauro, S. (1994) A splice junction mutation in a new myopathic variant of phosphoglycerate kinase deficiency (PGK North Carolina) *Ann Neurol 35*, 349-53.

[485] Yoshida, A., Twele, T. W., Dave, V. and Beutler, E. (1995) Molecular abnormality of a phosphoglycerate kinase variant (PGK-Alabama) *Blood Cells Mol Dis 21*, 179-81.

[486] Schroder, J. M., Dodel, R., Weis, J., Stefanidis, I. and Reichmann, H. (1996) Mitochondrial changes in muscle phosphoglycerate kinase deficiency *Clin Neuropathol 15*, 34-40.

[487] Ookawara, T., Dave, V., Willems, P., Martin, J. J., de Barsy, T., Matthys, E. and Yoshida, A. (1996) Retarded and aberrant splicings caused by single exon mutation in a phosphoglycerate kinase variant *Arch Biochem Biophys 327*, 35-40.

[488] Valentin, C., Birgens, H., Craescu, C. T., Brodum-Nielsen, K. and Cohen-Solal, M. (1998) A phosphoglycerate kinase mutant (PGK Herlev; D285V) in a Danish patient with isolated chronic hemolytic anemia: mechanism of mutation and structure-function relationships *Hum Mutat 12*, 280-7.

[489] Sugie, H., Sugie, Y., Ito, M. and Fukuda, T. (1998) A novel missense mutation (837T-->C) in the phosphoglycerate kinase gene of a patient with a myopathic form of phosphoglycerate kinase deficiency *J Child Neurol 13*, 95-7.

[490] Hamano, T., Mutoh, T., Sugie, H., Koga, H. and Kuriyama, M. (2000) Phosphoglycerate kinase deficiency: an adult myopathic form with a novel mutation *Neurology 54*, 1188-90.

[491] Aasly, J., van Diggelen, O. P., Boer, A. M. and Bronstad, G. (2000) Phosphoglycerate kinase deficiency in two brothers with McArdle-like clinical symptoms *Eur J Neurol 7*, 111-3.

[492] Morimoto, A., Ueda, I., Hirashima, Y., Sawai, Y., Usuku, T., Kano, G., Kuriyama, K., Todo, S., Sugimoto, T., Kanno, H., Fujii, H. and Imashuku, S. (2003) A novel missense mutation (1060G --> C) in the phosphoglycerate kinase gene in a Japanese boy with chronic haemolytic anaemia, developmental delay and rhabdomyolysis *Br J Haematol 122*, 1009-13.

[493] Noel, N., Flanagan, J., Kalko, S. G., Bajo, M. J., Manu Mdel, M., Fuster, J. L., Beutler, E. and Corrons, J. L. (2006) Two new phosphoglycerate kinase mutations associated with chronic haemolytic anaemia and neurological dysfunction in two patients from Spain *Br J Haematol 132*, 523-9.

[494] Martinov, M. V., Plotnikov, A. G., Vitvitsky, V. M. and Ataullakhanov, F. I. (2000) Deficiencies of glycolytic enzymes as a possible cause of hemolytic anemia *Biochim Biophys Acta 1474*, 75-87.

[495] Spiegel, R., Gomez, E. A., Akman, H. O., Krishna, S., Horovitz, Y. and DiMauro, S. (2009) Myopathic form of phosphoglycerate kinase (PGK) deficiency: a new case and pathogenic considerations *Neuromuscul Disord 19*, 207-11.

[496] Chiarelli, L.R., Morera, S.M., Bianchi, P., Fermo, E., Zanella, A., Galizzi, A. and Valentini, G. (2012) Molecular insights on Pathogenic Effects of Mutations Causing Phosphoglycerate Kinase Deficiency *Plos One 7,* e32065.

Editor Contact Information

Professor Mária Vas
Institute of Enzymology
Research Centre for Natural Sciences
Hungarian Academy of Sciences
H-1117 Budapest
Magyar tudosok korutja 2
Hungary
vas.maria@ttk.mta.hu
36 (1) 279-3152
Fax: 36 (l) 466-5465

Index

C

D

E

F

G

N

O

P

R

S